P9-DEM-000

QUINTESSENCE

ALSO BY LAWRENCE KRAUSS

Genesis: The Lives of an Atom (forthcoming)

Beyond Star Trek:
Physics from Alien Invasions to the End of Time

The Physics of Star Trek

Fear of Physics: A Guide for the Perplexed

A MEMBER OF THE PERSEUS BOOKS GROUP

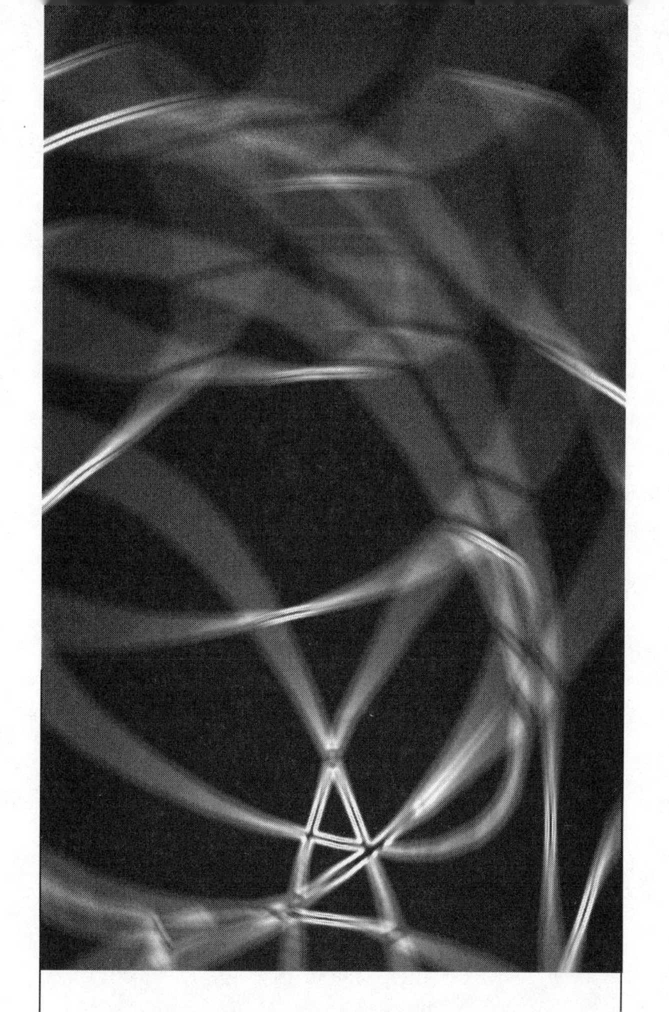

QUINTESSENCE

THE MYSTERY OF MISSING MASS IN THE UNIVERSE

LAWRENCE KRAUSS

Published by Basic Books,
A Member of the Perseus Books Group

For information, address Basic Books, 10 E 53 Street,
New York, NY 10022–5299.

Book design by Victoria Kuskowski

Library of Congress Cataloging-in-Publication Data
Krauss, Lawrence Maxwell.
Quintessence: The mystery of missing mass in the universe /
Lawrence M. Krauss.
p. cm.
Includes index.
ISBN 0-465-03740-2
1. Dark matter (Astronomy) I. Title.
QB791.3 .K74 1999
523.1′126—dc21
99-047676

FOR LILLI

DARKNESS THERE WAS AT FIRST BY DARKNESS HIDDEN. . . .

—Hymn of Creation, *the Rig Veda*

CONTENTS

* * *

PART III WHY THE UNIVERSE IS FLAT: THE BIG BANG, LARGE-SCALE STRUCTURE, AND THE NEED FOR SOMETHING NEW

* * *

PART IV THE NEUTRINO SAGA AND THE BIRTH OF COLD DARK MATTER

* * *

PART V THE CANDIDATES

* * *

PART VI DESPERATELY SEEKING DARK MATTER

QUINTESSENCE

PREFACE TO THE REVISED EDITION

The more things change, the more they stay the same. That was the challenge I faced as I considered rewriting a book originally penned over a decade ago about a subject that was then and still is at the very heart of modern physics and astrophysics. I received many requests from my colleagues and from lay readers to bring this book back into print, but it was clear I could not do so without playing proper homage to the remarkable past decade of research.

Today, as then, physicists wonder at the nature of unidentified dark matter—invisible to telescopes—that we are increasingly confident dominates the mass of our own galaxy, as well as essentially all other galaxies in the universe. Today, as then, teams of physicists are exploring new and ingenious terrestrial techniques for directly detecting this material, which we believe is likely to consist of some new type of elementary particle and thus is not only "out there" but "in here" as well. Today, as then, physicists are trying to determine how the presence of this material influenced the evolution of everything we see in the universe—clusters, galaxies, and stars. Today, as then, we are trying to use dark matter to give us an empirical handle on the earliest moments of the Big Bang explosion.

Yet, at the same time, everything has changed! The picture of cosmology we so carefully and optimistically developed over the 1980s, based on the fusion of ideas from particle physics and astrophysics, is now undeniably incomplete. Over the past five years it has become clear that "dark matter" alone is not abundant enough to eventually halt the observed universal expansion. More surprising still is the suggestion that the dominant energy in the universe may be associated with empty space, and moreover, that this energy is causing the expansion of the universe to accelerate with time! Exotic dark matter dominating the mass of the universe is a concept that may be hard for some to accept. But at least dark matter is "stuff," however exotic. How much harder is it to get used to the idea that most of the energy density of the universe may literally be associated with nothing at all? Moreover, if this picture is correct, the notion that we may one day know with certainty how the universe will end may have to be discarded.

Cosmology as a field has made remarkable developments over the past twenty years. It is now a mature science, ripe with new data. In the original version of this book I stressed the nascent observations that were at the basis of the ongoing transition of cosmology to a full fledged empirical science. Now that that transformation is essentially complete, I want to properly account for the plethora of new data while at the same time continuing to stress the many assumptions and remaining unknowns that continue to play a key role in our picture of the universe. I fully expect that in the next decade our picture will continue to evolve as swiftly as it has in the last, even if the fundamental tenets of Big Bang cosmology remain firmly grounded. There is more that we do not understand about the universe than there is that we do. Each time we open a new window on the universe I expect we will be surprised. Indeed, this is one of the most important reasons for unraveling the mysteries of dark matter. Perhaps nothing else can give us a window back to earliest moments of creation itself.

Thankfully, I felt that in spite of recent developments, the basic framework of the original edition of this book was sound. In spite of the great leaps in our empirical understanding of cosmology, most of the key issues that dominated our thinking at that time continue to remain central to the frontiers of current research. At the same time the past decade has not yet brought us substantially closer to solving the mystery of dark matter. Experiments that were only dim hopes ten years ago are now under way. Ideas that were only beginning to emerge have solidified. Yet at each stage we have been surprised, and in spite of a tremendous amount of activity and creativity, the search for the nature of dark matter remains one of the most important ongoing programs in all of physics.

The best reason for preserving the form of this book remains the one that motivated me to write it in the first place. Not only is the search for dark matter fascinating and thought provoking, but it presents in microcosm the state of the entire fields of particle physics and cosmology. By exploring the assumptions behind the endeavor, I can point out how far we have come in our understanding of microphysics and macrophysics, and at the same time how far we have yet to go.

I also still believe that we are on the verge of uncovering the nature of the dominant mass in the universe. Moreover, it is now overwhelmingly clear that there is a larger mystery out there to be solved, and that unraveling dark matter is the key.

A decade ago I coined the term "fifth essence" to refer to the dark matter we inferred to dominate the density of mass and energy in the universe. This was inspired by Aristotle's *quinta essentia,* the material of the heavens complementing the four elements earth, air, fire, and water. Aristotle's essence became the more modern aether, the material that was thought, right until the last century, to permeate empty space. A decade ago, it seemed to me that dark matter was a sort of modern "fifth essence." But even closer in spirit to Aristotle's heavenly aether or "quintessence" is the currently favored possibility that a nonzero energy exists throughout empty space. Thus the new title, *Quintessence: The Mystery of the Missing Mass*. The second half of the title has also changed and is particularly apt today. In the 1970s dark matter was often called "missing matter." This was a misnomer. It wasn't missing—we could identify where it was located by observing the motion of stars and gas in and around galaxies. It was simply invisible to telescopes. If we are now correct, and all the matter in the universe, including dark matter, is insufficient to account for the total energy density of the universe, then this new quintessence is truly still "missing." We will see. . . . But if anything, the search has gotten even more exciting in the past ten years.

I originally wrote *The Fifth Essence* with one goal in mind: I wanted never to appeal to authority, so that interested laypersons could decide for themselves what to accept and what not to. Moreover, I wanted to convey the satisfaction that can be gained only by puzzling through things for oneself. In this revised edition I have decided to add to the existing text only in those places where new developments have altered our thinking, and in these places I have attempted to explain why this is the case. At some points I have also removed material that is no longer of interest. I hope the resulting work remains true to my original intent.

When I wrote the first edition of this book, I at Yale University. This had been, at the end of the nineteenth century, the home institution of J. Willard Gibbs, perhaps the first great American theoretical physicist. In 1993, I moved to become Ambrose Swasey Professor and Chairman of the Physics Department at Case Western Reserve University, which was the home institution of the first truly great American experimental physicist at the dawn of the twentieth century, Albert A. Michelson. So I like to feel that I have continued to progress along with the field! In any case, I have had, thanks to the support of the administration at CWRU, significantly more resources at my disposal to build up a research group and to continue my own writing and lecturing than I had before. Three more popular books and a new line of research activities under my belt will no doubt color the presentation of new material here. I thank my colleagues at CWRU for their unfailing support of my efforts. It has been a joy being a part of the program we have all built over the past six years. I also have worked with many young people, who have taught me a tremendous amount since I dotted *i*'s in the last sentence of the first edition a decade ago. These include, but are not limited to, my research postdoctoral fellows and students over the past decade: Frank Accetta, Brian Chaboyer, Norman Cheng, Craig Copi, Andrew Delaix, Evalyn Gates, Junseong Heo, Peter Kernan, Scott Koranda, Hong Liu, Paul Romanelli, Todd Small, Raman Sundrum, John Terning, Mark Trodden, and Martin White. I am sure that many of the new insights I describe here are due in large part to the inspiration I gained from them. Finally, my family has had to endure an ever-greater drain on my time as a result of my writing and research activities, and their indulgence and support have continued to help make it all possible and worthwhile.

Lawrence Krauss

CLEVELAND, OHIO, SEPTEMBER 1999

PREFACE

A New Copernican Revolution?

You are a graduate student in physics. It's late Saturday night and you would much rather be at a party. Instead, you are a mile underground, in a cavernous enclosure, entertained only by the sound of a cooling fan whirring in the desk-top minicomputer that is monitoring pulses received from the gargantuan device located in the main chamber next door. It has been a boring eight-hour shift and you long to take the elevator ride up the mine shaft to the surface to breathe the fresh air and to watch the night sky, the stars twinkling, and the cool, evanescent glow of the moon bathing the earth's surface. You are, after all, studying to be an astrophysicist, not a geologist. When you forsook a lucrative programming job in order to return to graduate school, you envisioned working at a huge radio telescope aimed at the heavens, sensing the faint pulses emitted by quasars billions of light-years away. Yet here you are, deep underground, monitoring a new experiment built by a collaboration among four universities located on three continents. In order to pass the time you watch the calibration pulses appear with clocklike regularity on your monitor, noting how each exactly reproduces the last.

Suddenly, almost too fast to sense, you notice something momentarily different about the signal. You halt the online output on the computer and call up the program that single-steps through the data. While the program loads on the machine, your mind races. There is a small chance that the pulse you saw, or imagined, is the infinitesimally small signal from an elementary particle making up a totally new type of matter never before observed on earth that interacted in your detector. If so, this could be the first time this particle has so interacted in the ten to fifteen billion years since it was created in the fiery Big Bang. You may be looking at a signal from the beginning of time! Such particles may constitute one hundred times more material, by weight, than everything we can see put together, thereby governing the structure, evo-

lution, and eventual fate of the universe! Your discovery could affect the way we think about the universe as dramatically as had Copernicus's assertion that the earth moves about the sun. . . .

Or perhaps it is just a bit of noise in the detector. . . .

Over the past four centuries, we have come, in one way or another, to recognize the cosmic insignificance of humankind. The scientific revolutions associated with Copernicus and Galileo led to the realization that the Earth is not the center of the solar system. Later observations revealed that neither our solar system, nor our galaxy, nor our local neighborhood of galaxies is unique or special. Einstein's special theory of relativity taught us that we draw false security from the notion that the temporal milestones in human history have any absolute significance. Discoveries in quantum mechanics have further forced the realization that even absolute knowledge on our part does not lead to absolute predictive control over nature.

Indeed, as we have had the opportunity in each of the natural sciences to examine newly uncovered layers in the fabric of the physical world, we have found that the phenomena essential to our daily lives encompass a smaller and smaller piece of a much larger whole. The organic materials that form the basis for all living systems on earth involve only a small subset of the rich structure of modern chemistry. On a finer level, the discoveries in particle physics since the 1940s demonstrate that the fundamental constituents that make up terrestrial materials are themselves only a small fraction of all elementary objects.

In the midst of all this, modern astronomy has appeared to offer some solace. Part of its great success has been to demonstrate, through the use of photographic spectra taken by ever-larger telescopes, that the materials of the stars and earth are in fact the same. All visible objects, from the sun and moon to the distant stars, seem to be composed of elements that are abundant on earth. The nuclear burning that fuels the stars has, through cosmic time, cooked the same recipe throughout. We now know that the elements we find on earth were created in cosmic fireworks displays similar to the supernova we were privileged to witness in 1987.

But in spite of the fact that everything we can "see" appears to be made of the same fundamental constituents, that is, protons and neutrons, the actual number of these particles in the universe is minuscule compared to the number of "particles" of light, called photons, which we have detected as pervading the universe. For every proton in the universe today, there are roughly 10 billion of these photons! This ratio is one of the fundamental "observables" in cosmology, yet why it has exactly this value was completely without any proposed explanation until the 1970s. Even now, we still do not have a precise microscopic theory to predict this ratio of protons to photons unambiguously, although developments in particle physics have given us a great deal of insight into likely mechanisms by which this situation could come about.

The new particle physics has not only offered a possible explanation of why we see what we do, but it suggests the possibility that there may be more "out there" than meets the eye. If all that we can "see" is composed of particles that are relatively rare, then the relic photons in the universe today may not be the only other particles that are far more abundant. In raising the possibility that other unobserved backgrounds may permeate the universe, physics has returned to a theme that has resurfaced in many guises since the first recorded stories of creation in ancient times. For some reason, while striving to maintain our place at the center of the universe, it has seemed curiously comforting to imagine that the universe of our senses is not all there is. The developments of the past decade or so now give us reason to pause and rethink whether it is so natural that earthly materials should predominate throughout the universe. Perhaps this notion is merely one more extension of the fragile anthropocentric view of the universe which the Copernican revolution first began to undermine.

This book chronicles the process by which we have come to the brink of what may be the ultimate Copernican "revolution." Independent of these nascent rumblings from the field of particle physics, a series of remarkable developments in astronomy and astrophysics—some theoretical and some experimental, some carefully planned and some serendipitous—has shattered any remnants of the notion that what we see is all there is. It began rather innocuously. Working at Mount Wilson in 1933, Fritz Zwicky, an astronomer at the California Institute of Technology, noticed something unusual about the motion of a group of galaxies located tens of millions of light-years away. Their speeds relative to one another were so great that the gravitational attraction due to the visible material in the galaxies should not have been sufficient to hold them together. Yet held together they were!

Since the 1960s, the real significance of Zwicky's initial observation has begun to emerge. Systematic observations of systems as diverse as "dwarf" galaxies some 100,000 times smaller than our own, to giant clusters containing hundreds of galaxies as large or larger than ours, have established beyond a doubt that "something else" apparently resides in these systems whose mass is large enough to hold them together gravitationally. There is now overwhelming evidence that more than 90 percent of the entire mass within the visible universe is made of material that is invisible to telescopes. The gravitational pull of this "dark matter" therefore determines the motion of stars in galaxies, of galaxies in clusters of galaxies, and indeed of the universe itself.

It is simplest to assume that this dark matter is made of the same stuff as we are, but that, for one reason or another, it just does not shine. Yet when the recent parallel developments in the fields of particle physics and astrophysics are combined, the findings suggest that this is not likely. Moreover, there are powerful and persuasive theoretical arguments that arise from these findings which imply that even the inferred dark material around galaxies—some ten times as abundant as the visible material—is still only a small part of a vast cosmic "sea" of dark matter containing perhaps one hundred times as much material as is visible directly through telescopes. This would mean that almost certainly the dark matter and the visible matter in the universe are not made of anything even remotely alike.

How much farther can we fall? If these arguments are correct, then it seems very likely that we, and all that we can see, form an insignificant bit of "noise"—a cosmic afterthought, so to speak. Such arguments maintain that the dominant material governing the dynamics of the universe, from the formation of individual galaxies to the large-scale expansion of the universe, is made of something quite unlike the material we know on earth. Even if these ideas are not strictly correct, much of modern cosmology will still need revision. In either case, our picture of the universe will dramatically change.

It may seem profoundly arrogant for theoretical physicists to make such a claim—to assign a whole new physical character to the universe before we have even directly observed anything—merely to satisfy theoretical whims. But what makes this whole business so exciting is that we may be on the verge of discovering the identity of dark matter. Experiments are under way to close this final gap—to "see" this dark matter directly. And this should be possible if the dark matter is made out of "exotic" (that is, as yet undiscovered) particles, as predicted by particle physicists. If they are right, these exotic dark matter particles will be "in here" as well as "out there," traversing the earth, terrestrial laboratories, and even ourselves. The rewards of this search could be spectacular: a window on the universe at the earliest instants of creation, an understanding of its destiny, and finally, an understanding of the formation of all the structure we observe . . . for there is unquestionable evidence that the formation of everything we see is governed by that which we now cannot.

The book is about the search for dark matter. Because this search is inextricably linked to the dramatic revolutions in particle physics and astrophysics which have completely altered the way we perceive the universe on fundamental scales, the subject of this book is also much broader. Whether or not exotic dark matter survives the test of experiment, many of the ideas I describe in this book will form an integral part of whatever picture of the universe emerges in the future.

Nevertheless, because of the tentative nature of several of the conclusions presented here, I had some concerns about writing this book. I find distasteful those attempts, in some popular science writing, to interest people in modern science by portraying all new developments as intellectually profound, or as decisive as, say, the great revolutions in physics in the early twentieth century (and inevitably laying the mantle of Einstein on the physicists involved). I make no such claims here. The existence of dark matter in the universe neither presumes nor precludes the existence of new laws of nature. The intellectual baggage required to understand many of the results I shall describe is not too burdensome, and the theoretical ideas underlying much of what I will talk about are not necessarily as "deep" as those associated with the development of relativity or quantum mechanics, or even those associated with some of the leading research in particle and condensed matter physics. However, in importance and possible impact on our way of understanding the world, few fields of inquiry carry such potential as the search for dark matter. The interface between particle physics and cosmology makes for a unique combination of "simplicity" and significance, particularly amenable to popular description.

My discussion here is somewhat personal. I am not an unbiased observer. A fair portion of my own research has been devoted to the questions I shall discuss here. My own biases will be apparent, although I have made every effort here to justify them on the basis of experiment. Also, because the field of cosmology is of present necessity very speculative, I have made a special effort to emphasize not only the logical process by which the current wisdom has been derived, but also the *assumptions* that underlie this process. I have no doubt that certain of these assumptions may in the future prove to be wrong; but that is what makes part of the search so exhilarating.

Moreover, while it may be true that some of the conclusions in this book could be the pinnacles of some shaky houses of cards, the observations, arguments, and developments that lead to them are very much a part of modern physics and astrophysics. Their legacy will remain, no matter what form the answers to these questions ultimately take. I hope to convey the excitement of the recent convergence of "macrophysics" and "microphysics," which has captured the imagination of many practitioners in both fields. If the dark matter question can spur others' imaginations further to visualize and explore the ways in which particle physics and astrophysics are refining and unifying our vision of the physical world, or even spark people's creative energies in nonscientific directions, this work will have been worth the effort.

This book is an outgrowth in some sense of two "popular" articles I wrote on the subject of dark matter. The first was for the Gravity Research Foundation Prize competition in 1984, and the second was a longer piece for *Scientific American* in 1986. As the book now stands, however, it is really the product of many minds. It would not have appeared were it not for a remarkable group of individuals—friends, colleagues, and research collaborators—with whom I have had the pleasure to interact, and from whom I have learned much of what I relate here. The list of people I want to thank includes, but is not limited to, Larry Abbott, Frank Accetta, Blas Cabrera, David Caldwell, Sidney Coleman, Stirling Colgate, Marc Davis, Pierre Demarque, Savas Dimopoulos, Margaret Geller, Roscoe Giles, Sheldon Glashow, Alan Guth, Lawrence Hall, Gary Hinshaw, Craig Hogan, John Huchra, Rocky Kolb, Richard Larson, Don Morris, Gus Oemler, Jim Peebles, Joe Polchinski, John Preskill, Joel Primack, Martin Rees, Bernard Sadoulet, David Schramm, Irwin Shapiro, Ed Turner, Michael Turner, Alex Vilenkin, Steven Weinberg, Simon White, Frank Wilczek, Mark Wise, Mike Witherall, Ed Witten, and Michael Zeller. I am also grateful to the Harvard Society of Fellows and the Physics Department, as well as the Yale Physics and Astronomy departments and the Harvard-Smithsonian Center for Astrophysics, for their support of my research during the germination period of this work.

The entire first draft of this book was written while I was on leave from Yale. I would like to thank the university for a Senior Faculty Fellowship that allowed me to devote full time to this work and other research projects. I would also like to thank Tom Appelquist, chairman of the Yale Physics Department, for arranging this leave and for his support of my research. I also thank the following institutions and individuals for their hospitality and support: Jim Matthews and the Physics Department at Mount Allison University in Sackville, New Brunswick; Bernice Kelley of Amherst, Nova Scotia; the Aspen Center for Physics in Colorado; the Institute for Theoretical Physics at the University of California, Santa Barbara; the National Science Foundation; the Department of Energy; and NASA.

I am extremely grateful to colleagues from many institutions for taking time to provide me with graphic and photographic material for this book; the photo and figure credits list these individuals separately. In addition, I thank the Yale Beinecke Rare Book and Manuscript Library, the British Museum, the *Physical Review*, the *Astrophysical Journal*, *National Geographic*, *Nature*, *Science*, and the

Mount Palomar Observatory for granting me permission to reproduce material in their possession.

For a critical reading of all or parts of various versions of this manuscript I thank Blas Cabrera, Marc Davis, Feza Gürsey, Gus Oemler, and Frank Wilczek. In addition, I thank Suzanne Wagner at Basic Books for her marvelous production work and Adrienne Mayor for her copyediting of the manuscript. Of course, I take complete responsibility for any errors that remain.

Two people at Basic Books played an active role in bringing my writing of this book about. Martin Kessler, the president of Basic Books, first encouraged me to write about physics for a broad audience and then allowed me to put aside another project in order to complete this work. Richard Liebmann-Smith, my editor, continually provided support, advice, humor, and free lunches throughout this project.

Last, my wife, Kate, and my daughter, Lilli, through their love and support, influenced this work more than they know.

PART ONE

The Stuff of Matter

CHAPTER ONE

Making Something Out of Nothing

Quintessence: (ancient philosophy) quinta essentia, the fifth essence, the material of the stars, forming heavenly bodies and pervading all things; in contrast to the four elements (fire, air, water, and earth) in which all other matter was thought to exist.

The Fifth Essence of Aristotle was not the first manifestation of an all-pervading ethereal substance thought to permeate both the heavens and earth, nor was it the last. The notion of a universal background of invisible, ephemeral material, of which the dark matter and energy of elementary particle physics is the intellectual heir, extends as far back as some of our earliest written records. The history of this idea reveals some fascinating aspects of our changing concepts of the cosmos. From its beginnings in myth, through its rationalization in philosophy, to its utility in science, the idea of an all-pervasive essence appeared in many different places and times. Emerging knowledge and traditions have shaped each incarnation, providing subtle new guises. Indeed, speculation about the nature of the cosmos has gone hand in hand with our advancing knowledge of the "stuff" of matter here on earth. These advances taken as a whole trace out an amazing intellectual tapestry fashioned by the likes of Aristotle, Descartes, and Newton. Against this vivid tableau, the developments of recent years seem remarkably fresh in their empirical and theoretical bases and at the same time eerily familiar.

"Nothing can be created out of nothing,"[1] the Roman poet Lucretius declared. And indeed, creation from nothingness appears to have been anathema to ancient visionaries. Dispensing with nothingness, however, requires one to propose a replacement. Myths in cultures from ancient Babylon and Egypt to India and China all refer to an underlying, featureless, eternal substance from which all structure later emerged. These earliest notions about the origin of the universe, or cosmogonies, were inspired primarily by direct experience based on local geography. Civilization flourished where there was water—Egypt along the Nile, Mesopotamia between the Tigris and Euphrates, and so on—so it is not surprising that water plays a central role in creation myths. Witness the following excerpts from the Sumerian Enuma Elish (second–third millennium B.C.) and the Sanskrit Vedas (second millennium B.C.):

> When in the height heaven was not named,
> And the earth beneath did not yet bear a name,
> And the primeval Apsu, who begat them,
> And chaos, Tiamet, the mother of them both,
> Their waters were mingled together . . .
>
> (ENUMA ELISH)[2]

> Non-being then existed not, nor being.
> There was no air, nor sky that is beyond it.
> What was concealed? Wherein? In whose protection?
> And was there deep unfathomable water? . . .
> Without distinctive marks, this all was water.
>
> (*Hymn of Creation,* THE RIG VEDA)[3]

> When the great waters went everywhere, holding the germ, and generating light,
> Then there arose from them the breath of the gods.
>
> (*Hymn to the Unknown God,* THE RIG VEDA)[4]

For early cultures, water clearly held a sacred place as the origin of life. The annual flooding of the Nile valley brought a yearly cycle of activity for the ancient Egyptians, just as the alluvial silt carried by the Tigris and Euphrates, deposited at their mouths in the Persian Gulf, nourished the land and culture of the Mesopotamians. It was logical that water should play a vital role in early ideas about creation and the material nature of the universe. The Egyptians pictured the world as a bank of earth divided by the Nile and surrounded by the Great Circular Ocean. This mass of water had its origins in a vast Primeval Abyss of Waters, personified by the god Nu (or Nun), the source of all beings and things. The Papyrus of Nes-Menu (312 B.C.) describes a creation myth that probably dates back to before the third millennium B.C. (see figure 1.1):

> The sky had not come into being, the earth did not exist, and the children of the earth, and the creeping things, had not been made at that time. I myself raised them up from out of Nu. (version A)

> I am the creator of what hath come into being, that is to say, I formed myself out of the primeval matter, and I made and formed myself out of the substance which existed in primeval time. . . . I made all the things under the forms of which I appeared then by means of the Soul-God which I raised into firmness at that time from out of Nu. (version B)[5]

FIGURE 1.1

This papyrus, dating from 312 B.C., contains the Egyptian creation story, in which the world arises out of a limitless primeval sea, described in another text as "the infinite, the nothingness, the nowhere, and the dark." (Reproduced by courtesy of the trustees of the British Museum.)

According to the Egyptologist J. M. Plumley, this primeval abyss "was unlike any sea which has a surface, for here there was neither up nor down, no distinction of side, only a limitless deep—endless, dark and infinite." Another Egyptian papyrus expressed it as "the infinite, the nothingness, the nowhere, and the dark." Plumley points out that the "basic principle of Egyptian cosmology may therefore be said to be the Primeval Waters, which existed before the beginning and which would last forever."[6]

The creation story in Genesis is remarkably similar to these other early creation stories:

> In the beginning God created the heaven and the earth. Now the earth was unformed and void, and darkness was upon the face of the deep; and the spirit of God hovered over the face of the waters. . . . And God said: "Let there be a firmament in the midst of the waters, and let it divide the waters from the waters." And God made the firmament, and divided the waters which were under the firmament from the waters which were above the firmament; and it was so. And God called the firmament Heaven.

There are many reasons, beyond its intimate ties with the birthplaces of civilization, why water should have been pictured as the universal mother substance. Tasteless, odorless, colorless, it takes the shape of whatever contains it—in short, it seems to have no characteristics that appear to derive from anything else. Moreover, it is changeable and ubiquitous. It is the only substance to exist in all three of its phases (liquid, solid, and gas) under natural terrestrial conditions, and it can be observed under normal conditions to freeze or evaporate. In its gaseous and liquid forms it is omnipresent. It appears everywhere, in all living things, in rocks, in soil, and in the air.

Accordingly, in the sixth century B.C., when the Greeks first turned from creation myths similar to those of Egypt and Mesopotamia (for example, the "earth-encircling" river Okeanos, referred to by Homer) to rational attempts to explain the origin of the universe and determine its "raw materials," water again appeared as an underlying "principle."

Western philosophy apparently emerged in the Ionian coastal town of Miletus in the sixth and fifth centuries B.C. through the writings of three men: Thales, Anaximander, and Anaximenes. Although the records of their ideas exist primarily through secondary sources, it is clear that each of them in turn devoted powerful intellects to attempt a rational explanation of the world. Inevitably, then, each had to come to grips with two fundamental questions: What makes up the universe? and How did it achieve its present form?

The first of these philosophers, Thales, flourished in about 585 B.C. He was a consummate scientist and mathematician of his day, but he was also lucky. He is said to have been the first to predict a solar eclipse. If this is true, then one must assume on the basis of the astronomical observations available at the time (primarily Babylonian) that he would have been able only to foresee the likelihood of an eclipse within some broad time frame. The fact that it occurred during an important battle, and that it was visible in Ionia (now modern Turkey), was the kind of fortunate circumstance that has served to make many a latter-day scientist famous. Unfortunately, Thales appears also to have established a stereotype of scientists which remains to this day. According to Plato, "a witty and attractive Thracian servant-girl is said to have mocked Thales for falling into a well while he was observing the stars and gazing upwards; declaring that he was eager to know the things in the sky, but that what was behind him and just by his feet escaped his notice."[7]

It was Thales who first attempted to establish a rational cosmogony. He believed that there must be some fundamental substance out of which all structure emerged. According to Aristotle, Thales declared that

> the [underlying] principle is *Water* (and on account of this he also declared that the earth rests on water), perhaps coming to this belief by observing that all food is moist and that heat itself is generated from the moist and is kept alive by it . . . and because the seeds of all things have a moist nature, and water is the principle of the nature of moist things.[8]

Thales was no doubt influenced by his predecessors in arriving at this conclusion. Nevertheless, it is significant that he chose a familiar and observable substance as the germ of all structure. Whether such is in fact the case has been debated for more than two millennia.

It has been remarked that because our only vision of Thales' philosophy comes through the words of Aristotle and other later writers, we cannot say whether Thales actually believed that the visible world *is* water, or whether, in a view more characteristic of the Egyptian and Babylonian mythological antecedents, that the world arose out of an indefinite expanse of primordial waters. He might have believed that water is a hidden constituent of all things, as was proposed for other substances later.

The former interpretation, that the visible world is water, runs into logical problems, as was immediately stressed by Thales' younger contemporary, Anaximander, who pointed out that earthly materials have distinct properties, of which he named two pairs: hot and dry, cold and wet. Dry materials can be distinguished from moist materials by the absence of water. How then can the properties of these materials be explained in terms of the *absence* of a primeval substance? In a world made of water, all things should end up wet (an argument laid down at length later by Plato and Aristotle). By this reasoning Anaximander asserted that any fundamental substance must be neither hot, dry, cold, nor wet. He is said to have proposed instead that something different, which he called "boundless" or "indefinite," must be the source of all things (again we have no direct record of Anaximander's actual texts). Scholars disagree over Anaximander's actual terminology: some sources give *ἄπειρσυ* ("indefinite") while others use *ἀρχή* ("source"). Anaximander is said to have stated that this material

> is neither water nor any other of the so-called elements, but some different, boundless nature, from which all the heavens arise and the worlds within them; out of those things whence is the generation for existing things, into these again does their destruction take place, according to what must needs be.[9]

We will see many connections between the "indefinite" of Anaximander—yet another paradigm for a universal background medium—and the more clearly defined but no less ephemeral "vacuum" of modern physics. As we shall discover, one of the most popular theories in modern cosmology suggests that this "vacuum" may have been the source of all matter, observed and unobserved. Certainly there is something poetic about a featureless fundamental essence. A similar notion was introduced in about the time of Thales and Anaximander, in Lao-Tzu's Tao Te Ching philosophy:

> The way is empty, yet use will not drain it.
> Deep, it is like the ancestor of the myriad creatures.
> Blunt the sharpness;
> Untangle the knots;
> Soften the glare;
> Let your wheels move only along old ruts.
> Darkly visible, it only seems as if it were there.
> I know not whose son it is.
> It images the forefather of God.[10]

Yet Anaximander could claim a significant advantage in his theory of a neutral essence. His two sets of primary "opposites" in matter might result from the separating out of opposing qualities from this otherwise neutral material. The infinite nature of his "indefinite" material was also important, for it could explain why the earth was stable. If the earth was located in the middle of an infinite space, it would have "no more reason to go upwards than downwards or sideways . . . so of necessity it stays where it is."[11] The modern reader may find this argument less than compelling, but it was a vast improvement over Thales' "bed of water." In fact, the idea is not so different from modern ideas. By tying the motion of the earth to the global nature of the universe, Anaximander anticipated the arguments formulated some twenty centuries later by Ernst Mach, whose work had such influence on Einstein's derivation of the general theory of relativity.

Anaximander also probably established another precedent—the mathematical formulation of the laws of nature. His universe involved spherical rings; the innermost ring contained the stars, and the outer ring contained the moon and sun. The fact that he could reconcile this system with the fact that the moon obscured the stars as it passed in front of them indicates that the Greeks did not let empirical facts get in the way of a beautiful idea. On the other hand, he allowed the circles of the sun and moon to "lie aslant," which suggests that observations of the sun's yearly motion relative to the motions of the stars influenced his model building. In Anaximander's universe, the sizes of these rings were in fixed mathematical ratios to each other.

This symmetrical, or geometric, structure of Anaximander's theory is the first example of a mathematical style that would later be taken up by his successors: Pythagoras, whose "music of the spheres" would in turn be revived by Kepler; and Plato, whose geometric forms gave a mathematical framework to the universe. Of course, the mathematics of the ancient Greeks did not always correspond directly with empirical data. This concordance would require twenty centuries to evolve. The Greek "mathematics-for-the-sake-of-mathematics" attitude drove Aristotle to comment, "Philosophy has turned into mathematics for present-day thinkers, despite their claim that mathematics is to be treated as a means to some other end."[12] His complaint could easily be mistaken for a statement written today by a disgruntled physicist in this era of "superstrings" and "twenty-six dimensions." Nevertheless, we have come to recognize that it is only through mathematical formulations that a correct picture of physical reality can evolve, and it was this that Anaximander had in mind even then.

Between them, Thales and Anaximander produced the two fundamental opposing ideas that would set the general framework for future debates about a primordial essence. Is this essence like normal matter, or is it composed of something new? Their immediate successors refined those ideas, and in the process added new ingredients that would later play key roles in the modern versions of these notions.

Anaximenes, who followed Anaximander in the Ionian school, incorporated his predecessor's desire for a neutral, infinite base, but returned to Thales' preference for the familiar and the definite. Anaximenes chose air (or *pneuma*) as his universal source. Not only did this selection overcome one of the chief arguments against water, since it was not characterized by any specific opposite, but air was apparently just as abundant as water. Moreover, for the first time Anaximenes proposed methods for fabricating other components of the world from air: compression and rarefaction. It was claimed that when it is extremely thin or rarefied, air becomes fiery. Compression, on the other hand, first results in winds, then clouds, then water. Finally, further condensation produces earth, and then stone.

Anaximenes' *pneuma* was more than mere atmosphere; it had the germ of all creation and was therefore divine. As air gave breath to life, so *pneuma* maintained the stable pattern of existence. By assuming that different substances result from different configurations of the same material, he anticipated the atomic hypothesis. Nevertheless, his *pneuma* provided the continuum that held matter together; Aristotle and others who argued against a void would later reinstate this notion.

Although Anaximenes' *pneuma* theory had some of the virtue of explaining process as well as structure, it suffered from the same flaw that plagued the theory of Thales. The choice of air, or water, as a basic form of matter seems arbitrary. Why not choose earth, or fire, or anything else? Indeed, fire had been adopted by Heraclitus and earth by Xenophanes.

This consideration led in two different directions. Empedocles, who lived in the Sicilian town of Acragas in the fifth century B.C., is generally credited with originating the famous four-element theory, wherein earth, air, fire, and water are together considered as the fundamental, enduring elements of creation. He proposed that all materials and change could come about by the intermingling of these elements, and suggested two opposing "principles" responsible for such processes: Love and Strife. Love brings the elements together. Strife separates them. The terms for the principles are hardly scientific by modern standards, yet Empedocles' theory comes closest to suggesting some concrete methods by which one might understand both structure and change. His principles, which imparted "motion" to the universe, approach the modern idea of force. Moreover, his "cosmic cycle," during which the universe is alternatively dominated by Love and Strife, allowed a nonstatic but eternal universe. This possibility, first enunciated in Empedocles' work, has been a part of cosmological theory ever since.

His contemporary Anaxagoras, on the other hand, suggested what I view as a forerunner of the "bootstrap" theory of elementary particle physics which flourished in the 1960s: he suggested that "everything comes from everything." This implied that all materials were equivalent and any one substance could be extracted from any other, and thus that there were an infinite number of elementary substances (or none at all, depending upon how you look at it). He justified this idea by suggesting that all things were made of an infinite number of infinitesimal "particles" whose rearrangement produced all materials—a kind of primitive atomic hypothesis. Thus the characteristics of matter were not dependent upon the characteristics of some elementary substance but rather on the particular arrangement of these objects that themselves bore no resemblance to the objects they formed.

These men were as colorful as their theories. Empedocles was very likely a "medicine man" who became convinced of his own immortality. In what has to be considered one of the most extreme attempts at self-promotion, he supposedly met his end by leaping into the crater of the volcano of Mount Etna, in the hope of furthering his reputation for divinity. Anaxagoras lived in Athens where his students included the politician Pericles and the playwright Euripides. He was charged with (but acquitted of) impiety for making perhaps the first empirically based observation that the materials of the stars and the earth are the same. He had examined a meteor and concluded that stars and planets alike are merely burning rocks. This important observation probably influenced his belief in the "sameness" of all matter. Finally, in an act that should endear him to children everywhere, as he lay dying, he was said to have asked that the children of his town be allowed a holiday to commemorate his death every year.

In the philosophies of both Empedocles and Anaxagoras there seems to be no room for any material separate from and perhaps more pervasive than ordinary matter. Nevertheless, both philosophies contained a notion that had surfaced much earlier and that would later be elaborated in the philosophy of Aristotle. This notion would eventually play a pivotal role in the birth of twentieth-century physics. I refer, of course, to the "aether."

As far as I can ascertain, the term *aether* (αἰθήρ) first appears in the writings of Homer. There it refers to the "fiery sky," the upper atmosphere or celestial light, below heaven. It derives from the Greek root αἴθω, "to blaze." It is interesting that this vivid term would later come to epitomize an invisible medium, albeit one that propagated light. Some have argued that the aether of Homer refers to more than just a place, the fiery upper atmosphere of the sun and stars; it may also refer to celestial clarity from heaven which causes objects to be seen clearly. It is in this spirit that its original meaning was maintained through Aristotle and Newton, as we shall see.

The notion that the celestial bodies lay in a region of "fiery air" persisted after Homer's day and appears in the writings of all the early Greek philosophers. The observation that fire rises naturally suggested that the outermost regions of the universe should be the most fiery. It was no doubt this belief that led Anaximander to suggest, against the blatant evidence of the senses, that the circle containing the stars was located inside that of the sun and moon. Since the sun provides more light, it must contain a higher proportion of fire, and hence must lie in the region higher than the stars, which themselves must contain more air. Anaximenes made this line of thought an explicit part of his philosophy, maintaining that the thinner the air, the more fiery it must be. Although Anaximenes thought air to be the divine enfolder of all things, there is no evidence that rarefied fiery air, which he believed formed the heavenly bodies, had any special significance in his philosophy.

It is with Empedocles and Anaxagoras that aether begins to occupy a more special role, just as air (ἀήρ) itself evolved from its more limited meaning of "mist" or "haze" in the Homeric texts. Anaxagoras believed that all things were equivalent and interchangeable, but he proposed that the dominant components in the universe, and therefore in each thing, were air and aether: "For air and aether contained everything, both being unlimited. For these are the greatest items present in all things, both in quantity and in magnitude."[13]

Similarly, although Empedocles sometimes interchanged aether and air in his labeling of the divine atmospheric element (no doubt in part because of the earlier specific poetic meaning of the latter as mist), he treated aether in a special way in his cosmogony. The first element to be separated by Strife from the uniform whole is not air, earth, fire, or water, but the aether. Later aether separates into its components: air and fire.

It was Aristotle who finally and decisively took the step of elevating the aether to be his "quintessence," his fifth essence in the universe. In doing this he maintained its character as the material of the heavenly bodies. However, in a move important for its impact on the term's modern meaning, he no longer viewed aether as a kind of fire but rather as a "unique element, scarcely material in form, within which the operation of geometric law proceeds unclogged by any mechanical aid or impediment."[14] This view is much more in tune with the interpretation, mentioned earlier, of the Homeric aether as a "celestial clarity" from heaven which causes objects to be seen clearly. It also incorporates much that embodied the *pneuma* of Anaximenes while also returning in part to the "indefinite" of Anaximander. The main difference is that although Aristotle's quintessence permeates normal matter, it is fundamentally different from the elements out of which matter is made. In any case, Aristotle's hypothesis established firmly the notion of the aether, which later authors would borrow to describe an invisible background material through which the laws of nature are somehow transmitted.

One must recognize that in proposing such a quintessence, Aristotle was participating in an intellectual debate that raged in the fourth and fifth centuries B.C. about the nature and existence of a void. Though Anaxagoras proposed an atomic-like theory of materials, he was not an atomist. Nowhere did he insist upon a particulate description of matter, or a void between particles. As Empedocles had stated in his somewhat more flowery language:

> Fools—for they have no far-reaching thoughts—who fancy that that which formerly was not can come into being or that anything can perish and be utterly destroyed. For coming into being from that which in no way is inconceivable, and it is impossible and unheard-of that that which is should be destroyed. For it will ever be there wherever one may keep pushing it. . . . Nor is any part of the whole either empty or over-full.[15]

The true atomists, Leucippus and Democritus, for example, believed that if all matter were made of elementary indivisible particles, then the region between these particles must contain a void, that is, a region containing nothing. They also argued that without a void between particles, objects could not contract. Moreover, they, and later proponents of a void, maintained that motion was impossible without one. For, to quote Aristotle, "what is full cannot admit anything more into itself."

Aristotle argued that none of these conclusions was correct. He described how global rearrangements, as occurs in the observed bulk flows of liquids, allow qualitative alterations to occur in what is "full." Moreover, he claimed that contraction was possible by the expulsion of things contained within, such as air in water. In fact, Aristotle went further and declared that motion and change would be impossible if a void existed! He reasoned that in a void a body had no preference to move in one direction or another (recall the argument of Anaximander).

To understand Aristotle's reasoning we should recall that a key element of his thinking, and one to which much mental effort has since been devoted, was based on the question of the necessity of some agency or cause for change. Even if air can change to water, as the atomists might suggest, something must direct this change. How could this direction be communicated to the body? This is one reason why he argued that the motion of bodies depends crucially on the properties of the medium in which they are moving (an idea for which Aristotle has received a lot of bad press ever since).

This need for a controlling direction, which both provoked change and allowed the stability of the laws of nature, is what led him to posit a continuum in the universe, permeated by the aether. This fifth essence, then, not only formed the heavenly bodies, which after all exhibited the most elegant and stable motion of all, but pervaded all matter and in so doing ensured the harmonious operation of natural law. Aristotle further imagined a similar material, the *pneuma,* which he proposed as maintaining the stability of living things.

Aristotle's concern with the agency of change, a concern that in part motivated his philosophy, has been shared by many a scientist since. How can one object influence the motion of a distant object? Action at a distance has perplexed and piqued generations of natural philosophers. Newton, the first to formulate a law describing the results of exactly this process, was so troubled by this notion that he could not in good conscience entirely exclude a uniform background continuum material, atomist though he was.

Aristotle's ideas were to influence philosophy for the next two thousand years (although later philosophers such as Maimonides argued in favor of such anti-Aristotelian notions as a void). As spiritual and theological considerations began to become more important, starting with the Gnostics and the Stoics in Alexandria and continuing with the rise of Christian apologists and alchemists in the Middle Ages, philosophy blended once more with mystical notions: the search for spiritual salvation had begun. Thus, we find Aristotle's *pneuma* becoming a "Cosmic World-Soul" of divine proportions by the second century B.C. It was argued that the cosmos itself must have a soul, carried by a universal *pneuma* that "binds all the objects of heaven and earth in a common destiny."[16] This material of the highest grade, an extension of the quintessence of Aristotle, was identified by some to be the Deity itself.

One of the unfortunate by-products of this teleological philosophy was the resuscitation of ancient Babylonian and Egyptian astrology. If the same ethereal essence binds the destiny of heaven and earth, then studying the heavens for clues to one's personal destiny seemed to make sense. What is so remarkable and at the same time so disappointing is that while the philosophical and scientific basis of this ancient mysticism has long disappeared, astrology is still taken seriously.

Rational philosophy, which returned perhaps with Descartes in the early

1600s, accompanied by the physics of Galileo and Newton, which would lead to the toppling of much of the Aristotelian framework, inherited a great deal from its classical ancestry. Not least of this inheritance was the framework of debate among the Ionians of the fifth century B.C. What are the fundamental components of matter? Is there a void? What composes the heavenly bodies and what determines their motion?

Descartes proposed a cosmogony with certain features that bear a striking resemblance to Aristotelian ideas; other features are uncannily modern. Descartes rejected the notion of a void, insisting that space could be defined only by the existence of bodies. He imagined that motion was primordial, given by God, and that the "quantity of motion" was conserved. (The modern concept of the *conservation of momentum* derives from this early idea.) Since there were no empty spaces when God set the universe in motion, regions began colliding, which Descartes imagined led to the formation of large eddies, or vortices, moving in circles (see figure 1.2). At the interfaces of these eddies, adjacent regions were ground away, producing a vast quantity of aetherlike matter. The material inside the vortices aggregated to form observed matter. The tenuous aethereal material, which he supposed to have no weight, filled in the spaces in normal matter. In this way Descartes could argue that two bodies of the same size could weigh different amounts, having the same net amount of matter but different proportions of massless aether compared to normal massive material.

One element of Descartes's supposition of the existence of a uniform background of ethereal material is distinctly modern. It was designed to explain a specific physical observation: the difference in weight of various materials of uniform volume. This supposition is also distinctly classical in that it provides merely a *plausible* explanation of this phenomenon, not one that is testable or that follows irrevocably from some other line of empirically based arguments.

The modern concept of the aether, usually credited to Isaac Newton, was in fact developed roughly contemporaneously and published earlier (in 1690) in *Traité de la Lumière* (Treatise on Light), by the Dutch astronomer and physicist Christian Huygens, whose surname has defied proper pronunciation for generations of physics students ever since.

Among his achievements, Huygens discovered one of the moons of Saturn which he named Titan, invented the first modern pendulum clock powered by weights, and most important for our purposes, proposed a *wave theory* of light. Because Newton believed that light was made of particles, and because of Newton's eminence at the time, Huygens's wave theory did not immediately catch on. Nevertheless, the principle proposed by Huygens to understand the propagation

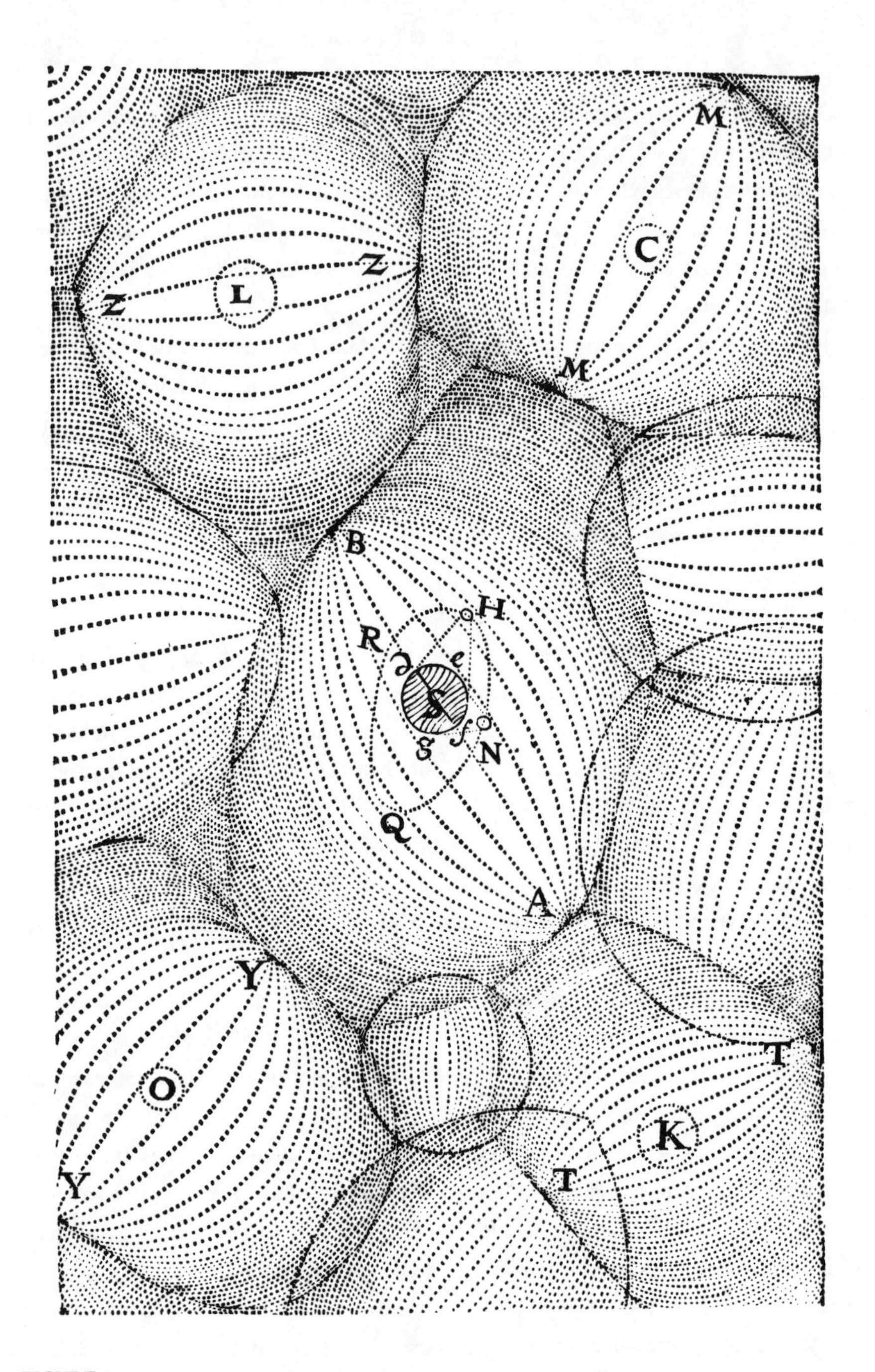

FIGURE 1.2

Descartes's vortices not only carried the planets around in closed orbits, as depicted here, but also led to the formation of large quantities of aethereal matter. (From R. Descartes, *Principia Philosophiae* [1644], courtesy of the Beinecke Rare Book and Manuscript Library, Yale University.)

of light waves is essentially that same principle used today to calculate the behavior of waves of all kinds. Moreover, in a very real sense, the mathematical method proposed by Huygens to explain the propagation of light led to the methods that would later form the basis of quantum mechanics and our modern understanding of the elementary structure of matter.

To Huygens it seemed inescapable that light was made of waves. His chief argument was that two light waves impinging upon one another from any direction will cross "without hindrance." Thus, in his words,

> One can well understand that when we see a luminous object, it cannot be due to the transport of some matter, which travels between the object and us, just as a shot or an arrow traverses the air: for assuredly this would impugn too greatly these . . . properties of light.[17]

Waves, whether they are sound waves or water waves, can be observed to pass through each other unscathed. If this were not the case, we should be totally unable to carry on a conversation across a crowded room. But waves were known to be caused by vibrations; therefore, in order to propagate they require some medium that can vibrate. This led Huygens to what appeared to be an inescapable conclusion. If light is a wave, there must be some medium permeating all space, even an apparent "vacuum," which could vibrate in response to a source of light and thereby allow the light signal to propagate. Here is how Huygens described his line of reasoning:

> Now if we examine what could be the material in which the movement coming from luminous bodies spreads, which I call Ethereal, one will see that it cannot be the same material which serves for the propagation of sound. For one finds that this material is really the air which we feel and which we breathe, which having been removed from some place, leaves behind the material which serves to propagate light. This can be proved by enclosing a body emitting sound in a glass vessel, from which one removes the air by the machine which Mr. Boyle has given us, and with which he did such beautiful experiments. But in performing this . . . it is necessary to take care to place the body on cotton or on feathers, so that it cannot communicate its vibrations to the glass vessel in which it is enclosed, or to the machine, a fact which has so far been neglected. For when all the air is then removed, one no longer hears any Sound from metal, even when struck.

One sees from this that not only our air, which does not penetrate through the glass, is the matter in which Sound spreads; but also that it cannot be this same air, but another material in which light spreads; since if the air has been removed from the vessel, the light does not stop traversing it as before.[18]

Since we see light from the sun and stars, Huygens had to presume that this "ethereal matter" was universal:

And I do not believe that this movement can be better explained, than by supposing that that which is inside luminous bodies which are liquid, like a flame, and apparently the sun and stars, is composed of particles which float in a material which is much more subtle, and which agitates them with great rapidity, and makes them collide with particles of the aether, which surround them and which are much smaller than they.[19]

Here, for the first time, the aether did not merely provide a plausible explanation of some phenomenon. Now it appeared *essential* if light were to propagate as a wave through the vacuum. Newton too was driven to this conclusion, for slightly different reasons. Whereas he believed that light was particulate, he proposed that light particles excite vibrations in a medium, producing waves that overtake the light and lead to "fits of easy reflexion and transmission" in order to explain apparent wavelike phenomenon such as the famous "Newton's Rings" that he observed. These alternating rings of color seen when a curved lens is placed on a flat surface are similar to the familiar bands of color observed when a thin oil slick floats on the surface of water. These patterns are reminiscent of "interference" patterns formed by sound waves or water waves. Observing that light also communicates heat, Newton described his motivation for the aether in a passage that has remarkable parallels to the work of Huygens:

If in two large tall cylindrical Vessels of Glass inverted, two little Thermometers be suspended so as not to touch the Vessels, and the Air be drawn out of one of these Vessels, and these Vessels thus prepared be carried out of a cold place into a warm one: the thermometer *in vacuo* will grow warm as much, and almost as soon as the Thermometer which is not *in vacuo*. And when the Vessels are carried back into the cold place, the Thermometer *in vacuo* will grow cold almost as soon as the other Thermometer. Is not the Heat of the warm Room convey'd through the *Vacuum* by the Vibrations of a much subtiler Medium than Air, which after the Air was drawn out remained in the *Vac-*

> *uum?* And is not this Medium the same with that Medium by which Light is refracted and reflected, and by whose Vibrations Light communicates Heat to Bodies, and is put into Fits of easy Reflexion and easy Transmission? . . . And is not this Medium exceedingly more rare and subtile than the Air, and exceedingly more elastick and active? And doth it not readily pervade all Bodies? And is it not (by its elastick force) expanded through all the Heavens?[20]

Newton's use of heat here was not strictly correct, because, as anyone who has used a glass vacuum thermos bottle to keep liquids warm knows, heat is communicated not just by light but also by the motion of air. Nevertheless, the fact that a thermometer in a vacuum heats or cools at all, regardless of how fast it does this compared to a thermometer in air, indicates that at least the heat communicated by light can travel in a vacuum.

His analysis may have been slightly flawed, but Newton was able to use his considerable mathematical ability, combined with his empirical good sense, to derive quantitative properties for this aether. Given the elasticity and the density of a medium, Newton could calculate explicitly the speed of a wavelike disturbance in it. By then comparing the known speed of sound with a reasonable estimate that he made for the speed of light, he calculated that the aether must be about 490 billion times more elastic, in proportion to its density, than air!

Newton had one more reason for wanting an aether to exist. In spite of the phenomenal success of his universal theory of gravitation, he was still bothered by the fact that his theory seemed to require "action at a distance." In other words, two bodies separated by a great distance must somehow "know" of the existence of each other and feel an attraction. This rather deep philosophical problem has been associated with a number of the major developments of twentieth-century physics, and Einstein himself said that what most aroused his admiration for Newton was his ability to see the flaws in his own intellectual edifice so clearly. In any case, the aether gave Newton a possible resolution of this problem. He proposed that the great elasticity of the aether might combine with a variation in its cosmic density to produce the gravity that attracts the moon toward the earth, and the earth toward the sun. This proposed use of the aether to produce the "harmonious" motions of the heavenly bodies is particularly reminiscent of Aristotle's quintessence.

If Newton's motivation for the aether was slightly less compelling than Huygens's, his authority was such that the notion of an aether became ingrained in physics. In the nineteenth century, after the experiments of Heinrich Hertz and the triumphant electromagnetic theory of James Clerk Maxwell, it was confirmed that light was indeed a wave phenomenon; and the aether hypothesis of Huygens seemed all the more compelling.

In one of the most elegant developments in theoretical physics, Maxwell showed how all the phenomena associated with electric charges and currents as well as magnetic fields in the laboratory could be predicted with complete accuracy using a succinct mathematical framework. The four "Maxwell's equations" represented a unification that was the culmination of literally hundreds, if not thousands, of years of investigation. Certainly, one of the most amazing aspects of his theory was a new prediction that it made. As the four equations show, if one "jiggles" an electric charge, Maxwell's theory predicts that oscillating electric and magnetic fields will be produced which travel outward together as a wave. Such an "electromagnetic" wave was in fact observed by Hertz. Even more remarkable is the fact that Maxwell could calculate the speed of this wave using only those fundamental constants of nature which could be measured in the laboratory and which give the magnitudes of (1) the force between two charges at a fixed distance, and (2) the force between two magnets at a fixed distance. The speed of the waves that Maxwell calculated turned out to be exactly (within the accuracy of measurement then available) that of a beam of light! Light was finally understood to be an electromagnetic wave.

While Maxwell's proof of the wave nature of light immediately reinforced the notion of an aether in which the electric and magnetic disturbances could propagate, it also led to the aether's eventual downfall. Einstein, while still in his teens, was motivated by Maxwell's analysis to ask the crucial question that not only led to his special theory of relativity but also irrevocably did away with the notion of an aether for light to propagate in.

Einstein noticed that the particular elegance of Maxwell's derivation of the speed of light also implied an apparent paradox. This derivation suggested that *as a law of nature* whenever a charge was jiggled, an electromagnetic wave should travel out with a speed that could be exactly calculated, given two fundamental constants of nature. Now, it had been universally accepted since the time of Galileo that the laws of physics should be the same for different observers in different places, or for observers moving at a constant velocity. For example, when you are traveling in an airplane moving very fast at a constant speed, as long as you are not looking out the window, nothing happening inside the cabin should alert you to the fact that you are moving: liquids pour into glasses the same way they do in your kitchen, and so forth. Hence, we expect that the constants of nature should be the same when measured in a laboratory inside the plane as they would be when measured on the ground. Thus, since Maxwell's calculation derived a speed for these waves which depended just on these constants of nature and nothing else, the speed of light should be independent of whether one is moving when one jiggles an electric charge, or whether one is standing still.

This reasoning is incompatible with the idea that light travels in a background medium. Imagine that you throw a stone in a pond to watch the waves emanate. If you are running when you throw the stone you would not expect the waves to have a different speed in the water just because you were running when you threw the stone. Then, there should be nothing surprising about the fact that the waves will have a different speed *relative to you* depending upon whether you are running or standing still. In the former case their speed relative to you will be the sum of their speed in the pond plus your speed relative to the pond. In the latter instance, since you are at rest with respect to the pond, their speed relative to you will simply be the speed of waves in the pond. Another way of stating this is to point out that it is easy to determine whether you are at rest or not: just measure the speed of water waves relative to yourself.

This behavior of water waves is exactly the opposite of what Maxwell's equations suggest for light. Instead, they indicate that no matter how fast you are running, when you jiggle a charge, you should see an electromagnetic wave move out with the same speed relative to you, independent of your motion, determined solely by the underlying constants of nature. We must either accept this or resign ourselves to having different constants of nature for every person in relative motion. Einstein rightly concluded that the latter alternative was unacceptable. He therefore made the bold assumption that the speed of light must have a constant value for all observers, regardless of their state of motion. To allow different observers moving relative to one another to measure the same light ray so as to have the same speed *relative to themselves,* Einstein was forced to allow the possibility that different people's measurements of time and distance could differ if they were in relative motion. Thus, his special theory of relativity was born.

This book is not the place to examine all of the startling implications of relativity. Nevertheless, there are two implications I want to discuss here that are of particular importance for our purposes. First, Maxwell's equations, and the theory of relativity which is based on them, are inconsistent with the idea that light propagates in some fixed universal background medium. Second, relativity suggests that it is impossible for two observers in relative constant motion to determine which observer, if either, is at rest. These two implications clearly go hand in hand. For if light travels at a speed v, in a fixed background medium, which is itself at rest (as the water in the pond), one could determine who among two observers in relative motion was at rest—namely, the observer with respect to whom the light was observed to be traveling at speed v.

Strange as Einstein's conclusions may have appeared when he wrote them down in 1905, or indeed still appear today, an experiment actually performed earlier, in 1887, by the great American experimental physicist Albert Michelson and his colleague Edward Morley, produced results consistent with Einstein's "prediction." (This work was done in laboratories at the Case School of Applied Science and Western Reserve University, which merged later in 1967 to form Case Western Reserve University, where I am now privileged to hold the position first held by Michelson.) They directed two light beams in two perpendicular paths of identical lengths. If one of the paths was moving along with the earth relative to the "aether," while one was perpendicular to its motion, the beams should arrive at their destination at calculably different times, well within the resolution of the scientific apparatus. They found no such effect. The light traveled at the *same* speed in both directions, no matter in which direction they pointed the two beams. There was no evidence that the light was propagating in any background material through which the earth was moving.

Since that time, Einstein's special theory of relativity has been tested innumerable times. Today many predictions of his theory, including the slowing of time for moving observers, are observed daily in the high-energy accelerators built by particle physicists to study the fundamental structure of matter. Since this theory is incompatible with the requirement that light travels in an aether, no large-scale effort has been exerted to extend the pioneering work of Michelson and Morley in probing for such a medium. One might thus date 1905 as the end of the aether story.

CHAPTER TWO

Filling The Void

The idea that a uniform background of "invisible" material might pervade the universe lay largely dormant for almost sixty years after the remarkable developments in physics at the turn of the century. In 1964, however, in Holmdel, New Jersey, a serendipitous series of observations would change the face of cosmology. The story of the discovery of the cosmic microwave background has been told in some detail in the excellent book *The First Three Minutes* by Steven Weinberg.[1] Quite by accident, two young astronomers, Arno Penzias and Robert Wilson, using an unusual radio antenna, discovered an unexpected nondirectional radio noise. They were unable to get rid of the signal, even by removing the pigeon droppings that had accumulated inside of the antenna horn. They did not know it at the time, but they had discovered the "afterglow" of the cosmic Big Bang. The radiation they were observing had been emitted when the universe was only about 100,000 years old, and since then had been traveling freely, waiting to interact with their antenna.

To understand the context and significance of their observation we need to backtrack a bit. The time will not be wasted, however, since many of the background ideas I discuss here will reappear in more detail later.

First, if we observe out in all directions (except along the axis of our own galaxy where our vision is obscured) the universe appears *isotropic*. That is to say, there are roughly an equal number of distant galaxies located in every direction. Next, the universe appears to be homogeneous. This implies that if we look at more and more distant galaxies, the density of galaxies appears relatively uniform. Finally, the visible universe appears to be expanding. By this I mean that all the galaxies we can see are receding from our galaxy. More than this, the velocity of their recession is on average directly proportional to their distance from us. (I will discuss in the next chapter how we measure distance and velocity in space.) This proportionality may seem mysterious, but it is a simple consequence of a system that is expanding. Consider a rubber box, one meter square. Fill it with dust particles and let the whole system expand uniformly, doubling in size in one second. Now, if the expansion is uniform, then two dust particles that are initially one centimeter apart will be two centimeters apart after one second. Thus the velocity of one relative to the other will be one centimeter per second. On the other hand, two dust particles initially one meter apart will be two meters apart after one second. Thus their relative velocity will be one meter per second. Therefore the relative velocity of dust particles will be directly proportional to their distance of separation.

The nature of this recession was first convincingly demonstrated through years of observation by Edwin Hubble at Mount Wilson Observatory in the 1920s and 1930s. The expansion implied by this recession is now called the *Hubble expansion,* and the parameter that describes how fast objects are moving apart is called the *Hubble constant*. This fundamental number in cosmology happens to be measured today to within a factor of 2. It turns out that objects separated from one another by roughly 3 million light-years (a light-year is the distance a light ray travels in a year, approximately equal to 1,000,000,000,000,000,000 [= 10^{18}] centimeters) are moving apart today at between 55 and 80 kilometers per second (see appendix A). I say today because the expansion rate of the universe is not constant over cosmic time-scales, and one of the chief challenges to cosmologists is to determine what the future course of the expansion will be.

Now, on large scales, since bulk matter is electrically neutral, the only force that affects the expansion is gravity. Galaxies that are very far apart from each other act like dust particles in the expanding box discussed earlier. It is fairly easy to calculate what will happen if one begins expanding a homogeneous isotropic volume of dust and then allows it to evolve only under the mutual attraction of the dust particles. As one can imagine, this attraction will cause the expansion to slow down. Whether it will stop depends on how many and how heavy the dust particles are. Indeed, if we are to determine the future evolution of our own universe we must determine these conditions for our dust particles—

the galaxies and clusters of galaxies we observe. (This issue, central to the question of "dark matter" in the universe, is discussed in chapters 3 and 4.)

In any case, knowing that the universe is expanding, and being able to calculate in general how it is expanding, means that we can deduce that at earlier times it was smaller. Next, by working the equations backward, we can calculate how much smaller and denser the universe was at earlier times. In some sense this is much easier than trying to predict its future evolution. Consider throwing a ball up in the air. It is likely to achieve a certain height and then fall back to earth. The eventual trajectories of balls thrown with different initial speeds can be quite different, but if we visually examine the early parts of their path, the trajectories appear quite similar. The differences in trajectories are enhanced at late times and diminished at early times. Thus, no matter what the future expansion of our universe may be, given the observed expansion today, we can estimate with good confidence what its past behavior was, at least as long as our approximation of a "dust-filled" universe remains valid.

How long does this approximation remain valid? This is where the cosmic microwave background comes in. One of the most remarkable features of this background, besides its amazing degree of isotropy (the radiation background is uniform over the sky to better than one part in a thousand; this notable isotropy is discussed in some detail in chapter 5), is the fact that the background is "thermal." We all know that when we heat something up, it glows: that is, it emits light. Light, as already noted, is one type of electromagnetic radiation (or wave). It was one of the great discoveries of the last century that whenever we heat something it emits electromagnetic radiation. Moreover, the "spectrum" of this radiation—that is, the energy emitted at each given frequency—depends in a fixed way on the temperature. As we heat something to higher temperatures, more of the radiation is emitted predominantly at higher frequencies. Now, visible light has a very high frequency (about 10^{15} cycles per second), so things have to be pretty hot before they emit waves significantly in this range. If an object is cooler, it emits waves of lower frequency. Since frequency is inversely proportional to wavelength, a lower frequency means a longer wavelength. In the electromagnetic spectrum, as an object emits progressively longer wavelength radiation, this radiation passes from the infrared (which we "feel" as heat) to microwaves and radio waves.

To visualize the relationship between the spectrum of radiation emitted and temperature, imagine a "black box" with walls that completely absorb any light incident upon them, no matter what the color or frequency, and that later reemit this energy as radiation just as "democratically." If one makes a little pinhole in one of the walls, any light emitted by any of the walls must bounce around a lot inside the box, being absorbed and reemitted many times, before it finds its way out of the hole. In this way, if one keeps the walls at a certain temperature, one can ensure that the radiation that escapes has settled to "thermal equilibrium" with the walls before it emerges. The resultant spectrum is called a "black body" spectrum. It peaks at a certain frequency, related to its characteristic temperature, and its intensity falls off on either side of this peak (see figure 2.1).

This observed form of the spectrum, with its characteristic peaking and decline, defied all theoretical explanation until about 1900, when a young German theorist, Max Planck, proposed an "absurd" explanation that worked. He suggested that light could be emitted or absorbed from the walls only in discrete packets, called "quanta," with energy proportional to frequency. The thermodynamics of such "quanta" in a black box resulted in the observed black body spectrum. With his proposal, quantum mechanics was born.

To return to our story, however, when a radio or microwave receiver receives a

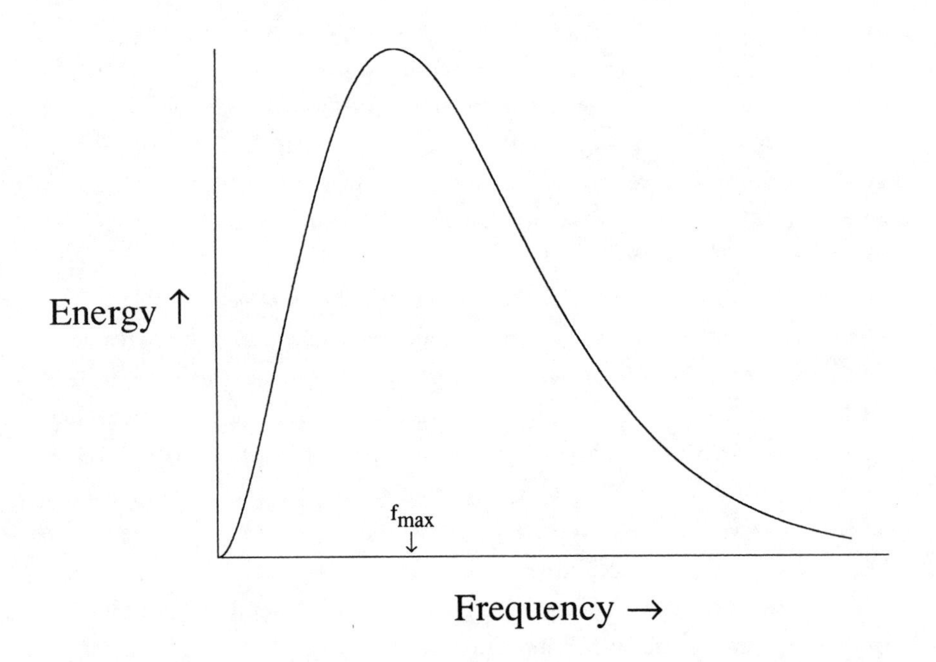

FIGURE 2.1 BLACK BODY SPECTRUM

The energy spectrum of radiation emitted by a "black body" at some temperature T. While I have shown no units here, the characteristic frequency f_{max}, at which the spectrum has its peak, is directly related to the temperature T in a fixed way. If one chooses a measuring system where frequency and temperature have the same units, then $f_{max} \approx 2.8T$ for this spectrum.

signal that resembles thermal noise, it is useful to imagine that the receiver is "inside" such a black box. One can calculate what temperature the walls would have to be to emit a spectrum with the observed energy at the frequency measured, and then one can "quote" a measurement of a certain "equivalent temperature." What Penzias and Wilson observed with their antenna in 1964 was a noise signal with an equivalent temperature of between 2.5 and 4.5 degrees Kelvin. (The Kelvin scale is the same as the centigrade or Celsius scale except that the zero point is more logically located at absolute zero, the lowest possible temperature, instead of at the freezing point of water. On the Kelvin scale, the freezing point of water has the value of 273 degrees K.)

Once it became clear that this noise signal was not spurious, it was recognized that their antenna was in fact inside a "black body," and that this black body (the universe) was radiating a thermal spectrum. This interpretation has since been confirmed by many different experiments measuring the spectrum at different

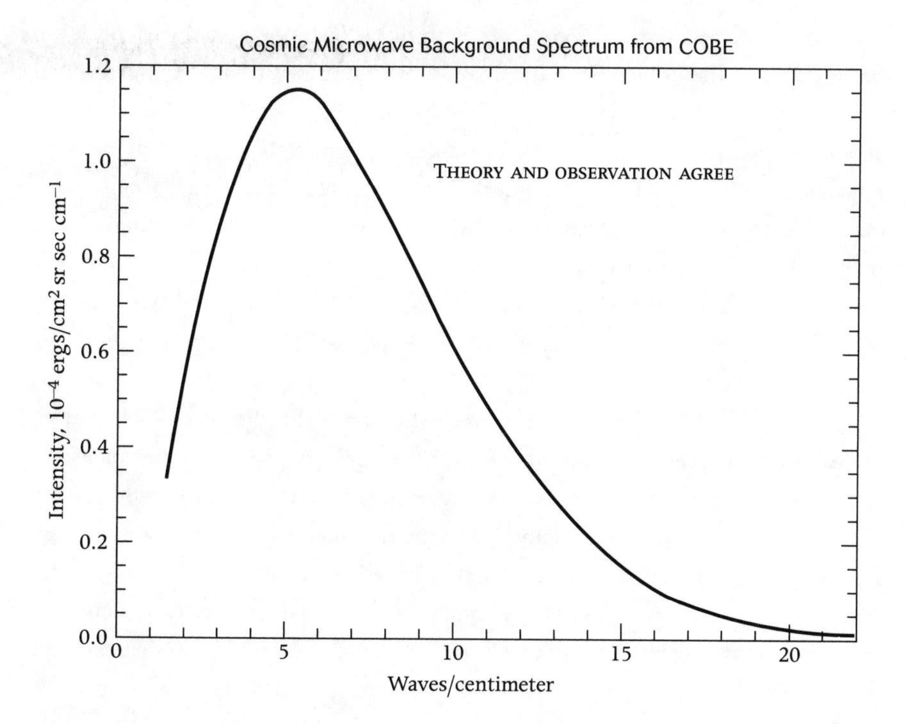

FIGURE 2.2

The observed CMB spectrum from the COBE satellite is compared to the theoretical black body curve. The size of the error bars on the observational points is much smaller than the thickness of the line displaying the theoretical prediction.

frequencies. The temperature associated with this spectrum is now measured to be about 2.7 degrees Kelvin. At various times, experiments have reported "non-black body" anomalies in this spectrum.

In November 1989, NASA launched a satellite called the Cosmic Background Explorer (COBE) Satellite, designed specifically to explore the Cosmic Microwave Background Radiation. It completely revolutionized this entire field. In the first eight minutes of operation, it measured the spectrum of the CMB more accurately than ever before. I display the measured curve compared to the prediction of a black body in figure 2.2. The agreement is incredible. In fact, it appears to be too good, because none of the predicted points lie outside the error bars of the observations. However, that is because the actual error bars are 10,000 times smaller than those shown. The COBE satellite was able to measure temperatures with an accuracy of a few millionths of a degree Kelvin, and in fact, the CMB agreement with a black body prediction makes it the *best measured black body in nature!* No more accurate black body emission has ever been produced in the laboratory.

Why should there be such a black body background, and what does it tell us? The answer to these questions comes from thinking about what happens to such a background as the universe expands. We know that in the case of dust, the dust specks merely become farther and farther apart as their density is diluted. What about in the case of light? Insofar as we think of light as being made of elementary quanta, *photons,* the same phenomenon happens. However, because the energy of these quanta is associated with a frequency, namely, because light behaves like a wave, something else happens. If we draw the crests of waves traveling in a box, and then double the box dimensions, the distance between crests will increase, just as the distance between dust particles did in our previous example (see figure 2.3).

The wavelength is the distance from crest to crest; this distance will increase, and thus the wavelength of the wave will increase, in proportion to the change in the size of the box. As long as the velocity of the wave remains the same (remember, the speed of light is a constant), then the frequency of the wave will remain inversely proportional to its wavelength. In other words, as the wavelength increases, the frequency decreases. Because red light has the lowest frequency of all visible light, this process by which the frequency of light waves decreases as the universe expands is an example of what is often called *redshifting.*

Consider the microwave background that has a black body spectrum with maximum energy at some frequency (f) associated with a temperature (T) of 2.7 K (see figure 2.1). *If this light has not been interacting with anything else that might have kept it at a fixed temperature,* then earlier it must have had a spectrum with

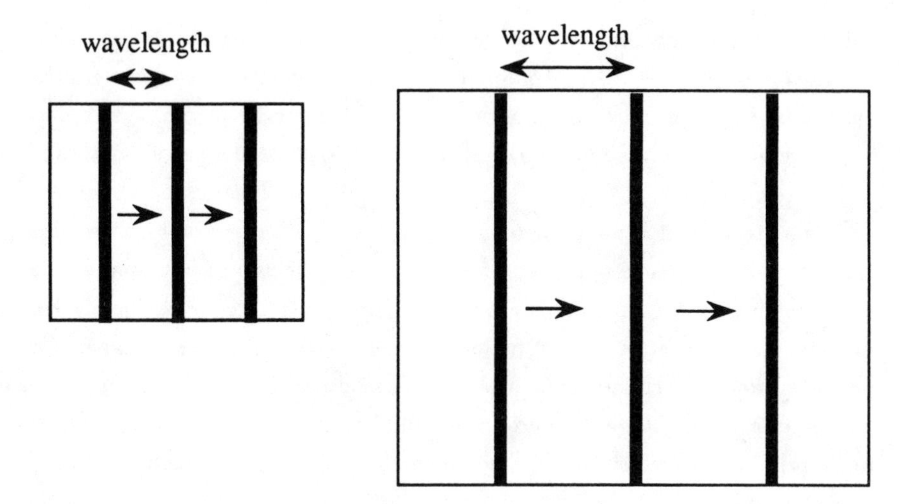

FIGURE 2.3 Redshifting Due to Expansion

Doubling all the dimensions of a box will double the distance between successive crests, that is, the wavelength of a wave propagating inside, as shown here.

maximum energy at some higher frequency (f'), since each light wave has been "redshifting" during the expansion. Such a spectrum would no longer be associated with a temperature T. However, if each light wave in a black body spectrum redshifts by the same amount, the shape of the spectrum remains the same. One can still associate a temperature with this spectrum, but this temperature would change in the same way as the frequency at which the maximum energy is contained had changed. This, in turn, as I have just described, decreases as the universe expands. Hence, we know that the "temperature" of this background would have been higher at earlier times.

This may sound almost mystical, but the same phenomenon is very straightforward when it occurs in less exotic contexts. Say we confine a gas to some volume and heat it to some temperature, and then let it expand freely. The gas will cool as it expands. This is the principle behind the modern refrigerator. In the context of the expansion of the universe, the "gas" in question is made of photons, which travel at the speed of light. We could say the same for any other gas of "relativistic" particles, that is, particles that travel at or near the speed of light.

Before I go any further, I want to explain why I italicized part of the chain of reasoning outlined in the previous paragraphs. What guarantees us that this background has not maintained a temperature of 2.7 K as the universe expanded, by interactions with matter which might have been fixed at that temperature? We can calculate, given our rough empirical knowledge about the density of charged particles in the galaxy and between nearby galaxies, about how long such a light ray would travel before being scattered or absorbed by a particle of matter. The mean distance turns out to be larger than the size of the visible universe today. Thus, we are safe in assuming that, at least at this time, the gas of photons is freely expanding.

What happens if we project back in time? The universe would have been smaller, and at the same time the temperature of the photon background would have been higher. We can calculate what the size of a given region in the present universe would have been as a function of time, based on the present expansion rate and also on the equations of gravity. We can also work out reliably what the temperature of the photon background should have been at any time, at least up to temperatures for which we believe that we understand the interactions of photons and matter. We can calculate that at a particular time, more than 10 billion years ago when the universe was only about 100,000 years old, the photon temperature would have been in excess of about 3,000 degrees Kelvin. At this temperature something important would have happened. Since the frequency of each photon would have increased as the temperature increased, and the energy of each photon increases with its frequency, the average energy of the photons in the background would also have been much higher at this temperature. When the temperature exceeded about 3,000 degrees K, the average energy would have been sufficient so that photons could have collided with hydrogen atoms, the most abundant element in the universe, and separated them into their two constituents, a proton and an electron. At this point the universe would no longer be transparent to light. This is because most of the protons and electrons in the universe then would not be bound together into neutral hydrogen, but instead would be moving about on their own. Thus the photons would encounter a background of charged particles rather than a neutral background of matter. Since charged particles easily interact with light, the mean distance a given photon would travel before interacting with matter would become very small, at least on a cosmic scale.

What happened when the light interacted with the matter? Since the light was thermal at a temperature T, the matter would quickly come into equilibrium with the radiation, at this same temperature. We must conclude, therefore, that further back in time not only the radiation alone but *both* matter and radiation would have been getting hotter together.

Note that this discussion outlined a procedure "backward" in time, instead of beginning with very early times, when the temperature was very high, and matter and radiation were in equilibrium. The time I refer to here, before which protons and electrons were separate, is usually thought of as the time *after which* they were combined. It is called the *recombination time* (at higher temperatures protons and electrons may have combined momentarily before being knocked apart almost instantaneously. At the time and temperature discussed here, they *recombined* again, for good).

We can thus think of the cosmic microwave background as a redshifted "snapshot" of what the universe looked like at the recombination time. The microwaves detected by Penzias and Wilson had last interacted with matter at the recombination time, when the universe was only about 100,000 years old, over 10 billion years ago. At that time the universe was hot, at a temperature of about 3,000 degrees K. After that time, the matter, which was now neutral, and the cosmic radiation "bath" no longer interacted. The thermal radiation background continued to redshift as the universe expanded, however, and now we are left with this mere "shadow" at 2.7 degrees K, reminding us of that early, violent, hot period of the Big Bang expansion.

In many ways, the cosmic microwave background was the first modern version of dark matter. It remained hidden until well after the age of telescopes had begun. Indeed, while photons are relatively easy to detect, such a low temperature bath is easily overwhelmed by other sources of "noise." Although the technology to detect this background probably existed ten to twenty years before the actual discovery in the mid-1960s, without a great deal of luck, or without some explicit reason to search for a signal at the appropriate frequencies, it could have remained undetected even longer. Beyond the difficulty of direct detection, the cosmic microwave background shares many of the prototypical features of the dark matter of the ancients. It is pervasive and apparently uniformly fills space. It has existed relatively unchanged since almost the earliest moments of creation, and it promises to remain that way. In a poetic sense, this cosmic microwave background evokes most of the qualities of the primeval "fiery light" of Homer, Anaxagoras, and Empedocles.

The microwave background plays another fundamental role earlier associated with the aether. It defines a reference frame against which we can compare our local motion, as well as the motion of nearby stars and galaxies. Though the temperature of this background is very uniform, there is a very slight observed increase in one direction, and a decrease in the opposite direction. This we understand very simply as a "Doppler" effect, like the change in pitch of the whistle of a moving train as it passes by. In this case, the frequency shifts reflect the fact that we are not exactly at rest with respect to the Hubble expansion. Our local motion includes the movement of the earth around the sun, the sun around the galaxy, and the galaxy around nearby galaxies. Along the direction of our motion, the photon background is slightly *blueshifted* (in other words, the characteristic frequency is higher), and in the opposite direction it is slightly more redshifted. In fact, this *dipole anisotropy* of the microwave background provides a very useful probe for cosmology. Our local motion is largely due, we assume, to the gravitational attraction of the dominant local mass concentration, whatever that may be. The magnitude and direction of our motion thus give us some indication of the size and distribution of this mass concentration. Comparison with visual observations can then offer a key to unraveling large-scale structures. As we shall see, it can also provide a test of models for the formation of such structures, which in a universe dominated by dark matter are determined by this distribution.

Although the use of the microwave background as a reference frame might at first glance appear to involve an aether-like violation of the axioms of relativity, it actually does not. Our motion relative to the frame defined by the microwave background is manifested by detection of differences in temperature, not velocity, associated with the electromagnetic signal received. As a result, while two observers in relative motion can determine and agree on who is at rest with respect to the microwave background frame (only one will see a uniform temperature), neither can assert, nor is it meaningful to assert, that this frame is itself at absolute rest.

The circumstances surrounding the origin and character of the microwave background motivate much of our speculation about the origin of dark remnants from the primordial Big Bang. We have seen how the photon background decoupled from matter early on, just when protons and electrons combined to form hydrogen. Could there have been other processes, perhaps occurring when the temperature of the universe was hotter, by which other more exotic remnants were produced and left to expand freely? As I stated in the preface, there are roughly 10 billion microwave photons in the universe today for every proton. The energy of these quanta is so small that they make up only a small fraction of the total energy contained in matter. It is logical to ask whether some other more exotic remnant might make up a substantially larger fraction of the total energy, or perhaps even dominate it. After all, if these unknown particles have even a small mass there need be nowhere near as many of them as there are photons today and they could still overwhelm normal matter as the dominant mass in the universe. As we shall see, these are not idle speculations, and in fact the CMB radiation provides crucial evidence that this may be the case.

We cannot close this brief history of cosmic matter, both dark and light, without discussing the modern implications of another major question debated by the early philosophers. Is there a void? In fact, the notion of the "vacuum," the modern terminology for a void, has evolved more than perhaps any other concept in science in the twentieth century. From antiquity until the early 1930s, this word denoted a space "entirely devoid of matter," or perhaps more poetically, "unchanging emptiness." While both notions persist in popular definitions, the vacuum of modern particle physics is teeming with activity. It is a bubbling, brewing source of matter and energy; it may even contain most of the matter in the universe!

Our concept of the vacuum began to change with the development of the theory of relativity. Einstein's famous equation relating rest mass and energy led to the realization that interconversion between these two was possible. The most well known version of this interconversion, the atom bomb, involves one direction, the production of large amounts of energy by the interchange of a minute quantity of mass. In this case, a heavy nucleus "fissions," or breaks up, into two lighter nuclei the sum of whose mass is less than that of the parent nucleus. But Einstein's equation opened the possibility for a more unusual interchange. If enough energy could be assembled in "the vacuum," could a chunk of matter emerge?

In one sense, this phenomenon can be observed in a commonplace process. When the light by which you are reading this book is absorbed by the atoms in your retina, the mass of those atoms increases. It was precisely this phenomenon that Einstein first predicted and which led to his most famous formula, $E = mc^2$, relating mass and energy. The more exotic application of his ideas, the conversion of "pure" energy into matter where none existed before, had to await the development of quantum mechanics.

Quantum mechanics, in its imposition of a wavelike nature to particles, and a particle-like nature to waves, has a staggering implication that intrigues philosophers of science. The so-called uncertainty principle, on its most grandiose scale, constrains humankind's ability to predict and control the future. In practical terms, it suggests the following limit: There exist pairs of quantities, position and velocity (more strictly, momentum), or energy and time, which cannot at the same instant be measured with arbitrarily good accuracy, *no matter how good the measuring apparatus.*

A standard explanation of one of these conditions is worth repeating here. Recall that Planck, in his original introduction of the "quantum," said that light comes in packets of smallest energy, related to its frequency. Thus, for a given frequency or wavelength of light, there is a smallest unit of energy that can be absorbed or emitted. This energy is that which is associated with a single "photon." Now, all we need to derive the uncertainty principle is one more fact about waves. Waves of a given wavelength will be "disturbed" by objects whose size is comparable to or larger than this wavelength. For example, a water wave will pass easily around a small pebble jutting out of the water, but it will be blocked—there will be a calm region of water—behind a large boulder. Similarly, sound waves, whose wavelengths are measured in meters, may pass easily around small objects in a room. Sound waves with the longest wavelength, the bass notes in a rock song, for example, pass around even larger barriers, remaining audible in places where the rest of the music is not.

Now let us imagine what happens when we send a very small particle on its way, moving through an experimental apparatus. If we want to measure the position of this particle, that is, "see" it, we must use light that has a wavelength small enough to "sense" this particle—by being reflected from it, for example. As we make the wavelength smaller, however, the frequency of the light increases. Planck informed us that the smallest packet of light we can use has an energy that increases with frequency. So, if we want to see the particle we must bombard it with photons of high energy. However, when such a photon bounces off the particle, allowing us to distinguish with arbitrary precision where it is, the photon will *alter* the motion of the particle by some amount. The minimum amount by which the reflection process *itself* will change the momentum of the particle is related to the minimum package of energy contained in a photon, which is in turn related to a universal constant of nature, appropriately called *Planck's constant.*

Because Planck's constant is so small, we do not distinguish this kind of uncertainty on the scales associated with everyday measurements. On an atomic, or subatomic, scale; however, such uncertainty is an everyday fact of life for experimentalists.

As I mentioned when introducing the uncertainty principle, another pair of quantities is subject to an uncertainty relation, namely energy and time. In this case, what is implied is the following: If we wish to measure the energy of a particle with arbitrary precision we must observe it for a very long time. If we have only a limited time to make the measurement, as is always the case, we can obtain only limited accuracy or knowledge of its energy. These two relations, between position and momentum and between energy and time, have profound implications for the notion of the vacuum.

Consider any physical process, say, the collision of two particles, which happens in some region of otherwise "empty" space over some interval of time. Since the region over which we can measure this interaction is limited in space and time, our knowledge of the exact energy and momentum associated with the particles during this interaction is limited. Although it is one of the basic laws of physics that energy and momentum are conserved—that the total energy of the particles before an interaction is equal to the total energy of all the particles after the interaction, and likewise for momentum—we can have no *direct empirical guarantee* that this need be the case at every point in between if there exists an inherent uncertainty in our knowledge of these quantities during the interaction. For example, one might imagine that during this period a new particle spontaneously appears out of the vacuum. As long as it disappears back into the vacuum on a time-scale that is small, we will not have time to measure any violation of energy and momentum due to the creation of something out of nothing. Because such a particle cannot be "real" in the sense that we can measure it directly, it is called a "virtual" particle. We can therefore imagine that surrounding every particle there might be a "cloud" of virtual particles burping momentarily out of the vacuum, carrying energies and momenta which are inversely proportional to the time and distance they travel before disappearing.

This may sound like some theoretical hocus-pocus contrived by physicists who have stared at blank notepads in rooms without windows for too long. Yet while virtual particles cannot be seen directly, their indirect effects *are* measurable. When an atom is sitting around minding its own business, quantum mechanics and relativity suggest that it can be constantly bombarded by virtual particles being momentarily burped out of the vacuum, only to disappear almost immediately. When the atom undergoes any physical process taking place for some limited time, it may be disturbed by such particles.

Imagine the process occurring in a light bulb, where an electron in an excited state in an atom relaxes to a lower energy state by emitting some light. One of the great early successes of quantum mechanics permitted us to calculate the "allowed" energy levels of electrons in atoms. In this way, we can figure out the difference in the energy levels, which should correspond to the energy of the light emitted when an electron relaxes from one level to another. But if we calculate this difference *directly,* we get the wrong answer! We are wrong by only a very little bit, but the error is big enough to measure. If, on the other hand, we include the fact that the atom is continually bombarded by virtual particles as it sits there, we find that the energy levels of the atom shift slightly. When we calculate the amount of this shift, we find that we obtain the right answer for the frequency of light emitted by the atom.

In fact, things are even stranger. If, during the time when one of these virtual particles happens to be "doing its thing," it absorbs enough energy by colliding with a real particle so that energy-momentum is no longer violated by its existence, then this particle can become "real." It need no longer disappear back into the vacuum after a short time. What will the outside observer then see? Initially, she might see a single photon of light traveling along in some material. At some point, however, she could suddenly see the apparent "creation" of a pair of particles accompanying the photon!

It turns out that when one combines special relativity and quantum mechanics, this process is not only possible, but *required*. The combination that results, called "quantum field theory," forms the basis of all theories by which processes involving elementary particles are presently understood. The formulation and application of quantum field theory is perhaps the greatest advance in physics in the second half of this century. Within its framework, not only can particles be "created" or "destroyed," but every particle has an associated "antiparticle" with which it can collide and mutually annihilate, leaving behind nothing but energy.

The notion of virtual particles resolves the philosophical problem of action at a distance which rightly bothered Newton. In the modern "quantum field theoretic" view, forces between particles result, not from some mystical action at a distance, but rather via the exchange of virtual particles. Imagine the following scenario: A particle such as an electron travels along spontaneously spitting out virtual photons. Some of these are reabsorbed by the electron as it travels. However, since the photon is believed to have no mass, as its wavelength approaches

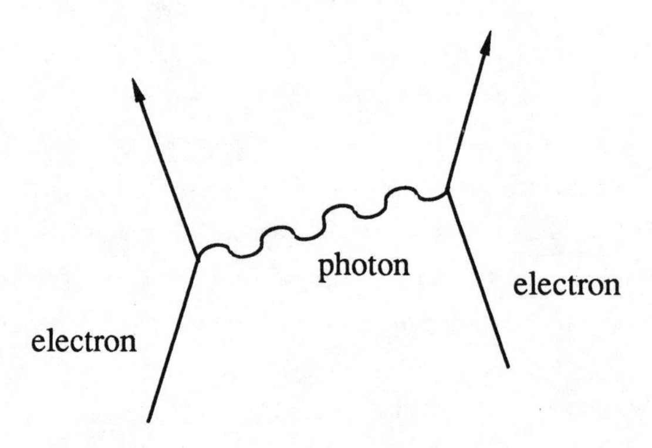

FIGURE 2.4

This Feynman diagram pictorially conveys the essence of our quantum field theoretic understanding of the electromagnetic force between electrons. An electron traveling along emits a "virtual" photon (*wiggly line*), which is subsequently absorbed by a nearby electron, thereby transferring energy and momentum between the two. The arrows represent the direction of travel of the electrons in space and time.

infinity, its frequency approaches zero. Therefore, so does its total energy, according to Planck. Thus, while a virtual photon of very long wavelength does carry some small amount of energy and momentum away from the electron, the uncertainty principle suggests that it can travel a large distance for some time before such a small violation of energy and momentum can be detected. The photon may encounter another electron on its journey and be absorbed by this new electron. In the process, energy and momentum can be transferred from one particle to another. We interpret this in classical terms as the force of repulsion of two nearby charged electrons. It is precisely because the photon is massless that a virtual photon can travel so far, and this is why electric forces are long range. If a photon were not massless, then even as its wavelength went to infinity, its energy would not go to zero. Forces transmitted by massive particles, of which there exist at least one, are therefore short range.

The brilliant and imaginative physicist Richard Feynman in the late 1940s first visualized the process by the image shown in figure 2.4. Such drawings have been termed Feynman diagrams.

One of the implications of this process is that forces are not transmitted instantaneously. Although virtual particles might involve short-term violations of energy and momentum, they are still constrained by relativity to travel forward in time at speeds less than or equal to the speed of light. It is only because this speed is so great that forces seem to be instantaneous on terrestrial scales. Nevertheless, if the sun, for example, were to disappear at this instant, then presumably (assuming gravity too is transmitted by massless particles) it would take about eight minutes (assuming a travel time at the speed of light) for the earth to know it should not be attracted to the sun anymore.

Another important byproduct arises from the quantum mechanical uncertainty relations. Since uncertainties in energy and time, or position and momentum, are related by the uncertainty principle, these quantities are in some sense interchangeable. For instance, if I want to measure processes that take place over very short distances or times, I need probes that involve very high energy or momenta. It is for precisely this reason that the "microscopes" of today are huge high-energy accelerators. Some may wonder why the particle physics community wants to build larger and larger machines involving ever-higher energies, such as the fifty-mile-in-circumference Superconducting Super Collider stupidly canceled by Congress in 1993. Without such high energies, which can only be generated in colossal particle accelerators, it is difficult, because of the uncertainty principle, to explore smaller and smaller distances in probing the fundamental structure of matter. This is an unavoidable consequence of the laws of nature. Of course, whether one believes that such explorations are worth the intellectual dedication and financial costs is a matter of personal prejudice.

These processes have great implications for cosmogony. I have argued that the temperatures of matter and radiation were increased in the primordial universe, and therefore the mean energy of particles was increased. As we go back in time, physics on smaller and smaller time and distance scales becomes more relevant, involving potentially exotic new particles. During very early times in the universe, such particles may have been created with abandon in thermal equilibrium. This could have been quite significant in the generation of particles that might eventually be identified as dark matter today. Moreover, this implies that the early universe can serve as a sort of laboratory for particle physics. In other words, by studying the "shadows" or remnants of the early universe, probing processes that may be inaccessible in terrestrial machines, we may learn a great deal about the fundamental structure of matter.

Because of these ideas, we now view the vacuum as anything but empty. Rather, the vacuum can be thought of as a vast storehouse of virtual particles waiting to appear. In a formulation introduced by Paul Dirac in about 1930, we can imagine the vacuum as "filled" with a "sea" of electrons with "negative" energy. (By convention, we usually think of a state with no observable particles as having zero energy. In this sense particles without enough energy to be measured on top of the vacuum state have negative energy. It turns out that for all considerations other than gravitational ones, which state we define to have zero energy is completely arbitrary. Of course, once we include gravity, we shall have to confront directly the question of whether the vacuum has energy or not.) Since these electrons with "negative" energy are not observable, their negative energy is not a concern. If the notion of energies that are less than zero is bothersome to imagine, think of it this way: All that is physically important is that the total energy contained in this "sea" is less than the energy of a state containing the sea *plus* one real particle.

Now, all we have to do to observe one of these particles in the sea is to "hit" it with something that causes a transfer of energy sufficient to make its net energy large enough so that it can appear as a real particle. What emerges is a real electron out of the vacuum. What is left behind in the vacuum is an empty spot or "hole" in the "sea." This hole, signifying the *absence* of an electron (with charge "minus") in the vacuum sea, appears to an observer as the *existence* of a particle of charge "plus" but otherwise with the same mass as an electron. This is the "antiparticle" of an electron, called a *positron*. Thus, one can create a "pair" of particles out of the vacuum where once there were none. When Dirac made this preposterous suggestion, no such particle as the positron had yet been observed. However, within two years of his prediction, positrons were actually observed among the cosmic ray particles regularly bombarding the earth. Once again, what seemed to be a totally ad hoc argument about the vacuum had a real and fundamental impact on our picture of the universe. As we shall see, this trend continues.

We have seen how energetic processes of the type discussed here can "kick" real particles out of the vacuum. Sometimes the vacuum can produce such particles all by itself! If the idea of a "vacuum" containing real particles seems like a logical contradiction, this merely reflects the fact that the popular definition of a vacuum does not exactly coincide with its definition in physics. We saw that because of the presence of virtual particles, it does not make sense to consider the vacuum as unchanging and empty. A better definition, from the point of view of a physicist, is to consider the "vacuum state" of a system, say, the universe (or perhaps a box in which we want to perform an experiment), as the *lowest energy configuration* of that system. In general it costs energy to create a real particle in the box, so one finds that the vacuum state usually contains no real particles; here the physics definition coincides more closely with the popular definition. But imagine the following situation. Say that one particle at rest is spontaneously created out of the vacuum. If this particle has a nonzero mass, then the energy of the system is increased. If energy is to be conserved, this particle must be virtual and must eventually disappear back into the vacuum. Now, however, imagine that two such particles are created at rest. If there exists an attractive force between these particles, then the total energy of the two particles will be somewhat less than the sum of their masses. This is true for any bound system. A hydrogen atom, for example, containing a proton and an electron, "weighs" less than the sum of the weights of a free proton and a free electron. This is just a consequence of the energy-mass relation discovered by Einstein. The binding energy of the proton and electron reduces the total energy of the system, and therefore the total mass.

One can imagine continuing this process. If things balance just right, one might imagine a situation in which, if enough particles are produced at rest, their mutual attractive force could result in a negative contribution to the energy which would cancel or exceed the contribution of their masses. In this case, the *favored,* that is, the lowest energy, state of the system would not involve a state with no real particles. Instead, one says that particles would "condense" out of the vacuum in a *zero momentum* state. The terminology here is appropriate. When water droplets condense out of the air, it is because the favored configuration of the water, due to changing conditions of pressure and temperature, no longer involves a gas of water molecules, but instead involves a state where water molecules bind together loosely into agglomerations that we recognize as water droplets in the liquid state. In the case of the vacuum, what would otherwise be virtual particles condense, under certain circumstances, into a zero momentum state containing real particles.

Analogies to this kind of behavior can be found in the physics of materials, called "condensed matter" physics. One example involves a ferromagnet such as iron. We can think of a macroscopic magnet as a large collection of microscopic magnets associated with the individual atoms in a material. Since at a reasonable temperature each of these atoms vibrates and constantly bumps into other atoms, the directions of these individual magnets tend to randomize, so that their magnetic fields tend to cancel on average. For example, a piece of iron that we pick up is not likely to be a magnet. But if we place the piece of iron in an external magnetic field, it becomes energetically *favorable* for the microscopic magnets contained in the iron to line up in the direction of this magnetic field, so much so that even the mean thermal agitation of the atoms cannot overcome this effect; the piece of iron then becomes a macroscopic magnet.

Now imagine the following situation. Say the atoms in the iron are in their standard disordered "vacuum state." (Since the iron is at a finite temperature it is not strictly true to call this a "vacuum state." One normally calls this a "ground state," but I hope the experts will allow me this misuse of terminology for the purpose of the analogy.) Suppose a random thermal fluctuation knocks a given atomic magnet so that it aligns in a certain direction. This is analogous to a "quantum fluctuation" spitting a virtual particle out of the vacuum. Just as the virtual particle will eventually be reabsorbed, we expect that thermal fluctuations will eventually tend to knock the atomic magnet into a different direction. Now, imagine that, instead of one atom, a fluctuation kicks several nearby atomic magnets so they point in the same direction. These together then act like a larger magnet. What happens to a nearby atomic magnet pointing in a different direction? Just as when the whole system is placed in an external magnetic field, in this case the odd magnet tends to want to reduce its energy by aligning with the field of the larger magnet created by its neighbors. Now an even larger magnet is formed. One can easily imagine a situation in which larger and larger regions align so that eventually all the "spins" spontaneously align in the same direction. Of course, this will happen only if the temperature is low enough so that thermal fluctuations cannot overcome the energy gained by having many or all of the atomic magnets aligned. Thus, while it costs energy to produce a large-scale magnetic field, the individual atomic magnet interactions with each other overcome this cost, as well as the thermal tendency to randomize, so that the favored state then involves the spontaneous creation of a macroscopic magnet with all atomic magnets aligned. Think of the "mean number of magnets pointing in a given direction" in the ground state as the equivalent of the "density of real particles" in a particle physics vacuum state. Then spontaneous magnetization can result in a nonzero mean value for this quantity, just as one might imagine that a nonzero density of real particles in the vacuum can result from the interactions of particles with each other.

In both cases, the macroscopic character of the lowest energy configuration of the system changes. We call this a *phase transition*. The change from water to ice is a phase transition, as is the formation of a macroscopic magnet in a previously unmagnetized piece of iron. Such phase transitions have come to play an increasingly important role in our understanding of matter on all scales: from boiling oatmeal to a boiling vacuum. So, too, they have naturally found a place in cosmology.

We have already observed in this chapter that matter and radiation cool as the universe expands. I shall describe later how it is possible that conditions can evolve so that phase transitions in the "vacuum" (or ground state) of nature can occur. In such cases, a finite density of real particles might "condense" in the new vacuum state. Such particles might later evolve into everything we see, or perhaps they could consist of something else that eventually would come to dominate the energy density of the universe, and so determine, by its gravitational effects, the formation of all observed structure. Both these possibilities emerge in modern cosmology.

The vacuum of modern particle theory is a strange place indeed. From an unchanging "void" it has become an active arena out of which particles might be created or into which they might be destroyed. Just as light was supposed to excite waves in the aether according to Newton, we now envisage elementary particles to be excitations out of the vacuum state. That vacuum might even be the "source" of all matter and energy in the universe. We may have dispensed with the classical aether of Aristotle and Huygens, but in the process have come to speculate about matters which may seem even more bizarre. Perhaps we have just come full circle. After all, how much closer can we come to the "indefinite" of Anaximander? Recall once again his words: the "indefinite"

> is neither water nor any other of the so-called elements, but some different, boundless nature, from which all the heavens arise and the worlds within them; out of those things whence is the generation for existing things, into these again does their destruction take place, according to what must needs be.

Anaximander's words now appear prophetically familiar. Although we have come a long way from the cosmogony of Anaximander and the earlier myths of the primeval ocean, and our explanations have become more sophisticated and more scientific, the basic questions driving our inquiries are the same. We are still searching for the fundamental constituents of the matter that we can see and touch, and we still wonder about the origin, nature, and existence of the dominant "stuff" in the universe. Throughout history these notions have been interrelated. As our knowledge of one realm has improved, so has our ability to speculate about and probe another. Curiously, the possibility that we may be immersed in a background of matter or energy, whose nature and constitution may differ substantially from that with which we are familiar, seems to form not just a natural part of our mythological tradition, but of our scientific one as well.

PART TWO

Weighing the Universe . . . and Coming up Short

CHAPTER THREE

First Light on the Darkness

This is the way the world ends
This is the way the world ends
This is the way the world ends
Not with a bang but a whimper.

—T. S. Eliot

Will our universe end with a bang or a whimper? To find out, we must first weigh it. This, as you can imagine, is a rather daunting prospect, and after more than half a century of dedicated effort we are still without a final answer. Nevertheless, the effort has already yielded at least one big surprise: there is a lot more "out there" than meets the eye. This chapter and the next explain how and why we believe that this is the case.

On the midway of any carnival, you may come across a booth where someone offers to guess your weight, for a price. The proprietor will size you up visually, and maybe even try to lift you once or twice before pronouncing an estimate. Guessing someone's weight seems a difficult task, so you go along. If the person's guess is wrong, you win a prize. If the guess is right, the person keeps your money and earns a little admiration. What stakes could we offer for the attempt to guess correctly the weight of our galaxy, much less the universe?

In chapter 2, I likened the visible universe to an expanding box filled with tiny dust specks. This of course was a cosmic understatement. The dust particles of our universe—the galaxies—are very considerable objects. There are on average about four hundred billion stars more or less like our sun in each galaxy. In the visible universe there are more than 100 billion galaxies. Each one is at least several thousand light-years across (see appendix A). The light emitted by my computer screen as I write this would take more than 100,000 years to get to the other side of our own Milky Way galaxy. Ours is a relatively typical spiral galaxy, with great majestic spiral arms filled with a "thin" disk of stars orbiting around a central dense bulge more than 5,000 light-years across (see figure 3.1). Spiral galaxies are only one type of galactic system. There exist huge elliptical conglomerations of stars, some of them ten times as big as our own Milky Way. Alternatively, there exist "puny" dwarf galaxies at least one hundred times smaller than our own—containing less than 100 million or so stars.

These awe-inspiring "island universes" are just that. With the exception of isolated collisions between galaxies in overpopulated regions of the universe, galaxies are few and far between. The closest galaxy similar to our own is more than 1 million light-years away from us. The average distance between two galaxies is about 10 million light-years, more than one hundred times larger than the visible size of an individual galaxy. Scattered about the universe are huge voids, regions 10 to 30 million light-years across, containing essentially no visible galaxies. There also exist enormous clusters of hundreds or thousands of galaxies, containing up to a million billion stars. More than half of the galaxies we can see are located in some cluster (see figure 3.2). Galaxies continue to be observed out to the furthest reaches of the visible universe, more than five billion light-years away.

The data needed to make each of the foregoing statements represent the life work of hundreds of dedicated observers. It is no mean task to observe extensively and categorize even a few stars with a telescope on a clear night, much less to attempt to determine the number of stars in our own galaxy, or the "billions and billions" of others we might see, and then weigh these stars.

The techniques used by astronomers to attempt this are not so different from those used by the weight-guessing experts at the fair. First, astronomers attempt to size up the universe visually. This itself must be done in a series of stages, and each stage entails an associated set of problems. Underlying all of these observational difficulties is a key theoretical caveat: we measure light, not mass, with telescopes. To estimate the density of matter in the universe we have to make some theoretical conversion equating a given amount of light to an equivalent mass. The more accurately we can make this conversion, the better our final estimate.

The first part of the task—measuring the light—may seem relatively straightforward. Astronomers have for some time scoured the sky with their telescopes,

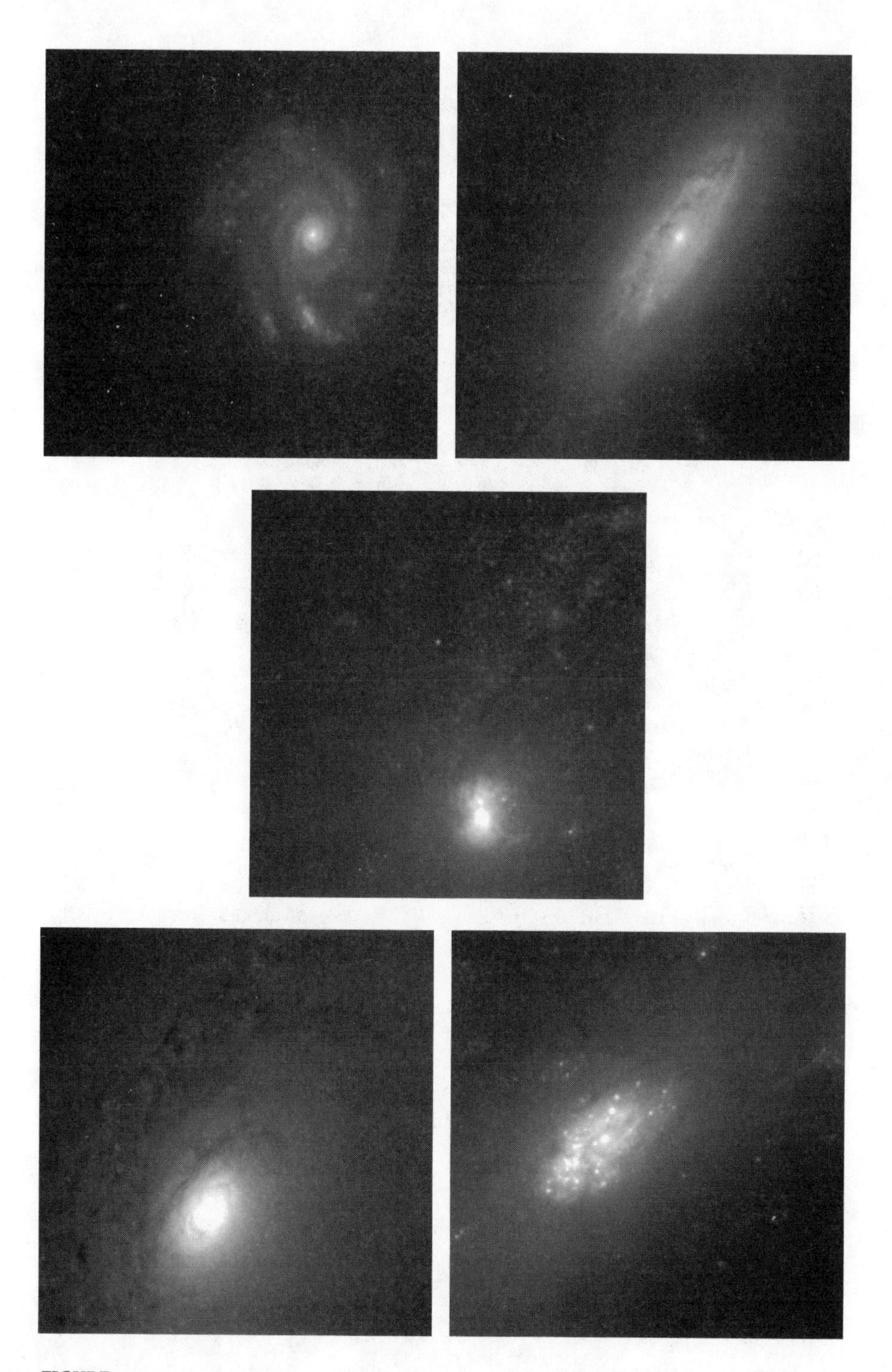

FIGURE 3.1

Various spiral galaxies. (Courtesy Torsten Boeker, Space Telescope Science Institute and NASA.)

FIGURE 3.2

Several hundred never before seen galaxies are visible in this "deepest-ever" view of the universe, called the Hubble Deep Field (HDF), made with NASA's Hubble Space Telescope. Besides the classical spiral and elliptical shaped galaxies, there is a bewildering variety of other galaxy shapes and colors that are important clues to understanding the evolution of the universe. Some of the galaxies may have formed less than 1 billion years after the Big Bang.

so one might imagine that determining the average amount of light associated with visible objects in the universe should be easy. Of course, even here, things are not as pat as they seem.

Say, for example, you see a light on a distant hill. How can you determine how large a wattage is powering the light bulb producing the light? If you know the distance to the hill, it is easy. You simply measure the intensity of light that your light meter registers. Then, assuming that the light spreads out uniformly in all directions and is not beamed at you, you can calculate the total power emitted by using the fact that the intensity varies inversely as the square of the distance to the source (since as the light spreads out, the same amount of light is spread over progressively larger spheres whose area grows as the square of their radii, that is, the distance to you). Once you know the intensity of light at the source, you know the power being emitted by it, or its luminosity.

Now, what if you do not know the distance to the light bulb? What can you do? Well, if you are an aficionado of light bulbs you can attempt to determine the "color" of the light bulb. From your experience you may be able to ascertain that it probably is a 100-watt "soft-white" bulb from company X, rather than a 50-watt "generic bulb" from company Y, because the light of "soft-white" bulbs is less yellow than generic bulbs, and 100-watt lights are also a little bluer than 50-watt lights. If you really know your light bulbs, you might then have some confidence in your prediction.

Similar techniques are used by astronomers. Measuring the distance to remote galaxies is very difficult to do directly. Whole books have been devoted to the "cosmic distance ladder," a chain of measurements that allows us to connect the distance of faraway galaxies to the distance of nearby stars that we can measure directly. However, if we accept the Hubble expansion velocity-distance relation discussed in chapter 2—that the speed of galaxies away from us is proportional to their distance—then within our uncertainty about the proportionality constant in this relationship we can estimate the distance of such galaxies if we can measure their velocity. As noted earlier, this proportionality constant is called the Hubble constant. This has been so notoriously difficult to measure that until recently there was a factor of 2 uncertainty in its value, even after over sixty years of observations! However, with the advent of the Hubble Space Telescope and new ground-based observing techniques, the uncertainty in this quantity has been vastly reduced to about 20 percent. Because the Hubble constant essentially sets the distance scale of the universe, it enters into all other cosmological quantities that one derives. Hence it is necessary to keep track of this uncertainty explicitly. This is done by writing the Hubble constant as some fixed number (see appendix A) multiplied by a "fudge factor," represented by the symbol h. This fudge factor is allowed to vary to represent our uncertainty in the Hubble constant. If we normalize the Hubble constant appropriately, h can be considered to have a value between about 0.6 and 0.8.

Measuring the recession velocity of distant galaxies is relatively straightforward, if time-consuming. We can use the fact that light from stars can be analyzed by its frequency components, or spectra. By measuring the light from nearby stars and galaxies, astronomers have known for some time that these objects contain elements, primarily hydrogen, just like the elements that exist on earth. They were able to determine this because each element, when excited so that it emits light, sends out that light at characteristic frequencies related to the differences in the energy levels of the electrons in the atoms of the element (see chapter 1). Since atoms of different elements have different energy levels, the "spectra" of light emitted by atoms of different elements act as fingerprints. If we measure the spectra of light from the sun, or nearby stars, we see exactly the same features in the spectra that we would see if we heat, say, hydrogen on earth to a high temperature. By matching the observed frequencies of the stellar spectra with the known spectra from elements on earth, we can ascertain the existence and the relative abundance of these elements in the stars (see figure 3.3).

In the preceding chapter I discussed how, in an expanding universe, light is "redshifted": its wavelength increases as it travels while the universe expands. I also mentioned that a very similar phenomenon occurs for light emitted by moving objects. This effect, called "Doppler shifting," occurs in all waves, whether sound or light, emitted from moving sources. I pointed out that it is manifested by the familiar change in pitch in a siren or train whistle as the source of the sound approaches you and once again as it moves away. As it approaches, a higher frequency tone is heard as the waves are compressed, and as it moves away one hears a lower frequency as the wavelength is stretched out. The same phenomenon occurs for light. As long as objects are moving slowly compared to the speed of light, even the formula for the Doppler shift is the same as for conventional waves. In both cases, the fractional change in the wavelength is proportional to the ratio of the object's velocity compared to the speed of the wave; in the latter case this is the speed of light.

Astronomers accordingly can determine how fast an object is moving away from us by observing the spectrum of that object, looking for familiar features due to the presence of hydrogen or some other abundant element, and checking to see how far these features are shifted from the frequencies at which they occur when that element's spectrum is measured in a laboratory (see figure 3.3). Note, however, that only the motion directly toward us or away from us can be measured in this way. Any motion *across* the sky yields no Doppler shift (unless the object is moving almost at the speed of light). However, if we are interested in measuring the Hubble expansion velocity of an object, we are only interested in the recession velocity, not motion across the sky. Any extra, peculiar velocity due to local motion adds some error to our estimate, but for objects sufficiently far away, their Hubble velocity dominates so that only small relative errors are introduced. Once one has figured the Hubble velocity, then a distance determination can be made, up to the ever-present uncertainty factor in the Hubble constant,

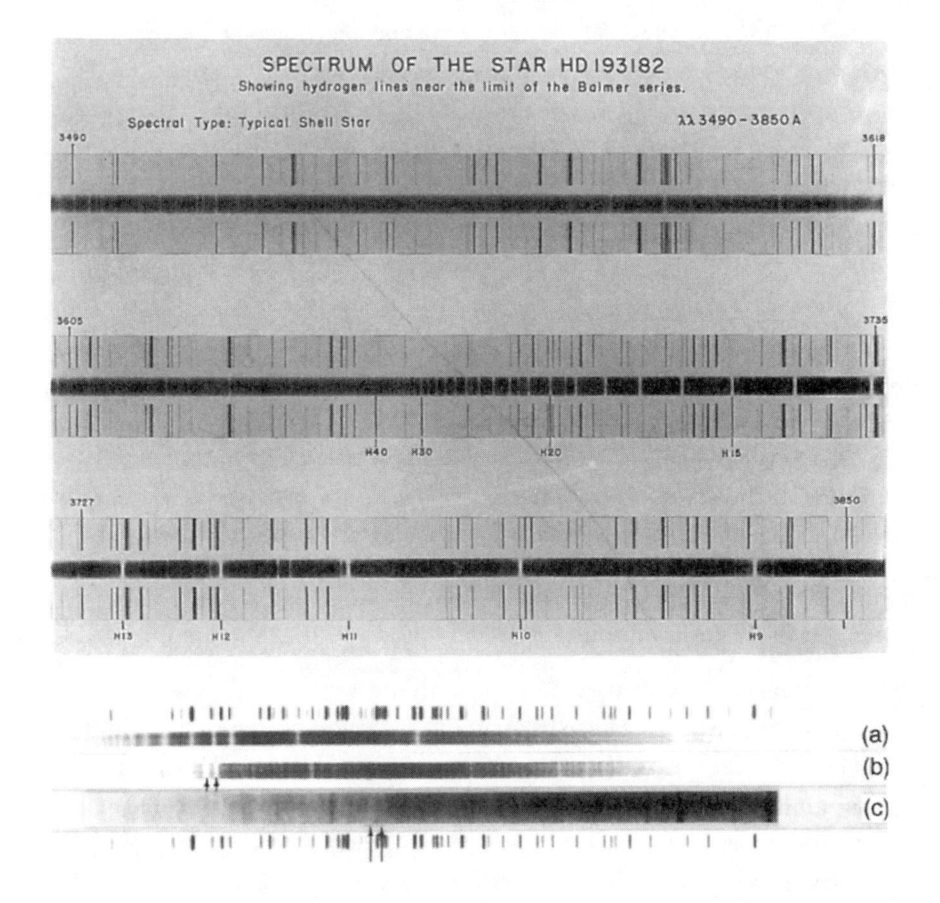

FIGURE 3.3

Shown here are spectra taken from objects at progressively larger distances. The same absorption line feature can be seen in each spectrum, but shifted in wavelength. This "redshift" allows a determination of the velocity with which these distant objects are moving away from us. If an independant distance measure to these objects is used, one finds that distant objects are moving away from us with a velocity proportional to the distance, the famous Hubble Law, indicating that the Universe is expanding. (Upper spectrum courtesy of the Observatories of the Carnegie Institution of Washington; lower spectra courtesy of A. Oemler © Yale University Observatory.)

represented by *h*. Once the distance is known, the actual intrinsic luminosity of an object can be deduced from the observed luminosity.

The "light-bulb" aficionado technique must be used to supplement this analysis, however. In the first place, the Hubble expansion velocity is not a useful distance measure for nearby objects, such as objects in and around our galaxy, which are bound to the Milky Way and therefore do not take part in the expansion relative to us. Second, we may measure light with a telescope in one range of the spectrum, say, blue light, and from this determine the luminosity of an object in the blue range of the spectrum. However, we must make some assumption about how the luminosity in the blue region of the spectrum represents the total luminosity, which in turn gives some idea of the total mass when we eventually come to "sizing up" galaxies. We do this by utilizing measured characteristics of the spectra which we can relate to presumably familiar objects.

We are aided here by the considerable body of knowledge concerning the structure and evolution of individual stars. Thanks to the ground-breaking work of A. S. Eddington, S. Chandrasekhar, and others in the first half of the twentieth century in applying the laws of thermodynamics and quantum mechanics to the interior of stars; to the later work of H. Bethe, F. Hoyle, G. Gamow, W. Fowler, and others in detailing theoretically and experimentally just how nuclear reactions power the sun and stars; and finally to the corps of individuals who use powerful computers to simulate in detail the dynamical processes in the interior of stars, we now have a very good idea of the life cycle of stars of all types. Supplementing this theoretical work, the painstaking observations made by astronomers over this century have allowed stars (and galaxies) to be catalogued in many different ways. When stars are observed, both their general color (related to their surface temperature by the same physics discussed in the preceding chapter) and their luminosity can be determined. When these are plotted against each other, a great deal of order emerges. On such a plot, known as a color magnitude or "Hertzsprung-Russell" diagram (after the two astronomers who first independently derived such catalogues), stars tend to fall onto restricted curves (see figure 3.4). The sun lies along what is known as the main sequence of hydrogen-burning stars. Other restricted regions are populated by Red and Blue Giant stars, and whitish blue small stars called White Dwarfs, which have primarily exhausted their hydrogen fuel and are burning other elements.

Not only do stars tend to cluster along curves in this diagram, but modern stellar evolution theory predicts extremely well the paths that stars will follow along these curves as they evolve in time. Our sun, for example, will burn along the main sequence for another 5 billion years or so (stellar evolution calculations

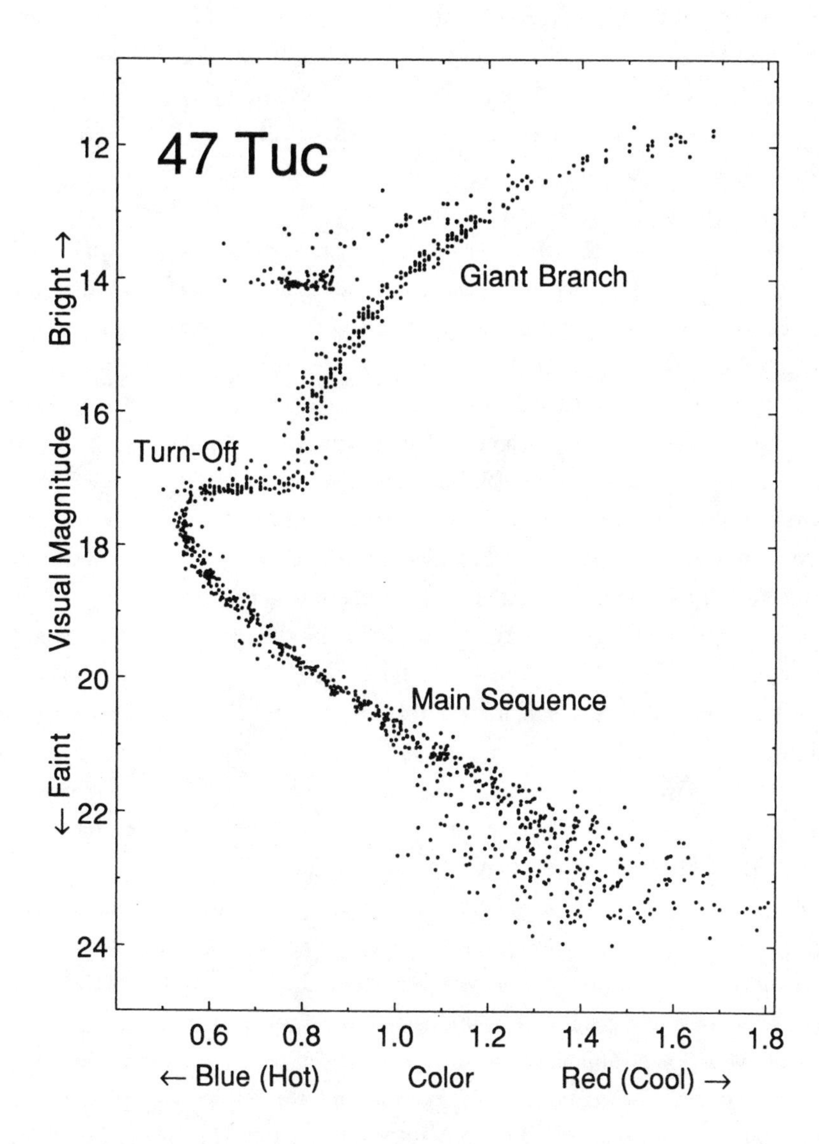

FIGURE 3.4

This Hertzsprung-Russell diagram has been obtained from observation of stars in the old globular cluster 47 Tuc. On one axis is measured the color of the star (obtained by observing the relative intensity of the light in different spectral regions, with larger numbers indicating redder stars); and on the other axis is the overall visual magnitude of the star (with larger numbers indicating fainter stars). It is clear that stars tend to fall only in certain regions of this diagram. Labeled on the diagram are those regions that are occupied by "main sequence" stars like our sun, the region where stars "turn off" the main sequence, and the region occupied by Red Giant stars. (Figure courtesy of D. B. Guenther, from data of J. Hessa et al.)

now carried out to an accuracy of better than 5 percent indicate that the sun is now about 4.5 billion years old), and then will move off the main sequence to the Red Giant stage. Eventually the sun will burn up most of its fuel, reduce in luminosity, and move to the White Dwarf stage, where it will exist for some time until, like a candle, it burns out. Other stars more massive than the sun will have a more dramatic end. As they end their life in the Giant stage, they collapse in a cataclysmic way, blowing off their outer layers in a supernova explosion. Supernovae are momentarily as bright as whole galaxies containing more than 10 billion normal stars. Interestingly, the evolution of the theory of stellar structure and evolution allowed astrophysicists to make many predictions in advance of direct empirical data, which were strikingly confirmed in the observation of the spectacular supernova of 1987. This explosion, called Supernova 1987A, was the first to be observed in the region of our galaxy in more than 400 years and it provided scientists with unprecedented data with which to test theory. In general, the agreement between theory and observation has been remarkable; for example, it was predicted and then observed that most of the energy emitted by supernovae tends to occur not by the emission of light but rather by the emission of very weakly interacting particles called *neutrinos* (which will be discussed at length in part 4).

Although the theory of stellar structure and evolution appears to work swimmingly, there remain two puzzles that continue to nag astronomers and physicists. First, the work on stellar evolution allows astronomers to predict the age of observed clusters of stars by determining how far stars of a given type have traveled along their evolutionary paths. Since smaller stars tend to evolve more slowly, one can look for the smallest stars that have begun to leave the main sequence, and from these observations assign an age to a given system. When such analyses were performed for what are believed to be very old entities called globular clusters of stars, the calculations originally assigned an age for these systems of about 15–20 billion years. Such age estimates sometimes exceeded other independent estimates of the age of the universe, for example, those which result if the Hubble factor h turns out to be in the range 0.6–0.8. The discrepancies here are not large, and there is ample margin for error on either end.

Most recently, in collaboration with my colleagues Brian Chaboyer, now at Dartmouth, Pierre Demarque at Yale, and Peter Kernan at Case Western Reserve, I have been involved in a comprehensive reanalysis of the age of globular clusters. Over the course of three years, we examined all of the uncertainties in the preceding analyses, evolved over 10 million separate stars on a fast computer in order to compare the predictions with observations. Our current conclusions are that the age of the oldest globular clusters lies between about 10 and 15 billion years old,

FIGURE 3.5

The large underground tank holding 615 tons (100,000 gallons) of liquid containing chlorine, located in the Homestake gold mine in Lead, South Dakota. The tank was built by a Brookhaven team, under the direction of Ray Davis, Jr., to search for the signal from neutrinos emitted by the sun. The photo shows the tank before the experiment was begun in 1967. This detector, now operated by the University of Pennsylvania, has operated for more than twenty years. (Photo provided by R. Davis, courtesy of Brookhaven National Laboratory.)

or up to 5 billion years younger than previous estimates. Whether this is young enough to accommodate our expanding universe depends on details we shall soon explore.

A second anomaly is even more annoying. It has to do with the astronomical object we like to think we know best—the sun. Based on all our visual observations, the theory of stellar structure applied to the sun appears to work remarkably well. Astrophysicists believe that they understand what the interior of the sun is like, right down to its fiery core, so that they can predict such quantities as the core temperature to an accuracy of better than 5 percent. But beginning in the 1960s a remarkable ongoing experiment undertaken by Ray Davis, Jr., and his colleagues began to produce some very surprising results. Using a huge vat containing a liquid that includes chlorine (essentially cleaning fluid) and located in a deep mine (see figure 3.5). Davis searched for those weakly interacting particles called neutrinos which are emitted in nuclear reactions and which fuel the sun, nuclear reactors, and supernovae. (I shall explain in part 6 why one must go underground to observe neutrinos coming from the sky.) Neutrinos interacting in Davis's detector can in principle initiate nuclear reactions that change chlorine atoms into argon atoms. However, neutrinos interact so weakly that in a few hundred tons or so of material, Davis expected about one to two such events per day induced by the highest-energy neutrinos from the sun. Instead, since the mid-1960s Davis consistently has seen on average less than one-quarter of the event rate predicted by the best solar models. Another experiment involving an 8,000-ton underground water detector in Japan was one of two experiments that observed neutrinos from Supernova 1987A, thus beginning the new age of "neutrino astronomy." This experiment and its successor involving a 40,000-ton water detector, the largest in existence, have also yielded information about solar neutrinos. They have confirmed that the rate at which the highest energy neutrinos are emitted from the sun is about 50 percent of that predicted by astronomers.

Since the nuclear reactions that produce neutrinos also power the sun, and since the sun is the fundamental object on which apparently successful stellar theory is based, the observations of neutrino behavior appear confusing. Astronomers are so convinced that they understand the structure of the sun that some have suggested that these anomalous results do not arise from the properties of the sun, but rather from the properties of the neutrino. Particle physicists have considered many models by which new physics, including a nonzero mass for neutrinos, might influence the neutrino signal that we observe from the sun. If this turns out to be the case, then astronomy, perhaps for the first time, will have given us information about realms of elementary particle physics well beyond the range of current accelerators. On the other hand, perhaps the standard solar model has to be "tweaked" or manipulated a little, or perhaps even the experimental results to date may be incorrect. Whatever the solution to the "solar neutrino" problem turns out to be, it remains one of the most puzzling issues in astrophysics today. I should point out, however, that any solution to this problem need not undermine the general body of knowledge on stellar structure or evolution. The standard solar model has to be adjusted only a bit to alter the rate of neutrinos produced by the sun. It is a measure of the confidence with which astrophysicists hold to their theories, and the degree to which other observations fall in line, that they are so reluctant to make even this small alteration. Indeed, further support for a particle physics solution has recently been provided by a new forty-thousand-ton Japanese detector, which has established evidence of a nonzero neutrino mass in observations of neutrinos produced by cosmic rays in the earth's atmosphere. If this is confirmed, a similar effect is likely to resolve the solar neutrino problem.

Returning to our original question of weighing the universe, the vast body of knowledge accumulated on stellar structure allows astronomers to predict in general the relative masses and ages (compared to the sun) of stars that we observe on the Hertzsprung-Russell diagram (see figure 3.4). This in turn allows us to convert "light" to "mass" when we attempt to weigh the universe. In the first place, being able to "recognize" different types of stars allows us to refine our distance measures by providing "standard candles." If we observe a system of a well-known intrinsic luminosity (as did our "light bulb" aficionado), then, from its observed luminosity, we can determine its distance, and therefore the distance to neighboring objects. We can continue this process out to larger and larger distances, in this way calibrating the Hubble law of expansion. As I indicated earlier, our calculation of the relation between distance and velocity is only accurate to within about 20 percent.

In addition to calibrating distances and luminosities, our knowledge of stellar structure allows us to convert the measured luminosity of a galaxy into data which are more useful for calculating its mass. When we observe a given system of stars in a galaxy, and determine its luminosity, we can also determine from its color and other characteristics the scarcity and abundance of different types of stars. Indeed, there is a "taxonomy" of galaxy types just as there is of stellar types. Knowing the abundance of stars of different types in a given galaxy type thus allows us to figure out the total mass in stars compared to the mass of the sun for such a system.

Before we can quote a number for the visible mass density of the universe, inferred from the total visible mass in galaxies and from the observed density of

galaxies, we must include one other component of visible matter. Stars form from the clumping of diffuse gas that heats up as it compresses, eventually igniting the nuclear reactions that power the stars. There still exists a great deal of such embryonic diffuse gas—primarily hydrogen—inside galaxies and among clusters of galaxies. Some of this diffuse gas is hot enough to emit light. From its spectrum we can estimate the amount of such material within our galaxy and within neighboring galaxies. We must then include this contribution when making the conversion between light and mass in the universe.

One final bit of common-sense intuition is helpful in sizing up the total luminosity of galaxies. Bigger galaxies are likely to be intrinsically more luminous. Luckily there is a way to make an independent empirical estimate of how big a galaxy is. The stars in more luminous galaxies tend to move around these galaxies with greater velocities than stars in less luminous galaxies. There is a theoretical reason for this relationship which I shall make a great fuss about in the next chapter, but here it suffices to say that the relationship has been observationally established for both major types of galaxies, spiral and elliptical. Like other well-established phenomenological results, it even has a name, or in this case two names. For spiral galaxies the relationship between galaxy luminosity and stellar velocity is called the Tully-Fisher relation, and for elliptical galaxies it is called the Faber-Jackson relation. Once these relationships are quantitatively established for galaxies whose luminosities could be determined from other measures, they can be used to verify less firmly established luminosity estimates for other galaxies.

When all of the "dust" has settled we can tally up how much light the galaxies of our universe are emitting to arrive at a mean luminosity density of the universe, in some appropriate frequency band. It appears that there is an *average "light equivalent"* of about 200 h million stars as luminous as our sun in each cubic volume of roughly 1 million "parsecs" (1 parsec $\approx$ 3 light-years) across on each side.

Note that the ubiquitous factor of h, the "fudge factor" representing our uncertainty in the distance scale of the universe, enters in this estimate, as we might expect, given that our ability to determine luminosity depends crucially on our ability to estimate distance.

When we wish to convert this number into a mean visible mass density, so that we can complete step one of "weighing the universe," we need to estimate how much mass in a given galaxy results in a luminosity about equal to the sun's luminosity. Since a great deal more mass in ionized diffuse gas is required to produce the same luminosity as that produced by gas compressed in the form of a star, and since there are a lot of old stars in galaxies which are relatively dim compared to our own sun—which is in the prime of its life—it turns out that the average luminous mass corresponding to the light emitted by the sun is more than a solar mass. It turns out that *a light equivalent to a solar luminosity is produced on average by about* 25 h *solar masses of galactic material.* (This number varies somewhat for individual galaxy types and also varies depending upon such conventions as the frequency range in which the luminosity is tabulated.) Thus, the mean visible mass density of the universe is roughly 4.5 h billion solar masses per cubic megaparsec (1 mega-parsec = 1 million parsecs).

Let me emphasize once again that this amount only represents the "luminous" mass density of the universe. As such, this number provides merely a *lower limit* on the total mass density. One could easily imagine ways in which material could be missed in this counting procedure. In most cases, however, other measurements constrain the possibilities. For example:

1. Ordinary diffuse gas that has not been excited to emit light would not be included in our tally. Even if such material does not *emit* light, however, it can *absorb* light from more distant objects, such as quasars. Quasars are the extremely bright point sources as luminous as a whole galaxy (we do not yet know for certain why) which are visible to distances as far as we can see, perhaps 5–10 billion light-years away. When the spectra from quasars are examined, one would expect significant absorption in certain regions if the intervening space contained significant amounts of neutral hydrogen. Such absorption "troughs" have indeed been observed, but the hydrogen gas indicated by their presence is such that this material constitutes significantly less than the luminous material that we do observe.

2. Gas that is extremely hot does not emit visible light, rather it emits X rays. In fact, X-ray emission is observed from galaxies and clusters, as I shall discuss in the next chapter. The total amount observed suggests the mass in such hot gas located outside of and between galaxies in clusters in fact exceeds that in the stars associated with the galaxies by perhaps a factor of 2. In the 1980s there appeared to be a diffuse isotropic X-ray background that could have been associated with a density of hot gas as large as 100 times the amount of matter in galaxies (although no one could think of a mechanism to keep such a vast quantity of gas hot). However, subsequent X-ray observations have resolved this background in terms of point sources, indicating that an additional significant density of hot gas does not exist on large scales.

3. The one possibility not strongly observationally constrained is that a lot of

mass is contained in very compact dark objects, such as "intermediate mass" black holes (if they were too big then we would see the light emitted by objects falling into them, and if they were too small they would have to be so abundant that we would see the gravitational effects of nearby black holes), planetary objects the size of Jupiter, or dead stars that have not produced supernovae. None of these possibilities can be ruled out, but none is very compelling either. With the exception of dead stars, which some scientists have proposed may be abundant in our galaxy, no explanation for the formation of galaxies suggests that such objects should form in great abundance. But this prejudice alone does not exclude the possibility that these entities could be a significant or even dominant component of the mass in galaxies.

A recent set of experiments begun a decade ago have demonstrated that in fact a nonnegligible fraction of the mass of our galaxy exists in such objects, which go by the acronym MACHOS (Massive Astrophysical Compact Halo Objects), to distinguish them from the elementary particle WIMPs (Weakly Interacting Massive Particles) that will be featured in the latter half of this book. On the surface, these observations seem impossible to perform. If a significant fraction of our galactic mass is made of MACHOS, such objects will periodically pass in front of more distant visible stars. As I shall discuss shortly, the gravitational effect of a massive object passing between us and a more distant source allows this object to act like a lens, literally magnifying the light from the more distant object. (This takes place because the light from the more distant object bends as it passes by the intervening mass, due to Einstein's General Relativity.) The lensing effect causes the background object to "twinkle," or brighten for a short period, in a very predictable fashion. By continually monitoring the light from over 10 million distant stars on the outskirts of our galaxy each night and searching for a handful of such gravitational microlensing events over the course of a year, two different teams of researchers have established that up to 2–4 times as much matter may exist in such MACHOS as exists in all the visible stars in our galaxy.

Beyond these three specific alternatives and constraints, some theoretical arguments place an upper limit on the mean density of material in the universe made up of protons and neutrons—the components of normal matter. These arguments suggest that no more than about ten times the matter accounted for in visible stuff, as estimated earlier, can be present as normal matter in the universe today! I am sure you are salivating to know how we can make such a bold assertion.

However, right now I want to review the data only. Later I will discuss theory. Suffice it to say that these arguments, combined with the observational constraints discussed here, suggest that we are not missing too much ordinary material in our visual count.

Uncertainties about unaccounted-for matter aside, our weight estimate for luminous matter in the universe comes down in the end to knowing the mass of our own sun. Once this is determined, we can factor that mass into the expression given earlier and arrive at a number in pounds or kilograms or whatever. It is just as if the barker at the weight-guessing booth sized up a small boy as being equivalent to about five sacks of sugar. Unless the man knew how much a sack of sugar weighed, he would be no richer at the end of the day.

Weighing the sun is one of the most elegant and useful measurements we can make in physics. Describing how it is done requires returning to the seventeenth century when Isaac Newton unveiled his Universal Law of Gravity.

What made Newton famous was not his discovery of the law of gravitation, but rather his application of this theory to predict correctly the time it would take the moon to orbit around the earth. When all the relevant numbers were factored into his universal formula, he was able to claim that if the moon were pulled to the earth by the same force that apparently pulled Galileo's cannonballs down from the Tower of Pisa, and the same force that apparently made the moons of Jupiter orbit around that planet, then our moon should take about twenty-eight days to orbit the earth. Of course, this is indeed the case. Explaining the value of one of the most well known, but otherwise arbitrary, numbers in astronomy trumpeted for all the world that the heavenly bodies did not act capriciously, but that there existed order in the chaos, and that order was described by the laws of physics. Some two centuries later another young scientist would gain instant fame, not just by writing down a beautiful theory, nor even for having revolutionized his field some years earlier. What catapulted Einstein's name to the headlines was his correct prediction of a remarkable observation made during a special expedition to observe a solar eclipse. As he had claimed must be the case, photographs taken at the time of the eclipse displayed that even light itself must bend inward as it travels in the gravitational field of the sun. As we shall see, these two predictions, made by these two great physicists, form the key to probing for dark matter hidden among the stars.

How did Newton come to his celebrated conclusion about the moon? By "standing on the shoulders of giants." A generation or so earlier, the great Danish astronomer Tycho Brahe, followed by his German assistant Johannes Kepler, devoted the better part of their adult lives to classifying the motions of the planets in the sky. From the painstaking observations made by Brahe without the aid of telescopes, Kepler derived three celebrated empirical laws of planetary motion:

1. The planets move in elliptical orbits, with the sun as one of the foci of the ellipse.
2. The radius vector (that is, a line from the sun to the moving planet) sweeps out equal areas in equal times.
3. The square of the orbital period of each planet is proportional to the cube of its mean distance from the sun.

Newton was to depend crucially on each of these in deriving his Universal Theory of Gravitation. Earlier, he had codified his famous Second Law of motion, establishing the magnitude of the force that was required to govern the motion of a given object of known mass. Before him, Galileo had brilliantly established the idea that, left to their own, objects will continue to move in a straight line at constant velocity. While all physical objects on earth appear to slow down unless we continue to push on them, Galileo deduced that this was just an artifact of the existence of frictional forces. With no friction, a puck sliding on the floor should continue to slide forever, just as a puck at rest should continue to remain at rest. Zero velocity—an object at rest—is just a special case of constant velocity. To change the speed or the direction of motion of an object requires some force.

Newton correctly determined, using simple geometric arguments, that Kepler's second law of planetary motion mathematically reflected the fact that a force must be being directed radially along the line connecting the sun and each planet. The arguments that Newton used can be followed by anyone with high school mathematics, and for the interested reader I recommend the especially lucid discussion by Richard Feynman in his *Character of Physical Law*.[1]

Kepler's third law provided the key to determining the precise form of this cosmic force. This law can be simplified to yield a fixed relationship between each planet's orbital velocity around the sun and its distance from the sun. In this form, Newton was able to explain Kepler's third law based on a new "universal" law of gravity. Newton showed that if one posits a force to exist between the sun and the planets which depends on the product of their masses and inversely on the square of the distance separating them, then one could derive Kepler's third law. In addition, Newton showed that the proportionality constant in Kepler's law was in fact *directly related to the mass of the sun*.

But Newton went further: he also suggested that the same law of force would also be responsible for the motion of objects falling toward the earth and for the moon orbiting around the earth. Comparing the distance of the moon from the earth's center with the distance for objects located at the earth's surface, he was able to compare the predicted orbital velocity of the moon with the known motion of falling cannonballs and the like. From this he predicted the time it would take the moon to complete an orbit, and came up with his celebrated answer.

Having demonstrated the validity of his theory, Newton could use the new relationship that he had shown to exist between planetary orbital velocity and the mass of the sun in order to "weigh" the sun. He needed only to determine exactly the overall constant relating orbital velocity, distance, and mass. This could be done by applying this same relationship to the earth-moon system, where not only the velocities and distances were known but the mass of the earth as well. By inserting all the measured quantities into the relationship that he had determined should exist between them, Newton could extract the value of the

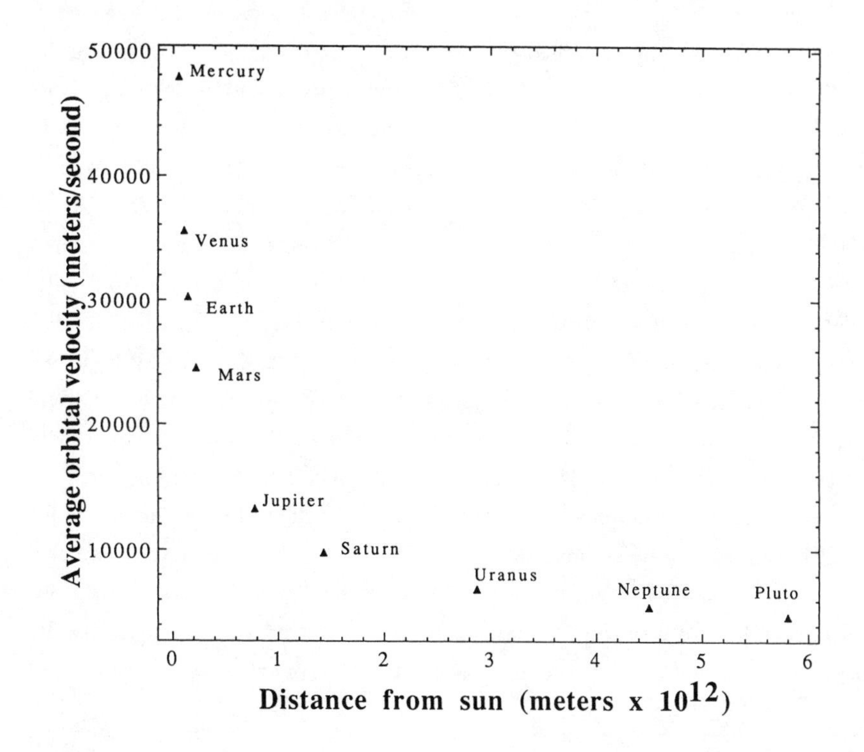

FIGURE 3.6

The average orbital velocity of each planet plotted against its average distance from the sun.

overall constant which was necessary to make the left- and right-hand sides of his relationship numerically agree. He then claimed that this constant was universal, in the sense that it was independent of the objects involved and dependent only upon the inherent strength of gravity.* He could then use this constant in the relationship he argued should exist among the planetary orbital velocities, their orbital radii, and the sun's mass in order to determine the mass of the sun.

Now in science, when one measurement is good, more than one is always better. For example, instead of applying this relationship once using just one planet, more accuracy is gained by trying to determine the sun's mass using the data simultaneously for all nine planets. Just for fun, I have plotted the average value of the orbital velocity of each of the nine planets versus their average distance from the sun (see figure 3.6).

Next, I draw a curve determined by Newton's relationship—with the mass of the sun adjusted so that this curve, *whose shape is fixed by this relation,* will fit these points as closely as possible (see figure 3.7).

In my entire career in physics, I have only once otherwise seen such good agreement between data and theory (and that was for the agreement between the COBE spectrum and the curve predicted for black body radiation, shown earlier). Newton's law of gravity works! From data such as these, we find that the mass of the sun is about 2×10^{30} kilograms. The accuracy of this value is limited by our uncertainty in Newton's constant, G. Were it not for this uncertainty, the planetary data would in fact allow one to determine the mass of the sun to better than one part in a billion.

We can imagine how this calculation might have gone astray, however. Suppose, for example, that there is a significant amount of dust in the solar system. Not enough to be noticeable, but say that out to the orbit of the earth, that is, in a volume that is about a billion times the solar volume, there exists about one solar mass worth of dust (in other words, the density of the dust is about one-billionth the mean solar density). Suppose also that the density of dust rapidly diminishes for larger distances, as, say, the inverse square of distance. What should one then predict for the orbital velocity of the planets? Well, Newton's relation would still hold, but now instead of utilizing the mass of the sun in this relation one must use the mass of the sun *plus* the dust contained inside the orbit of any given planet.

---

*This constant is called Newton's constant and given the label G (for "gravity"). We now know the value of this fundamental constant of nature to a part in ten thousand. This accuracy may seem impressive, but it is rather crude by modern standards, making G one of the poorest known fundamental constants and reflecting the difficulty of doing detailed gravitational experiments.

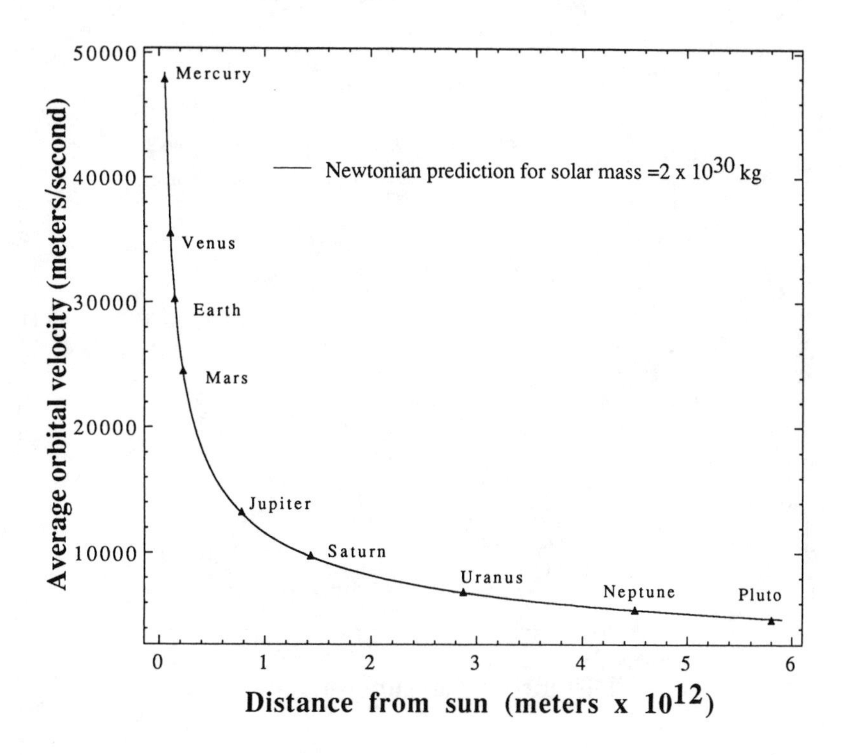

FIGURE 3.7

The average orbital velocity of each planet plotted against its average distance from the sun. Superimposed on the data is the predicted curve for orbital velocity versus distance obtained from Newton's analysis.

Let's forget the sun for the moment and just consider the effect of the dust alone. (It is easy to put the sun in again afterward if we wish.) If the density of dust diminishes inversely as the square of the distance, then it is straightforward to show that the total mass of dust enclosed inside any sphere centered on this distribution will continue to increase in direct proportion to the distance from the center. Using this result in Newton's relation among orbital velocity, distance, and mass, we would predict a planetary orbital velocity that is constant, independent of distance, for planets located outside the earth's orbit. This prediction is shown by the curve in figure 3.8.

If these conditions had been true, the history of science would have been radically changed. Kepler would have determined from observation that the orbital velocity of the planets approached a constant value instead of falling inversely as the square root of distance. From this relation, Newton would have derived a force law for gravity which varied inversely with the radius, not the *square* of the radius. He would have then applied this result to the moon, gotten the wrong

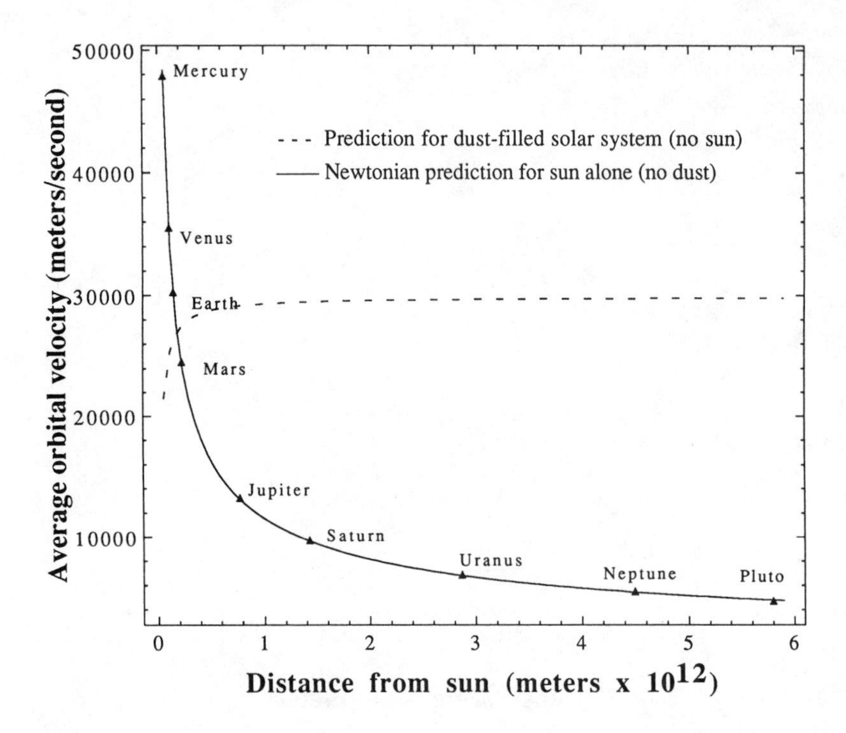

FIGURE 3.8

The average orbital velocity of each planet plotted against its average distance from the sun. Superimposed on the data is the predicted curve for orbital velocity versus distance obtained from Newton's analysis (assuming no dust), and also the prediction in the case that the solar system has a dust component of mass equal to that of the sun whose density decreases as the inverse square of the distance from the sun. In the latter case, only the contribution of the dust has been taken into account in deriving the curve.

answer, and perhaps would have retreated completely into mysticism and alchemy.

In any case, because the agreement between Newton's law of gravity and observation is so good, we can say with some confidence that there in *not* enough otherwise invisible dust in the immediate region of the sun to yield a total amount in the solar system which would constitute even a small fraction of a solar mass. One can place a limit on the amount of such dust in the solar system to be less than one-millionth of a solar mass.

Because we have a value for the sun's mass, the value presented earlier for the mean visible mass density of the universe can now be written in terms of kilograms per cubic mega-parsec if one wishes. But I have gone into such detail for a more important reason. This approach provides a second methodology for weighing the universe, akin to the man at the fair who hefts his customer before making a final estimate. *We need not rely on visual estimates of stars and gas to weigh the galaxy; we can use gravity to do it directly.* Moreover, while determinations based on visual material can always miss things, gravity probes all massive matter, regardless of its luminosity.

As I mentioned at the beginning of this chapter, our Milky Way is a classic example of a spiral galaxy. Such systems make up about 70 percent of all known galaxies. We are fortunate that one of our neighboring galaxies, the Andromeda galaxy, is also a spiral galaxy so that we may get a good idea what our own system looks like from outside. The COBE satellite in fact provided a recent picture of our own galaxy seen edge on, viewed in microwave and infrared radiation (see Figure 3.9). This image demonstrated that we indeed do live on the edge of a spiral galaxy. The Andromeda galaxy was first recorded on a celestial map by Abd-al-rahman al Sufi in A.D. 964, but it was not recognized as a separate galaxy until almost 1,000 years later. Andromeda is about twice as massive as our own galaxy. Because of the approximate circular symmetry of the disk of our galaxy, the sun orbits around the galactic center just as the planets orbit around the sun. As a result, Newton's arguments can be directly adapted to considerations of the sun's orbit around the galaxy.

Once again, by measuring the orbital velocity of the sun around the galaxy, we can estimate the mass enclosed inside our orbit. If we do this, the calculation appears to work remarkably well. We find that the mass enclosed is roughly equal to 10^{11} solar masses. Luminosity measurements also suggest that there are about this number of stars in our galaxy if each star, on average, is as luminous as the sun. Since we are on the outskirts of our galaxy, and the sun's orbit is determined by the gravitational pull of all the stars inside its orbit, it is reasonable, and reassuring, that these two numbers should agree.

Following the great success of measuring the mass of the sun, we can attempt to do better, by measuring the velocity of objects yet farther out, and then fitting these data to the curve predicted by Newton; one expects to find the curve shown in figure 3.10.

While most objects lie inside the orbit of our sun as it travels around the galaxy, there are certain objects which have more outlying orbits and which we can use as test probes in Newton's relation. Large "clouds" of diffuse molecular carbon monoxide (CO) gas can be measured somewhat farther out from the center of the galaxy. Still farther out can be seen small orbiting groups of stars called globular clusters (GCs). Even farther are the satellite systems of stars and gas

FIGURE 3.9
False-color image of the near-infrared sky as seen by the DIRBE. The image is presented in Galactic coordinates, with the plane of the Milky Way Galaxy horizontal across the middle and the Galactic center at the center. The dominant sources of light at these wavelengths are stars within our galaxy. The image shows both the thin disk and central bulge populations of stars in our spiral galaxy. Our Sun, much closer to us than any other star, lies in the disk (which is why the disk appears edge-on to us) at a distance of about 28,000 light years from the center. The image is redder in directions where there is more dust between the stars absorbing starlight from distant stars. This absorption is so strong at visible wavelengths that the central part of the Milky Way cannot be seen. DIRBE data will facilitate studies of the content, energetics, and large scale structure of the galaxy, as well as the nature and distribution of dust within the Solar System. The data will also be studied for evidence of a faint, uniform infrared background, the residual radiation from the first stars and galaxies formed following the Big Bang. (Courtesy of NASA Goddard and M. Hauser, T. Kelsall, D. Leisawitz, J. Weiland.)

called the Magellanic Clouds (MCs). This bright patch in the sky observable in the Southern Hemisphere was first recorded for Europeans by Ferdinand Magellan during his travels in the sixteenth century. (Recall that the Large Magellanic Cloud was the site of Supernova 1987A.) Even farther away we can observe small satellite galaxies orbiting in the gravitational field of our own Milky Way. When we measure the apparent orbital velocity of each of these systems we obtain the values shown in figure 3.12.

What has gone awry? Apparently, the orbital velocity of these objects around our galaxy remains roughly constant out to distances of almost ten times the distance of the sun from the center of the galaxy. At these distances the density of

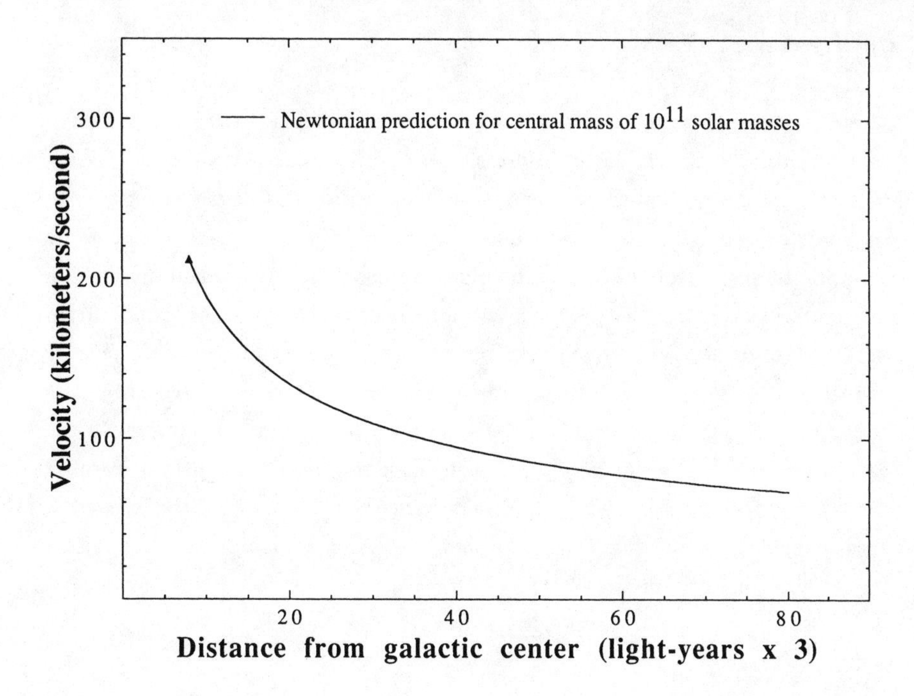

FIGURE 3.10 OBSERVED GALACTIC ORBITAL VELOCITY OF THE SUN AND PREDICTED CURVE FOR OUTLYING SYSTEMS

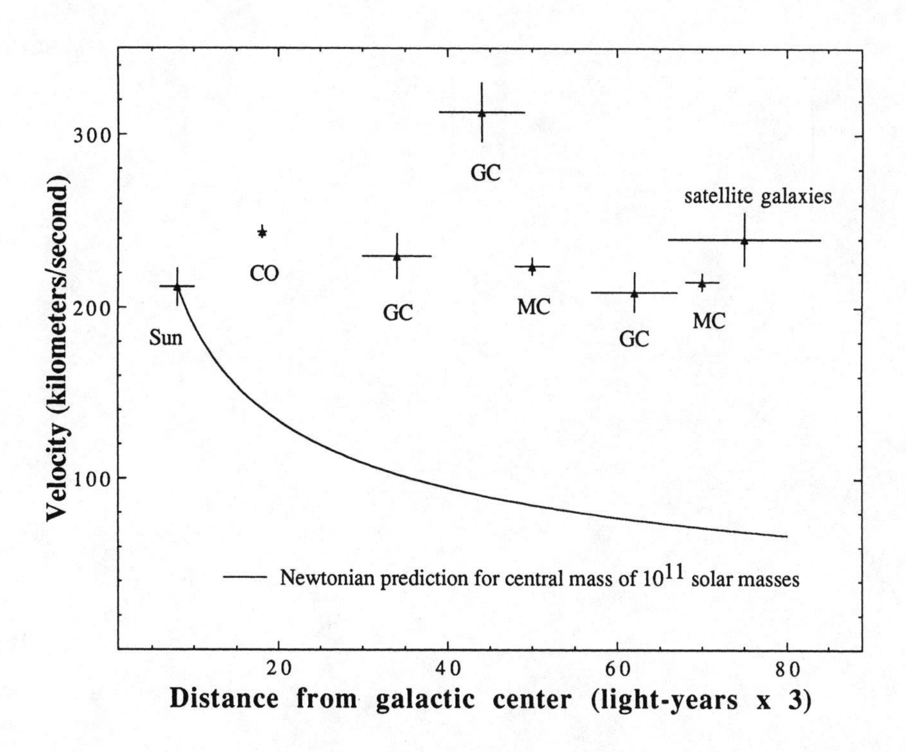

FIGURE 3.11 ACTUAL ORBITAL VELOCITIES IN GALAXY

stars has dropped off to almost nothing. How could such a situation arise? Well, as my cunningly chosen earlier digression on a dust-filled solar system demonstrated, such a situation can happen if, instead of the mass density dropping to zero outside the region of visible stars, it falls roughly as the inverse square of the distance as one travels outward. In this way, the mass enclosed in any region increases linearly with distance. Now, since the observed constant orbital "rotation" curve for our galaxy continues out to distances ten times that of the solar orbit, then the mass enclosed must apparently be ten times as great as the mass contained within this inner orbit. But almost all of the luminous matter in the galaxy is contained within this orbit. This leads us to conclude that at least *ten times the mass associated with all visible matter in the galaxy can be inferred to exist in the region of our galaxy. And this mass is distributed out to distances at least ten times as far as the visible matter.*

CHAPTER FOUR

Beyond Our Island in the Night

Not only does "dark matter" exist in our galaxy, it apparently dominates it! Such a surprising result drives us to investigate whether our galaxy is anomalous, or whether in fact this strange situation is generic. In some ways it is easier to measure the general properties of other galaxies than to measure our own. This is because most of our galaxy is obscured from our sight by dust. It is remarkable that while supernovae are perhaps the brightest objects in the universe during their short climax, we are unable to observe 90 percent of those that occur in our own galactic system. As noted earlier, Supernova 1987A, which occurred in a neighboring satellite galaxy, was the closest supernova observed in more than four centuries. Nevertheless, we believe, based on observations of supernovae in many other galaxies, that a supernova explosion happens almost once every thirty to fifty years in our own galaxy. These would be much closer to us than Supernova 1987A was, so they should be much brighter. Unfortunately, because we live at the edge of our galactic disk, a great deal of "junk" obscures our vision in the direction of what we see in the sky as the band of the Milky Way. We can only see clearly when we look away from this plane in perpendicular directions. In this way, we can look outside our galaxy at other galactic systems much farther away. Because we are located outside those systems, much of their detailed structure is visually more accessible.

As I described in chapter 3, when we observe distant galaxies we see that they are moving away from us in a manner described by Hubble's empirical velocity-distance relation. This relation is determined by measuring the amount by which their light is redshifted. We can observe literally millions of galaxies through visual and radio telescopes. Of these, the spectra of perhaps ten thousand have been measured with enough sensitivity to obtain a redshift estimate. (The ambitious Sloan Digital Sky Survey is currently under way. It will measure the redshifts of over 1 million galaxies and will completely revolutionize the details of our current understanding of large scale structure.) For a subset of these galaxies, perhaps a few hundred or so, we have done even more.

Since galaxies can be resolved as more than "points" of light on photographic plates, in principle we can observe their structure. If we observe a spiral galaxy that happens to lie at an angle to our galaxy, then as it rotates, one side of the galaxy will be rotating toward us and one side away from us. In this case, as we scan across the galaxy, light emitted from different parts will be redshifted by different amounts. In general, the Hubble expansion velocity away from us will dominate, so that lines will all be redshifted compared to their positions in terrestrial spectra. However, the side that is rotating toward us will have a slightly smaller velocity away from us, so that its spectral lines will not be as redshifted. Similarly, the other side of the galaxy should produce lines that are redshifted by a greater amount. Since the amount by which lines are redshifted is directly proportional to the velocity of the light-emitting object, as we scan across a galaxy the relative displacement of a given line compared to the mean galactic Hubble redshift should map out directly its relative velocity compared to the center of the galaxy. In this way we can determine how the orbital velocity of objects around the galaxy varies with redshift. If this velocity falls to zero, then the lines from emitting regions on the left and right sides of the galaxy should converge.

This kind of redshift analysis of light emitted in spiral galaxies has been carried out extensively by Vera Rubin and her colleagues at the Carnegie Institute. In practice, it is difficult to make direct spectroscopic measurements of the light from individual stars, so instead the light from clouds of gas surrounding certain very hot stars is examined. In figure 4.1 several spiral galaxies are displayed, along with spectra—obtained by Rubin and collaborators—of light emitted in a certain frequency range as one scans across the visual image. The two brightest lines represent the emission of light from hydrogen and nitrogen in the gas surrounding hot stars. As can be seen, on one side of each galactic center the spectral lines are shifted to a lower frequency and on the other side they are shifted higher. Note that these shifts quickly approach their maximum values and then remain constant out to the edge of the luminous disk, well past the region where most of the light is. On the right-hand side of the figure are rotation curves inferred from the observed shift in the spectral lines. Constant rotational velocity appears to be the norm, not the exception, out to a distance that in some cases exceeds 50,000 parsecs.

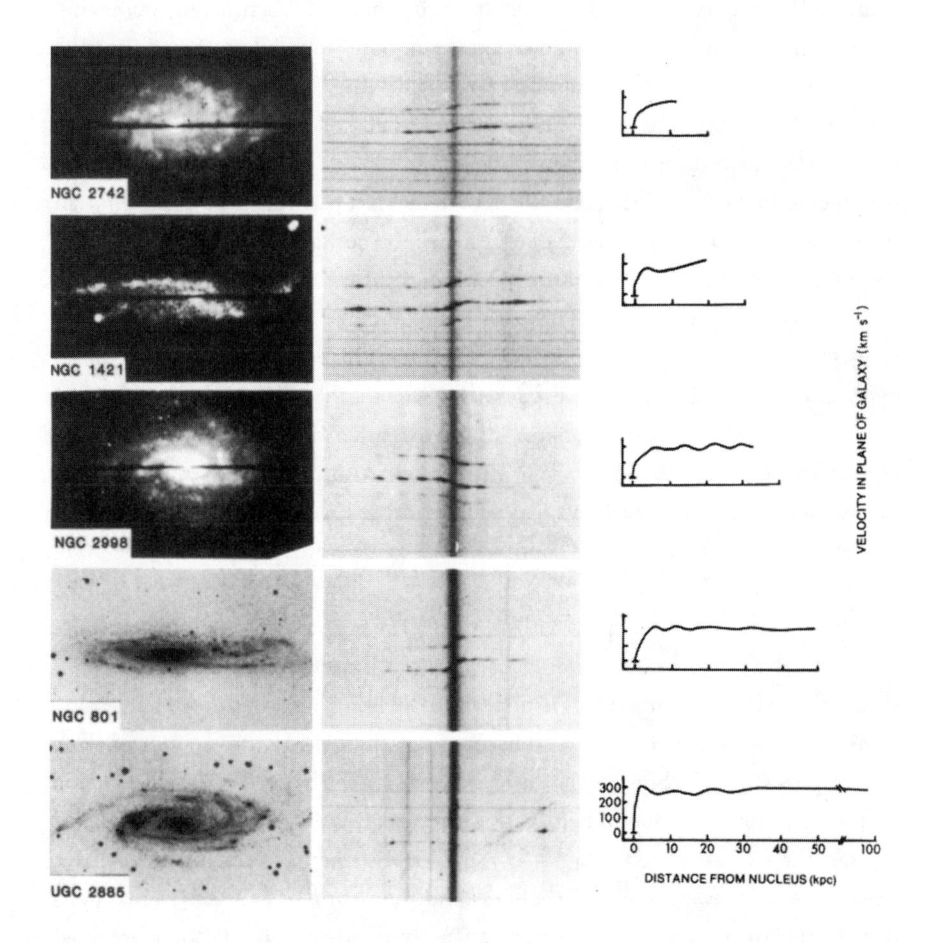

FIGURE 4.1

Shown here are five spiral galaxies inclined at an angle to us, along with spectra of light emitted in a certain frequency range and rotation curves of orbital velocity (in kilometers per second) as a function of distance (in kiloparsecs) inferred from these spectra. The two brightest lines in the spectra are from the emission of light by hydrogen and nitrogen in the gas around very hot stars. The Doppler shift of frequency in different directions on either side of the galaxy is clear. It is the fact that the shift remains constant out to the edges of the luminous disks which leads to the inferred flat rotation curves, suggesting a linearly increasing mass well outside the region in which most of the light is emitted. (Figure courtesy of Vera C. Rubin, from "The Rotation of Spiral Galaxies," *Science* 220 [June 24, 1983]. Copyright © 1983 by the AAAS.)

This kind of analysis has been performed for numerous spiral galaxy systems and the results always remain essentially the same: rotation curves are constant out to the limit of the luminous disk—in some cases almost four times the size of the region where the majority of the light is emitted. This appears to be true no matter what the size of the galaxy or its environment. Unfortunately, observing stars, or the gas around stars, allows one to explore only out to the edge of the luminous disk of a galaxy. Researchers have made efforts to circumvent this problem by probing the Doppler shifts in the frequency of radiation emitted (not in the visible range of the spectrum) by diffuse clouds of neutral hydrogen gas which extend out to well beyond the visible disk. In this way, measurements can be extended out to over twice the optical size, or perhaps eight to ten times the size of the region where most of the galactic light is emitted. Once again, rotation curves appear to remain flat. These results, combined with the optical results, suggest that at least 75 to 80 percent of the mass of these systems is "dark." Moreover, almost all of this material is contained outside the region in which most of the light is emitted. These estimates can be extended by examining binary systems: two galaxies that orbit around each other. Although these results are less firmly established, in binary systems separated by 100,000 parsecs, the rotation curves suggest that dark matter continues to provide a linear increase in the overall galactic mass.

These measurements imply that five to ten times more dark matter than visible stuff exists in galaxies. But perhaps more significant is the fact that this dark stuff is distributed differently. It appears to exist in an extended "halo" surrounding a galaxy. Inside the visible region of the galaxy, the amount of dark matter probably roughly equals the magnitude of the visible material. Outside this region, however, the mass to light ratio can grow by over a factor of 100 at the outskirts of measured rotation curves. This is one of the many puzzles we must solve if we want to claim to understand the nature of this material. Why is dark matter distributed so much more diffusely than stars? In addition, we must explain the apparent "conspiracy" that results in nearly equal levels of dark and light matter in the luminous cores of galaxies, so that as luminous matter drops off, the rotation curves remain constant, without increasing or falling off.

At this point the skeptical reader might complain that the interpretation of a constant rotation curve in terms of a spherical halo whose total mass increases with increasing radius may be unjustified from the data. After all, the disks of spiral galaxies are not spherical, merely circular. How can we assume that dark matter is distributed uniformly in a spherical volume when the luminous matter is restricted primarily to a disk?

To answer this we can rely on a wonderful property of nature. Anything that *can* exist usually *does*. Indeed, the observational astronomer scanning the heavens today with a telescope must feel somewhat like Darwin did as he roamed the Galapagos Islands in search of new life forms. As if designed to answer our question, a special type of spiral galaxy has been observed: here, in addition to the spiral disk, a "polar ring" is oriented perpendicular to the disk. Comparing the velocity of objects in the polar ring with those in the disk at the same radial distance from the center of the galaxy allows one to learn whether the gravitational force objects feel is the same in both places, and thus whether the underlying mass distribution is spherically symmetric. Measurements of at least four of these systems indicate, within an accuracy of 10 percent, that the systems seem spherically symmetric.

Theoretical support exists not only for a spherical distribution of dark matter, but also for the existence of spherical halos. In fact, some of that support comes from work performed well before these halos were first inferred from rotation curves. In 1973 P. J. E. Peebles and J. Ostriker pointed out that spiral disks, when left on their own, are gravitationally unstable. They will collapse over time into rotating bar-shaped objects. Roughly one-third of spiral galaxies do have central bars. However, Peebles and Ostriker showed that the spiral disks of galaxies could be stabilized if they were embedded in a spherical distribution of matter of comparable mass. Thus, the fact that we see spirals at all could have been taken at that time as a priori evidence that halos must exist. In any case, since these calculations were performed in 1973, other dynamical evidence, such as the existence of warping at the edge of some spiral disks, continues to point to the necessity for the existence of spherical halos.

The evidence for dark halos surrounding spiral systems is overwhelming at this point, but there is also evidence, although less well established, for dark halos in elliptical galaxies. Elliptical systems vary widely in mass and size; some form the largest known galaxies, with stellar envelopes extending out to more than 300,000 light-years. Because of the ellipticity of these systems and because we can only see a two-dimensional projection as we look out at the sky, it is difficult to use rotation curves directly to probe the gravitational potential. Some assumption about the shape of the galaxy must be made. Nevertheless, for the most spherical of the elliptical galaxies, rotation curves again appear constant, indicating the presence of a dark halo.

A very exciting potential direct probe of the mass distribution in elliptical galaxies involves the use of X-ray observations. As I noted in chapter 2, hot gas emits X rays. The energy distribution of the X-ray photons depends on the temperature of the gas that emits them, in the way that Planck derived in 1900. Assuming the hot gas is in pressure equilibrium (a reasonable assumption as long as the system does not change shape in the time it takes a "pressure" wave to travel across the system, which is the case for most galaxies), then the gas automatically will be spherically distributed about the galaxy. Moreover, because the gas particles frequently collide, one can be assured that their individual orbits are isotropically distributed. Now, in such a spherically symmetric system the density and temperature of the gas at any point is very simply related to the total mass distribution, and thus to the gravitational attraction at that point. Since the X-ray luminosity of the gas is related to its density, if we can measure both the X-ray luminosity at different points along the galaxy and also the temperature of the emitting gas throughout, then it is possible to determine unambiguously the underlying mass distribution.

Because the earth's atmosphere blocks out X-ray radiation from space, to measure the X-ray emission from galaxies we need to use probes outside the earth's atmosphere. Such an effort began with high-altitude balloon flights during the 1970s. Later, in 1984, the Einstein X-ray Satellite Observatory was launched; it provided unprecedented mapping of the X-ray sky. While excellent spatial imaging was possible, the range of photon energies to which the Einstein satellite was sensitive meant that only relatively poor temperature determinations could be made. Over the past decade several X-ray satellites have been launched—notably the ROSAT satellite launched in the early 1990s—that have improved tremendously our ability to probe systems on the scale of clusters of galaxies, as I shall describe later, although not yet with resolution on subgalactic scales. The Chandra satellite, launched in 1999, will raise our ability to explore X-ray sources to a new level.

Nevertheless, some data already exist. For one particular elliptical galaxy, M87, the data is good enough to allow a preliminary estimate of its total mass. Within a distance of about 600,000 light-years of this galaxy, the total mass enclosed is more than 200 times that which would arise from stars, each of about one solar mass, emitting the luminosity of this system. From this, we can infer the existence of about ten times as much dark matter as luminous stars and gas. This result confirms earlier provisional estimates based on the motion of satellite galaxies around M87.

Thus, dark matter halos appear to be relatively ubiquitous. Of course, further analysis will be needed to verify the early results for elliptical galaxies, but all evidence so far points to the existence of from one to ten times as much dark matter as light matter in and around systems varying in size by more than 5 orders of magnitude.* The evidence for dark matter in the smallest of these, the so-called dwarf spheroidal galaxies, is only now being amassed; the data could prove to be very telling. In the first place, these galaxies have very low luminosity, perhaps pointing to some mechanism that drives gas out of the system, such as supernova-driven winds. If this is the case, and if dark matter is made of something other than gas, there may be an excess of dark matter in dwarf galaxies compared to other galaxies. Early estimates of mass to light ratios in a few candidates support this idea. In addition, it is difficult to pack material that tends to be diffuse into a system as relatively small as a dwarf galaxy. If dark matter estimates hold up for these objects, this could place a very strong constraint on the nature of the material that makes up the dark matter. I shall return to this theme later.

Once we infer the existence of halos that cause the mass around galaxies to increase linearly with distance, the next question is, do these halos terminate? Does the linear increase continue out to fill in the mega-parsec or so gaps between galaxies? One way to determine the answer is to try to weigh much larger systems, enclosing many galaxies. As I stated earlier, about half the galaxies that we can see are parts of such systems, called clusters, or superclusters. How can we weigh these?

As we attempt to measure larger and larger systems, the process becomes more and more difficult. Observational data are more scanty and of poorer quality, and more theoretical assumptions need to be made. To be sure, these difficulties have not stopped astrophysicists from using existing data to try to answer the questions just posed. In fact, the first dynamical evidence for the existence of dark matter came from a consideration of one such system in the 1930s by Fritz Zwicky of the California Institute of Technology, an astronomer responsible for many seminal developments in the first half of this century. Zwicky examined the relative redshifts of many objects in the Coma cluster of galaxies. To his surprise he found that these redshifts varied by a large amount for different galaxies in the cluster. This implied that the relative velocities of these objects were large: that is, they were moving fairly rapidly relative to one another. On the other hand, when he estimated the total mass of the system by adding up the luminous galactic material, he found something very disturbing. Apparently the individual galaxies in the cluster were moving so fast that their velocities tended to exceed the velocity that would allow them to escape the gravitational pull of the system. The Coma cluster should not be stable. After the length of time since its forma-

*See appendix A for a discussion of orders of magnitude.

tion, it should have "evaporated," yet it had not. All evidence appeared to show that the cluster was a stable amalgam of galaxies.

Zwicky precociously concluded, and we now concur, that a resolution of this apparent paradox was possible if the cluster was a lot heavier than he had surmised by adding up the luminous matter. In this case the escape velocity from the system would be increased and the cluster would remain dynamically stable. In arriving at this conclusion Zwicky set an important precedent. He showed that dark matter can be detected indirectly by its *gravitational* effects, even if it can not be seen directly. To date, all demonstrations of the existence of dark matter, on structures ranging from dwarf galaxies to superclusters, have been based on this idea.

We use essentially a minor variation of Zwicky's method to weigh clusters and superclusters today. We are not able to apply Newton's relation among orbital velocity, distance, and mass directly because these systems are less coherent than the solar system, or the galaxy. In the first place, superclusters are not well-defined, symmetric, orbiting systems. They are just what the name implies: a cluster of galaxies jumbled together, each moving in the gravitational field of all the other galaxies in the system, but in no uniform way. In such a situation we rely not on measurements of individual motions but instead on statistical arguments to relate velocities to mass. The basic idea goes like this: in "self-gravitating" systems—systems held together by their own gravity—the average velocity of objects relative to each other increases as the total mass of the system increases in a prescribed way. Specifically, the square of the average relative velocity of objects in the system depends directly on the total mass in the system.

Newton's relation demonstrated that this result is explicitly true on an individual basis and not just in a statistical sense for the special case of objects in orbit around a central mass such as the sun or the galaxy; therefore, it is not so hard to believe that this result can be generalized. In fact, it derives from a very general theorem in mechanics called the *virial theorem*. Closely related to both Newton's Second Law and the conservation of energy, it states that if any self-gravitating system settles into dynamical equilibrium, its total energy will be balanced in a fixed way between the "kinetic" energy of motion of its constituents and the "potential" energy that is stored due to their mutual gravitational attraction.

This general result is intuitively sensible. Imagine that I drop a handful of marbles into a well whose sides and floor are very hard so that the marbles do not lose any energy when they bounce off them. As they fall, the marbles speed up. The deeper the well, the greater will be their final speed before they hit the floor. After they hit the floor they will rattle around the well, bouncing against each other and the walls. Since no energy is lost to the outside, their total energy of motion will remain the same. Thus, if we measure the average speed of the marbles after they have rattled around for awhile, we can get a good estimate of the depth of the well. The same holds true for galaxies, or clusters of galaxies. Stars first form as diffuse gas particles "fall" together under their own gravitational attraction. Galaxies form as stars "fall" together, and clusters form as galaxies "fall" together. It is reasonable that if no forces other than gravity are relevant and if the system has come into equilibrium, the relative velocities of the galaxies in a cluster should reflect on average the "depth" of the gravitational "potential well" into which they first fell.

This statistical variation of Zwicky's original analysis has since been performed on many different clusters of galaxies (as well as on individual galaxies), and all results are pretty consistent. About ten to twenty times the mass exists in these largest systems compared to the luminous mass associated with the galaxies and gas. This result is very important. It suggests that while dark matter does indeed dominate these structures, the linear growth in total dark mass seen around galaxies does not seem to continue at the same rate, at least when we probe out to scales associated with clusters of galaxies—approximately 3–30 million light-years across.

Applying the virial theorem can be a very tricky business. It is clear that if applied to systems not acting under their own mutual gravitational attraction the theorem could yield nonsense. Two unassociated objects flying past one another would in this analysis suggest that there was a great deal of mass associated with the system in which one presumes they are a part. However, as their motion is unrelated to their gravitational attraction for one another, this conclusion is not warranted. How, then, can we tell whether objects are responding to gravity—whether they form part of a "self-gravitating" system—or whether their relative motions are purely fortuitous? It is not always so easy, and some well-known mistakes have been made.

One such example is the case of the Cancer cluster. When observed against the plane of the sky this cluster of galaxies appears to be the model test system. It looks spherical and not too lumpy—an ideal candidate for a system of closely associated galaxies in dynamical equilibrium. When the virial theorem was first applied to this system, it yielded a result consistent with what one might expect from a continuation of the flat rotation curves of galaxies out to larger distances. This cluster appeared to contain almost fifty times as much mass as could be associated with the galaxies within it. The total mass in the halos around the galaxies appeared to continue to rise linearly as one traveled outward, so that the cluster mass was five to ten times more massive still than the mass of the galaxies contained within it, even when the nearby dark halos supposed to surround these galaxies were included.

This result stood until Gregory Bothun and colleagues at the Smithsonian Center for Astrophysics reanalyzed the Cancer cluster in 1984. Although the Cancer system appears homogeneous when viewed on the plane of the sky, new data on the individual redshifts of each galaxy in the system made it possible to conclude that the galaxies are not distributed homogeneously. Remember that objects that are farther away have a larger redshift than nearby objects, due to the Hubble expansion. If a series of closely associated objects in the cluster has a systematically greater redshift than other objects, one can investigate to learn whether this subgroup actually forms an isolated cluster that is spatially separated from the rest, but only appears to be associated with them because we are using a two-dimensional projection to look at a three-dimensional system. This was the flaw in the case of the Cancer cluster. With the aid of computer-generated projections along other directions, Bothun and colleagues were able to show that the Cancer cluster in fact contained five separate self-gravitating subgroups which were spatially removed from one another. Within each subgroup, the spread in velocities was much less than the spread in velocities between subgroups. What had earlier been interpreted as a large virial mass could now be seen to be just a reflection of the Hubble expansion.

This example reminds us that we must be very cautious when drawing conclusions from observations of large-scale structure. The field is very young; "reasonable" assumptions may be incorrect. It is useful to keep this in mind as we proceed in this book. Much will be built on a theoretical edifice, which is at times primarily based on plausible assumptions and a limited amount of data. Although it is very unlikely that all of the assumptions are wrong and that the entire theoretical framework is misguided, it is likely that as new data are obtained certain currently accepted ideas may have to be altered. This is not a cause for despair, but rather for excitement. Observational cosmology is in its infancy and we are no doubt in for some surprises.

There exist other independent ways of measuring the mass of clusters and superclusters, and it is significant that they all tend to give similar answers. Two methods, in particular, have over the past decade provided the best constraints on cluster masses and dynamics. The first method makes use of Einstein's General Relativity theory, as we shall discuss later in this chapter. The second method has already been briefly introduced. Clusters are known to have substantial hot gas in them. In fact, the amount of hot gas in between galaxies in clusters far exceeds the total mass in the galaxies themselves, by at least a factor of two. This gas is distributed in such a way as to suggest that the systems are often in hydrostatic equilibrium—that is, that the gas pressure balances the gravitational forces in the systems. Recently, with the EINSTEIN, GINGA, ASCA, and, most importantly, the ROSAT satellites, resolution has been obtained that allows the mass of these systems to be explored with unprecedented precision. In fact, these observations give the best limits on the mass of these, the largest gravitationally bound systems in the universe, and also on the total baryonic matter in the universe. Limits suggest dark matter to luminous matter ratios in the range of 10–30.

To date, the virial theorem has yielded evidence of dark matter in individual systems ranging from dwarf galaxies containing millions of stars to large clusters containing perhaps a million billion stars. Since it is primarily a statistical measure, the virial theorem has the virtues and failings conventionally associated with statistics. If the sample studied is "unbiased" and the assumptions used to relate various average quantities are valid, then as long as the number of data points is large enough the viral theorem can yield an accurate answer. Of course, it is not always easy to determine that these two prerequisites are met. We have seen one example where the second requirement—validity of the underlying assumptions—was not met, and the inferred results were false. The first requirement—that of an unbiased sample—is generally much more subtle and difficult to ensure. A great deal of probability theory is devoted to measuring just how "unbiased" a sample can be. The sensitivity of the virial theorem estimates to this aspect of statistics is nowhere more apparent than in the ambitious "cosmic virial theorem."

Jim Peebles, at Princeton University, in the 1960s and 1970s pioneered the use of statistical measures of galaxies to probe for cosmological structure in the universe. Peebles demonstrated the existence of an intuitively believable generalization of the virial theorem as described here. If the mean velocity of pairs of galaxies in the "potential well" of a cluster should reveal the depth of this well, and hence the mass of the cluster, then the mean velocity of pairs of galaxies taken all over the sky, whether in clusters or not, might give a good idea of the mean mass density in the universe. This, after all, is one of the things we are after here. Moreover, one might hope in this way to reduce the impact of the difficulties associated with individual systems such as the Cancer cluster, discussed earlier.

In the cosmic virial theorem, one relates a statistical estimate of galaxy clustering on different scales with a statistical measure of the relative velocity of galaxies. If clustering is caused by the gravitational attraction of galaxies, then

Peebles showed that the probability of finding one galaxy within a given radius of another should be related to the average value of the square of the relative velocity of galaxies separated by this same radius. The constant of proportionality in this relation is, not surprisingly, the average mass density of the universe.

Statistics have been compiled both on the clustering and on the relative velocity of a large number of galaxies on scales ranging from about 50,000 light-years—when galaxies are nearly touching—to more than 10 million light-years, in excess of the average distance between two well-separated galaxies. The mean mass density of the universe so derived turns out to be once again between ten and thirty times the mass density of luminous matter.

This result should, in principle, settle the issue of dark matter abundance. Not only does this "cosmic" result suggest that dark matter exists universally, but the measurements should give a "system independent" estimate of just how abundant on average it is. The cosmic virial theorem agrees numerically with essentially all the other independent dynamical measures I have described. Apparently between ten and thirty times the visible mass in the universe is dark. This material is located around galaxies of all types, from the smallest to the largest, and it is distributed such that the mass around galaxies continues to grow linearly out at least as far as distances of a few hundred thousand light-years. Beyond that, the measures of binary galaxies, clusters of galaxies, and the cosmic virial theorem all suggest the dark matter contribution may level off.

However, just when it seems safe to accept this result as a given, and go on to discuss its implications, I want to discuss why it might be wrong. The cosmic virial theorem relies on one key assumption: that galaxies provide good "tracers" of mass in the universe—namely, that they provide an "unbiased" sample. But why should they? After all, I have just spent most of this and the last chapter showing that the visible matter associated with galaxies probably makes up less than 5–10 percent of all the matter in the universe. If dark matter is the dominant matter in the universe, it is reasonable to expect visible material to cluster around the places where the dark matter is dense, rather than where it is not. But it does *not* follow that the dark matter should always cluster where the light matter is. Since there is much more dark matter than light, there may be huge concentrations of dark matter where there are no galaxies. I will describe in the next chapter why, if galaxies are relatively rare occurrences, one could expect this latter situation to occur often. If galaxies turn out to be biased tracers of mass, then the estimate we obtained for the mean mass density in the universe goes out the window. The mean mass density cannot be much smaller than the value I have quoted, but it could be larger—five to ten times larger. Further probes are required in order to further explore this possibility, and we shall discuss these in subsequent chapters.

Now, readers may point out that the cosmic virial theorem was only one of several different dynamical arguments to probe the dark matter abundance which I have described, *all of which* appeared to lead to the same estimate. Well, the problem is that all of the arguments that I discussed so far are based on probing the abundance of material within a few tens of millions of light-years of galaxies at best. If dark matter or energy exists in empty spaces of the universe, some of which are also tens of millions of light-years across, in a way that is uncorrelated to its abundance near galaxies, then *none* of these methods would be sensitive to this material. This issue has recently taken on renewed significance.

Only in the past decade or so have we really begun to learn about the structure of the universe on such large scales. We do have some provisional measures, but they are inconclusive. One of the better established of these involves our "infall" toward the center of the nearest supercluster of galaxies. This system, called the Virgo supercluster, is centered on the Virgo cluster of galaxies, about 45 million light-years away from us (assuming $h = 3/4$). If we measure the abundance of galaxies in the direction of Virgo out to this distance, compared to that in other directions, we find it to be about twice as large. This suggests that there should be a net gravitational attraction in the direction of Virgo, that is, our galaxy should be falling toward the center of this system. If we can find out how fast, then we can estimate the mass enclosed in this region.

How can we measure our local motion toward Virgo? There are two ways. First, if we accept that our local motion is superimposed upon the uniform Hubble expansion velocity, one way to extract this local component would be to look for anisotropies in the Hubble velocities of all galaxies in our neighborhood. If we are all falling in toward Virgo, then galaxies that we measure in a direction away from the Virgo supercluster should have systematically smaller velocities away from us than galaxies measured in the direction of Virgo. If we assume a spherical infall, and measure this anisotropy, we find that on a scale of about 30 million light-years around us, the total mass enclosed appears to be about twenty to thirty times that accounted for by visible material. Once again, this accords with the earlier estimates made over smaller regions. Unfortunately, however, this figure depends crucially on the assumption of spherical symmetry. And if this assumption is relaxed, the number can vary significantly.

There is another way to measure our local motion, which takes us back to the notions of the aether discussed in chapters 1 and 2. Recall that attempts to measure the earth's absolute motion failed, demonstrating that there was no possibility of defining any absolute frame that all observers agree is at rest. Nevertheless, recall as well the discovery of the cosmic microwave background. This bath of microwave radiation permeates space as an afterglow of the Big Bang. We believe that this background originated from a time when the universe was very young. Hence the microwave photons, which we observe coming from different directions, have traveled billions of light-years. This light has been redshifted by over a factor of 1,000 since its emission. Effectively, the sources of these photons are much farther removed from us than any of the galaxies we can see. The microwave background, therefore, *itself* emanated from an average frame that was independent of any motion in our local "neighborhood." In this case, we can use the microwave background to probe for such local motion.

Although the Michelson-Morley experiment yielded a null result, indicating that the speed of light is the same in all directions, I have described how, if we are moving relative to the source of a light ray, this light will be redshifted or blueshifted by the Doppler effect. Now, if we are moving relative to the frame defined by the network of sources of the microwave background, then this background will be redshifted more in a direction *opposite* to our motion than in the direction toward which we are moving. Such a situation is called a dipole anisotropy, since its redshift varies from one "pole" to the other (see chapter 2). Since we imagine that the sources of the microwave background are scattered over a very large sphere with a radius on the order of 10 billion light-years, it is reasonable to expect that these sources define a frame that is at rest relative to the Hubble expansion. Any small-scale deviations from this background expansion should be wiped out in the general microwave background signal. Hence, if we can measure the dipole anisotropy of the microwave background as seen from the earth, we should be able to extract out our local motion from the background Hubble flow.

This measurement has been done with great accuracy by the COBE satellite. The result implies that our local velocity compared to the frame defined by the microwave background is almost three times as large as our apparent infall toward Virgo, as measured by fitting the anisotropy in the Hubble flow of galaxies to a spherical infall toward Virgo. Furthermore, the direction of our local motion is *not* in the direction of the Virgo supercluster.

This situation has caused a great deal of confusion. Are we falling toward Virgo or are we not? Is our local motion due to the gravitational pull of nearby objects, or is it due to some primordial "kick," which would be very difficult to explain at present? Some steps have been taken to understand this discrepancy. One group of seven astronomers from almost as many institutions, nicknamed the Seven Samurai, claimed to have measured a net infall not to Virgo but to another, possibly much more massive system located in another direction. This system has become known as the "Great Attractor." If this theory holds up, it might suggest a great deal of mass in this faraway object. On the other hand, given current models for galaxy formation that I will describe later, it would be extremely baffling to understand how this large mass distribution could have accumulated. An alternative explanation of this situation was proposed by Marc Davis, at Berkeley. Based on data from a satellite measuring infrared radiation, Davis and collaborators made an infrared map of the galaxies over the whole sky. From this map they claimed to be able, at least in principle, to determine the "potential well" in which we sit, due to the gravitational attraction of all of these systems. They suggested that our local motion compared to the microwave background might be due entirely to the net gravitational attraction of the known massive systems that their survey probes. This implied a mean mass density out to very large regions, which is perhaps twice as large as suggested by the cosmic virial theorem. Unfortunately, more recent analyses by Davis and his colleagues suggest a smaller value, in disagreement with their initial estimates. In any case it is clear that one way or another, dynamical measurements such as these, even if they confuse us in the short term, may eventually resolve the controversy over just how much matter exists in the universe—and where it is.

So far, all the dynamical tests for dark matter that I have described involve only Newton's laws, and Newtonian gravity. This fact has caused at least one set of researchers to suggest that a failure of the universal law of gravity may be responsible for the apparent excess of dark matter over light matter. It has been suggested that if the force of gravity deviates from its Newtonian form at large distances, one might try to explain the flat rotation curves without resorting to dark matter. Unfortunately, this appears to be one of those cases in which the solution is uglier than the problem. It may be slightly radical to suggest that galaxies are dominated by dark matter. But to suggest an alteration in one of the four known forces of nature to explain these observations seems to me excessive. From an elementary particle perspective, which is, after all, how we must understand microscopic forces in a quantum world, there is no present theoretical or experimental justification for this proposal. Nevertheless, our recognition of the dependence of all of these tests on Newtonian gravity reminds us that nowhere in this analysis have we utilized the most modern theory of gravity: Einstein's general theory of relativity.

In order to describe a number of features of the universe as a whole, we require the curved space-time of general relativity. Yet because general relativity

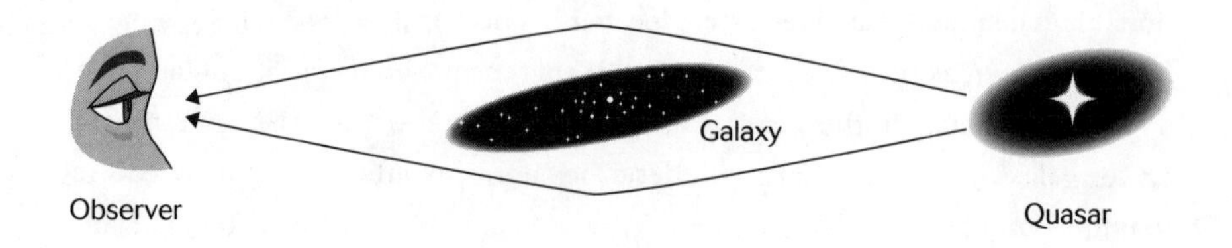

FIGURE 4.2
Light rays emanating from a distant quasar may be bent in the vicinity of an intervening galaxy so that they can be "refocused" for an observer on earth.

becomes equivalent to Newtonian gravity when masses are not too large and speeds are nonrelativistic, for many cosmological purposes we can approximate space-time as flat. This may come as a surprise, but one need only recognize that the dynamical velocities in even the largest superclusters are only a small fraction of the speed of light and one can show that the general relativistic corrections to the dynamics of these systems are miniscule.

Nevertheless, it is perhaps fitting that the observation which made Einstein famous can also be appropriated to probe for dark matter, just as Newton's relation has been used so exhaustively in this search. General relativity theory predicts that space-time is curved near massive objects. One of the direct implications of this prediction is that light, which travels locally in straight lines, will appear to bend around massive objects because the space-time in which it is traveling is curved. One can try to mimic this action by using Newtonian gravity and merely allowing light to be attracted by gravity, just as matter is. This is certainly a very logical thing to expect, given the equivalence between matter and energy exposed in Einstein's special theory of relativity. Yet if we do this, we arrive at a different answer than we get by assuming that space is curved. It turns out that, in the latter case, light bends by exactly twice as much as we would calculate in the former case. It is this bending that was measured by Eddington in 1919, and which helped make Einstein's name a household word.

Because light bends around objects, one can imagine the rather fantastic following scenario: If enough mass is located in a region, then light rays emanating from some source far behind this mass distribution may be bent sufficiently on both sides to be refocused at the point where an observer is located. Imagine the situation shown in figure 4.2. In this case, the mass distribution will act as a "gravitational lens." Not only might intervening masses magnify distant objects in this way, but they might also produce multiple images of these objects. Indeed,

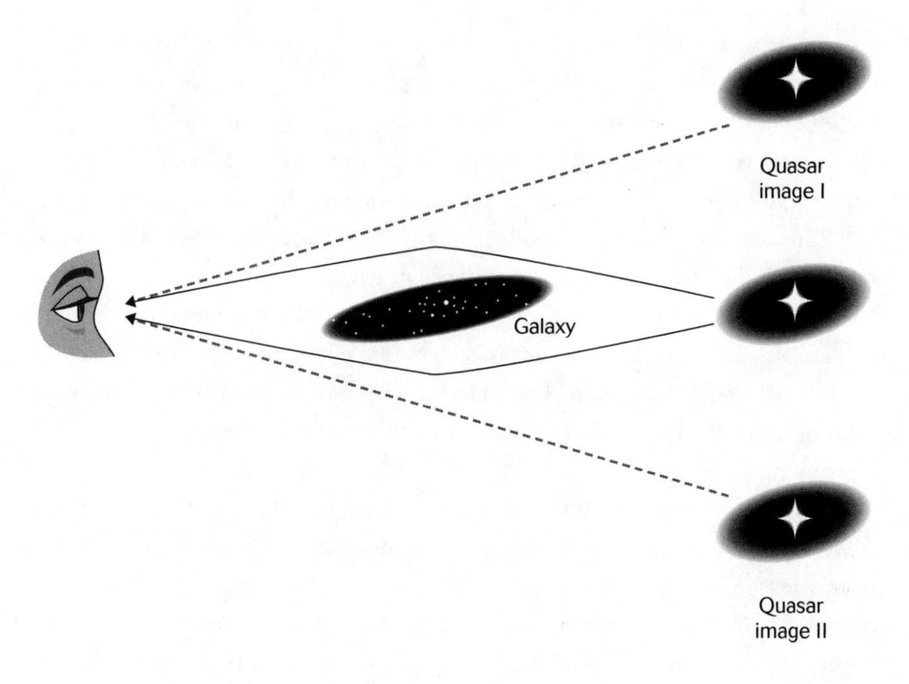

FIGURE 4.3
Double image of a quasar, as might be seen by an observer, due to gravitational lensing.

observers will perceive two images, because they will imagine that the two rays shown will have traveled in straight lines. Extrapolating backward, they will see, instead of the original object, an image in each of the locations shown in figure 4.3.

Einstein himself realized this peculiar possibility in 1920, but quickly dispensed with it, saying that the effects were so minute that they could not be measured. Yet sixty years after he dismissed the problem, such effects were observed! Earlier, I mentioned the existence of quasars: unbelievably bright, compact objects seen at the outer reaches of the observable universe. Perhaps these are primordial galaxies, perhaps they are supermassive black holes gobbling up whole galaxies. No matter, we see them, and like all luminous objects, each quasar produces a spectrum of light that is essentially unique. And since 1979, observations of quasars have yielded many candidates that appear to be multiply imaged.

Since the first observation of a candidate multiply-imaged quasar by D. Walsh and his collaborators in 1979, dozens of groups around the world have studied visible and radio images of thousands of quasars looking for gravitational lensing candidates. Over the past five to ten years, the Hubble space telescope has been of great use in searching for multiply-imaged quasars. First one searches for quasars that are very close together on the plane of the sky. Next one determines whether quasars close together on the plane of the sky have nearly identical redshifts. This is relatively unlikely to occur naturally, since this would imply that the quasars were very close together in three-dimensional space, and quasar densities are not that high on average. Next, one examines the spectra of the quasars. If the spectra are identical, the odds strongly suggest that they are not two separate objects, but rather two images of the *same object*. Finally, one searches for an intervening galaxy or cluster of galaxies. From the observed mass distribution, one can attempt to model the kinds of images that might be produced if such systems are acting as gravitational lenses. In several cases the intervening system has been observed and the predictions can be compared with the observations (see figure 4.4).

Now what does all this, interesting as it may be, have to do with dark matter? A lot. In the first place, for a gravitational lens candidate to produce a given set of quasar images, the total amount and the distribution of matter in the system is then fixed. In order to match the observations for the very few "gold-plated" events with a known gravitational lensing candidate, one can determine unambiguously what the matter distribution is in this system. And once again observers have found evidence for dark matter. Studies of the gravitational lens candidates suggest that the lensing galaxies can contain more than ten times the total visible mass. What is surprising from this growing set of observations, however, is the *variability* of the amount of dark matter needed to explain the observations. Some galactic systems appear to require no significant amount of dark material in the central galactic regions. The cases where models can be compared with observations also confirm that where dark matter is required, this material cannot be distributed like the luminous material. More interesting perhaps, there are also cases where it appears that the dark matter distribution need not have the same center of mass as that of the luminous material, that is, the two distributions can be "offset" from one another. Also of great interest in using lensing to constrain the total clustered mass on the scale of galaxies and beyond is the case of an "Einstein Ring." As Einstein first noted, by symmetry arguments, an intervening mass located directly along the line of sight to a distant object will produce not separate multiple images of that object, but rather a continuous ring. The radius of this observed ring is proportional to the mass of the lensing object. Several cases of near-perfect Einstein ring images of more distant galaxies have appeared in radio observations. Of course, from a few gravitational lenses one cannot infer general statements. Several sky surveys are under way in search of more lenses; we await the results with some eagerness.

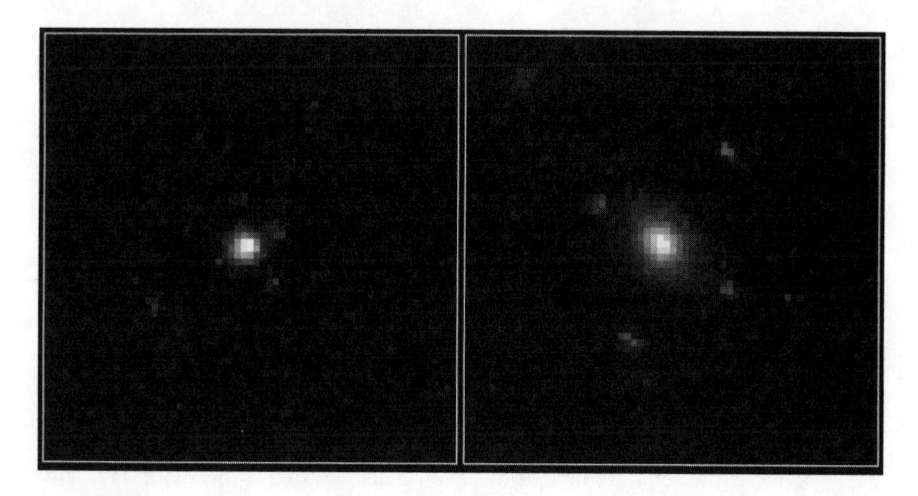

FIGURE 4.4
These two objects represent a new distant class of quadruple, or cross-shaped, gravitational lenses which might eventually provide astronomers with a powerful new "magnifying glass" for probing the cosmic mass distribution. Depending on the alignment between the objects and the mass distribution of the foreground lens, the more distant object can be smeared into arcs or split into pairs, triples, or even quadruple images. The model for the observed lens configuration shows that the mass of these galaxies consists of predominantly dark matter in a very elliptical distribution. (Kavan Ratnatunga [Johns Hopkins University, Baltimore, Md.])

Another result of some import for dark matter considerations may come from searches for gravitationally lensed quasars. Although the two or more images of a given quasar may be separated on the plane of the sky by less than a thousandth of a degree in a given lens, the light that produces each image has traveled on a path that may differ from that of a neighboring image by several light-years. This is true because the quasar is likely to be several billion light-years away from us, so that even a minute change in path can result in a significant difference in distance. If the light from two images travels different lengths, then it will take different amounts of time to reach us. If the path-lengths differ by one light-year, it means that the light from the images will be one year out of synch. One can then imagine the following scenario. Say that one of two quasar images is observed to "flicker," that is, the light output of the quasar varies slightly on some time-scale, say days. Observers then monitor the companion image scrupulously for the next two years, and during this time the companion image "flickers" in exactly the same way. Not only would this finding provide additional evidence that the pair is in fact due to a single quasar that is gravitationally lensed, it could also provide, at least in principle, a quantitatively accurate distance measure for objects that are very far away. This is because from the observed image splitting, and the redshifts of the quasar and the lensing galaxy, we can deduce what the expected difference in path-lengths should be and the expected time delay between the images. Observing such a time delay would then pin down accurately the actual distance from us of these objects, and allow us to correlate redshift and distance very accurately. One of the main problems in weighing the universe is the uncertainty in the Hubble constant relating these two quantities of redshift and distance. Such a measurement would pin this down, at least in principle.

In fact, the time delay between separate images of a lensed quasar—in fact the very first multiply imaged quasar, 0957 +561, discovered in 1979—has now been very well measured. In a remarkably convincing set of observations, the time delay between the separate images has been demonstrated to be about 417 days, with an uncertainty of only about a day or two. Nevertheless, like many cosmological observations, what in principle allows a simple unambiguous extraction of the Hubble constant is in practice more ambiguous. This is because two crucial factors in our ability to relate the measured time delay to a distance are the gravitational potential of the galaxy and the mass distribution in which the galaxy sits. Our uncertainty in this mass distribution, while now greatly reduced, is still responsible for the remaining uncertainty in the Hubble constant deduced from these observations. Nevertheless, the value deduced from these observations is now completely consistent with other measurements, suggesting the Hubble "fudge-factor" h is in the range 0.65–0.75.

We might learn still more from observations of gravitational lensing. Once a large data base of lenses is amassed, we can begin to undertake statistical research. On the basis of the observed galaxy distributions in size and in space, we can make predictions for what the frequency of lensing and what the mean separation of quasar images should be, for example. If the observations do not agree with our expectations, this could be a hint that something else is contributing to the lensing of distant quasars. Already there are several confusing features of the observations. Of the handful of good lensing candidates, only three have visible galaxies that could act as lenses. In at least two cases one can infer that the object or objects that are doing the lensing should be large enough to be visible, but nothing is seen. One is tempted immediately to take this as evidence for some massive dark object, such as a "dark galaxy." On the other hand, I have learned after a decade or so of false alarms in my own field of particle physics that the statistics of rare events can be very misleading. With only one or two observations, many anomalous factors can contribute to some very perplexing observations.

One must await more and better observations before jumping to any such conclusions, exciting though they may be.

In fact, several theoretical factors argue against the idea that these events are due to large systems composed primarily of dark matter. Initial theoretical models developed by Ed Turner and his colleagues at Princeton suggested that if there were enough such isolated systems to alter our estimates of the total dark matter content of the universe, then we should see many more lensed quasars than we do. One might hope, accordingly, to turn this suggestion around to place a limit on the amount of dark matter in the universe. Unfortunately, the existence of dark matter and the existence of dark matter distributions that can produce lenses are two different things. Indeed, work done in 1985 by Gary Hinshaw and myself suggests that since halos in observed galaxies are diffuse, one would expect a "dark galaxy" consisting primarily of dark matter to be too diffuse to produce multiple images of distant quasars. Thus, regrettably, we cannot probe for such systems as effectively as one might have wished by using lensing statistics. Also, this provides further ammunition for believing that our present observations are not likely to be due to dark matter. One other way to discover whether dark matter is present in huge quantities, even if it is not clumped and thus does not directly cause a lensing phenomenon, would be to see how the number and type of lensing events increase as we look out to farther distances. In a universe with four or five times more dark matter than is indicated by the virial estimates of galaxies, the number of galactic lenses seen as a function of distance should be measurably different, because the geometry of space-time would be different.

In fact the statistics of gravitational lensing of quasars can do more than merely tell us about the geometry of space-time. It can also tell us about the composition of the dominant matter and energy in the universe. There is a relation, given by general relativity, between the actual distance to distant objects and the observed speed of redshift of these objects. This relation varies as the dominant energy governing the expansion of the universe varies. For example, in a universe in which the character of the expansion is dominated by the energy associated with the virtual particles in the vacuum, as opposed to the energy associated with real particles of matter, the physical distance between us and an object at a given redshift is larger. In this case, one would expect to have more intervening galaxies between us and that object, and thus a higher probability of lensing. If one had a sufficiently large database of lens surveys, then one could use observations to constrain the nature of the dominant energy of the universe in this way, as well as the total amount of matter in the universe. Unfortunately, the limits that are currently obtainable in this way are still somewhat crude. However, this will become an important probe of cosmology in the years to come.

One other type of lensing statistical analysis has been performed which may have a profound impact on our interpretation of what forms the dark halos around galaxies. Anthony Tyson, of AT&T-Bell Laboratories, has used sophisticated imaging technology to search, not for lensing of quasars by galaxies, but rather for lensing of distant galaxies by foreground galaxies. He has estimated that if the halos around galaxies continue out to very large distances, in excess of 100,000 light-years, then the images of distant galaxies should be distorted by having passed with high probability at some time near such a distribution on the way to the earth. As I shall discuss later, most elementary particle models of dark matter suggest that the diffuse halos should continue out indefinitely. Scott Tremaine, of the Canadian Institute for Theoretical Astrophysics, has argued, based on the dynamics of satellite galaxies near the Milky Way, that the halo of our own galaxy drops off at large distances. If this analysis holds up, we may have to rethink some of the current wisdom about dark matter.

Tyson's techniques, however, have formed the basis of a far more interesting probe of dark matter on the scale of clusters and beyond. Tyson and colleagues recognized that "weak lensing"—that is, distortions of background galaxies, rather than multiple imaging of these objects—could be used to literally "map out" the distribution of matter in the intervening space. For example, stretching of galaxies into arcs or ellipses is expected if these objects are located behind a distribution of galaxies in clusters (see figure 4.5). Recently, Nick Kaiser, at the University of Hawaii, and colleagues have used this technique to provide definitive observations of the mass distribution in several clusters. There results confirm the results from virial analyses and X-ray measurements. The clustered mass in the universe appears to be between 10–30 times the mass associated with luminous objects.

As I earlier described, if the dark matter in galaxies is in the form of very compact astronomical objects such as black holes or Jupiter-sized planets, and not in the form of diffuse elementary particles, one might hope to probe for the existence of these objects through gravitational lens effects, too. These objects should produce "microlensing" events. As a compact object moves across our line of sight to a star or galaxy, or supernova, it should momentarily magnify it. Several groups exploring our own galaxy have produced remarkable images displaying precisely the pattern of magnification predicted if distant stars are magnified due to the presence of a compact object passing between these objects and our

FIGURE 4.5

This NASA Hubble Space Telescope image of the rich galaxy cluster, Abell 2218, is a spectacular example of gravitational lensing. The arc-like pattern spread across the picture like a spider web is an illusion caused by the gravitational field of the cluster. The arcs are the distorted images of a very distant galaxy population extending 5–10 times farther than the lensing cluster. Also seen are various multiple images of galaxies. Abell 2218 has an unprecedented total of seven multiple systems. The abundance of lensing features in Abell 2218 has been used to make a detailed map of the distribution of matter in the cluster's center. (Figure courtesy of W. Couch [University of New South Wales] and NASA.)

FIGURE 4.6

Another example of lensing by a cluster of galaxies, showing multiple distorted images of a single faint blue galaxy located far behind the cluster. (Courtesy W.N. Colley and E. Turner [Princeton University[, J.A. Tyson [Bell Labs, Lucent Technologies] and NASA.)

telescopes. The implications are not yet definitive, but the characteristic mass of these dark objects appears to be between 1/10 and 1/2 of the mass of our sun, which is in the range of small stars. It is clear that the abundance of these objects is not sufficient to account for all the dark matter in the halo of our galaxy, but it probably accounts for some of it. As we shall describe in the next chapter, this is not that surprising. We expect some dark "baryons" in the universe; indeed, perhaps more dark normal matter than visible normal matter exists in nature.

Whatever results the ongoing search for gravitational lensing brings, it is clear that this emerging field holds great promise for our under standing of what makes up the universe and how it is distributed.

The general theory of relativity could in principle offer one more test for the total amount of matter in the universe, this time on the grandest scale we can measure. Because space-time is curved in the presence of matter, the standard geometric relationships appropriate to flat space often break down. There are many textbook examples of this kind of alteration in geometry. My favorite involves drawing triangles on a sphere. We learn in school that the sum of the three angles inside a triangle always equals 180°. It is very clear that we cannot draw a triangle with three angles each of which is a right angle, because we always end up with two lines parallel to each other (see figure 4.7).

When Euclid invented geometry one of his five axioms was that parallel lines never meet. Hence the two sides of the triangle will never join. However, if I draw a triangle on a sphere, all of these rules are tossed aside. Lines that start out parallel often end up meeting. Consider the lines of longitude on the globe. Each of them starts out at the equator at an angle of 90°. Thus, nearby lines of longitude are parallel at the equator, yet we know that they all meet at the North Pole (see figure 4.8). On a sphere, it is easy to draw a triangle that has three angles of

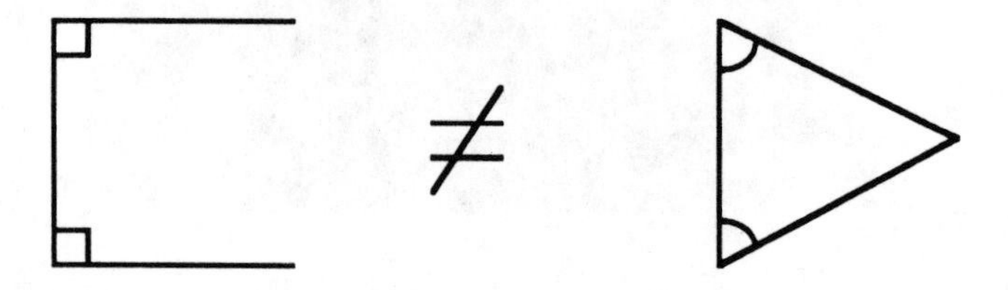

FIGURE 4.7

Three lines joined by two 90° angles, as shown on the left, cannot form a triangle if the figure is drawn on a flat surface.

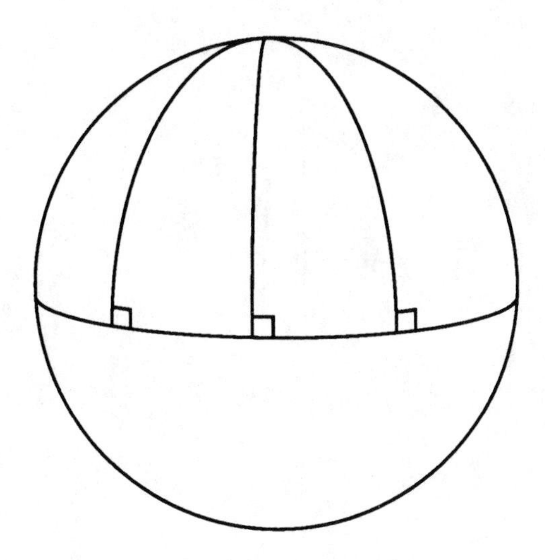

FIGURE 4.8
On a sphere, lines that start out parallel to each other can eventually meet.

90° inside it. Simply join up two lines of longitude, which meet at the North Pole, with the line along the equator.

As you might imagine, other familiar rules regarding triangles are altered when we draw them on a sphere. For instance, the area of a triangle drawn on a flat piece of paper is one half the length of its base times its height. Now consider the triangle on a sphere obtained by taking as its base a line almost completely around the equator, with length almost equal to the circumference of the earth—given by π (≈ 3.14159) D, where D is the earth's diameter.

The height of the triangle is the distance from the equator to the North Pole. This is one-quarter of the circumference of the earth. The rules of plane geometry tell us that the area of the triangle should be one-half the base times the height, or one-eighth times the square of the circumference of the earth. By looking at the triangle, however, we can see that its area encompasses one-half the area of the earth's surface. The area of a sphere is equal to the circumference of the sphere times its diameter, or about one-third the square of the circumference. The area of the triangle must then be about one-half of this, or about one-sixth the square of the circumference. Thus the flat geometry estimate of the area is about 30 percent smaller than the actual area (see figure 4.9).

This kind of alteration in the relationship between geometrical quantities persists when we consider a curved four-dimensional space-time versus a flat space-time. Just as by measuring the area enclosed by triangles on a sphere we can

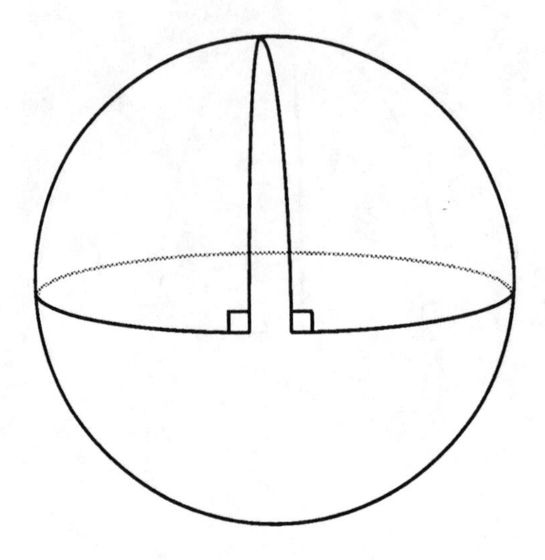

FIGURE 4.9
A triangle drawn on a sphere whose area deviates by 30 percent from that which one would calculate by using the formulas of plane geometry.

prove that the earth is round and not flat, we can try to measure the relationship between, say, the volume and radius of a large sphere in the universe to probe the average curvature of space, and from this the average mass density. We cannot measure volume directly, but if we assume that the number of galaxies in a given volume is roughly constant throughout the universe, which is certainly supported by observation, then we can try to count the galaxies in progressively larger volumes and see how the volume contained in a spherical region scales with its radius.

Just such an effort was undertaken by two young Princeton astronomers, Earl Spillar and Ed Loh, in the mid-1980s. When they completed their study, they announced a stunning result: the mean curvature that they inferred suggested an average mass density of the universe four to five times larger than virial estimates. Their announcement caused a momentary stir, and their analysis was soon attacked on several fronts. Aside from experimental problems, the facts that galaxies change their characteristics in time and that galaxies sometimes merge together compounded the uncertainty of their analysis. It is generally accepted that this approach cannot, at the present time, give a definitive probe of the geometry of the universe.

More recently, and probably for the first time in the history of cosmology, a set of observations is being made that will, within the next decade, directly pin down the geometry of the universe. These observations are associated with the most direct probe of the universe on large scales, the Cosmic Microwave Background Radiation. As I have described, the Cosmic Background Radiation (CBR) provides us with a "picture" of the universe when it was about 3,000 degrees in temperature and about 250,000 years old. Since this epoch was well before the time of galaxy formation, observing the spatial structure of the CBR gives us invaluable information on the truly primordial structures, built into the early Big Bang itself, which would later collapse to form all large-scale structure we see in the universe today. For this reason, I shall devote considerable time later to discussing how the COBE observations and subsequent CBR measurements have literally revolutionized our ability to compare theoretical predictions of structure formation to observation, and with that comparison, to constrain the nature of dark matter. However, for the purposes of this chapter we need only explore one facet of the physics affecting the CBR.

Just before the observed CBR formed, matter and radiation were in equilibrium, because the matter was ionized and interacted strongly with radiation. Because radiation travels at the speed of light, it exerts a large pressure on anything it interacts strongly with. Thus, the pressure in this radiation-matter bath was great. This implies that any primordial lumps of matter would not have been able to collapse due to gravity, because the pressure of radiation would have stopped such collapse. In fact, the dissipation of energy by the scattering of matter and radiation would have served to smooth out any such lumps in matter. Of course, such lumps could only have been smoothed out on distances smaller than light could have traveled since the Big Bang. This is because light provides a cosmic speed limit, and thus no material could have been moved around on larger scales than a light ray could have traveled.

At the time the CBR formed, however, matter became neutral, and then it was free to start collapsing, as it no longer interacted strongly with radiation. Since lumps smaller than the distance a light ray could have traveled since the Big Bang—called the horizon distance—would have been smoothed out before this time, the first scale on which primordial lumps could have collapsed would have been the horizon size at the moment matter became neutral. This sets a characteristic size scale that would have been imprinted in the CBR at that time. On smaller scales the radiation-matter mix would have been largely smoothed out. On larger scales it would not have.

How could we resolve such a scale? We could observe the CBR on small angular scales in the sky and look for small temperature deviations in this background. The characteristic lumps at the horizon size at recombination should be represented by temperature fluctuations on the angular scale associated with that horizon size. Here is where geometry comes in. As we look back at the CBR, we are looking back at a surface located about 10 billion light years away in all directions. The characteristic angular scale associated with a fixed physical distance at that time depends upon the geometry of the universe. For example, if the universe is curved, we can see that the angle subtended by a fixed distance is different, depending upon the curvature (see figure 4.10).

Primordial fluctuations in the temperature of the CBR have been searched for almost from the time this background was discovered. As of 1988, no such fluctuations had been discovered at all. As far as experiments could tell, the CBR was uniform in temperature across the sky to better than 1 part in 10,000! Why this is the case was completely inexplicable until about twenty years ago, as I shall describe in part III of this book. However, for the moment it is worth pointing out that, finally, in about 1992, the COBE science team reported the discovery of small fluctuations in the CBR temperature at a level of about 1 part in 100,000, if one compared points on the sky separated by about 10 degrees, the resolution of the COBE detectors. Since that time ground-based observations have been made on even smaller scales in order to explore the nature of the primordial fluctuations in the CBR, which in turn can tell us a tremendous amount about the early universe. Preliminary observations just made on the scale associated with the horizon size at recombination, about 1–2 degrees on the sky, suggest something remarkable. The geometry favored by the data suggest a very particular geometry for the universe, and one that involves 4–5 times as much mass or energy density in the universe as can be accounted for by even the dark matter in clusters of galaxies! But more than that, it implies that there is far more dark matter or

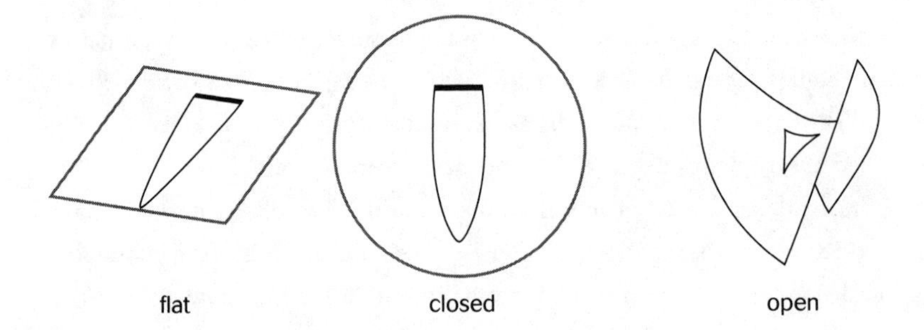

FIGURE 4.10

A fixed length such as that of a ruler will subtend a different angle for observers located at a fixed distance in universes of different geometries. This is because light rays converge in a closed universe and diverge in an open universe, as show in this figure.

energy out there than even our gravitational probes of large scale structure have indirectly inferred.

The foregoing discussion hints at the real reason for worrying about exactly how much dark matter exists in the universe. Because the geometry of space responds to the matter content, the structure of the universe as a whole is determined once we know the distribution of matter (and energy) and its motion. The general theory of relativity distinguishes between three different kinds of expanding universe: closed, open, and flat. Only in the first case is the universe of finite spatial extent—there is enough matter present to curve space sufficiently so that the universe closes back upon itself. The universe is then like the surface of a large sphere, except that it is spatially three-dimensional rather than two-dimensional. In the latter two cases, open and flat, the universe is of infinite spatial extent. Geometrically these two cases are distinguished by their "curvature," an important term in Einstein's equations for an expanding universe. A closed universe has positive curvature and an open universe has negative curvature; as you might guess, a flat universe has zero curvature.

Certainly these descriptions both tax and excite the imagination, but, in a universe dominated by matter, a much more practical and more relevant distinction exists among an open, closed, or flat universe which does not rely on one's ability to visualize four-dimensional space-time. In a matter-dominated closed universe, the density of matter will eventually halt the expansion due to its gravitational attraction. In a matter-dominated open universe, the expansion will continue forever at a finite rate. In a flat matter-dominated universe, the boundary between these two cases, the expansion will slow down but not halt in any finite time (see figure 4.11).

One of the most remarkable aspects of Einstein's equations for a matter-dominated expanding universe in the general theory of relativity is that they can be derived exactly by using purely Newtonian reasoning. One merely adds up the total energy of matter as the universe expands. There are two contributions to this energy. The energy of the motion of objects in the expansion, called kinetic energy, is always positive. On the other hand, the gravitational potential energy due to the gravitational attraction of objects is usually considered to be negative. This is because we usually consider the potential energy of an isolated object under no force to be zero. In order to reduce the gravitational force of attraction between two masses to zero, we would have to separate them infinitely far apart. But it takes energy to do this. Hence the potential energy of the two objects when they are only separated by a finite amount must be less than zero.

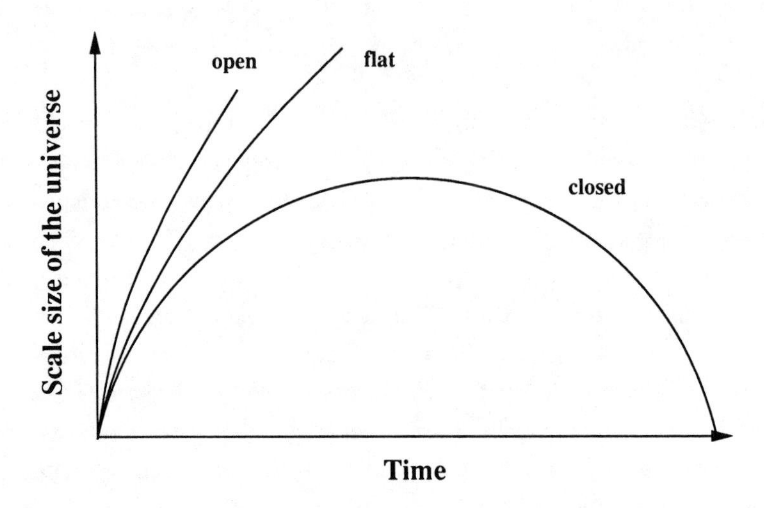

FIGURE 4.11
The distance between galaxies will continue to increase indefinitely if a matter-dominated universe is either open or flat; if it is closed, eventually the universe will stop expanding and recollapse.

The total gravitational energy of an object or objects is just the sum of these two contributions, so that we may write in a straightforward way:

KINETIC ENERGY + POTENTIAL ENERGY = TOTAL GRAVITATIONAL ENERGY

This "equation" is very useful here on earth. If we consider any object, for example, a ball or the space shuttle, we can determine whether its positive kinetic energy due to motion overcomes the negative contribution due to the attraction from the earth. If it does, so that the total energy is positive, then the object will escape from the earth. If the total energy is negative, so that the potential contribution outweighs the kinetic contribution, the object will remain bound: if we send it up, it will come back down. In the borderline case, where the total energy is exactly zero and where the two contributions cancel, the object can just barely escape to infinity, but its velocity will approach zero as it does. The minimum kinetic energy, which just balances the potential energy, comes when the object's velocity is just great enough so that the sum of the two terms on the left-hand side of the equation equals zero. This velocity is called the escape velocity. When NASA decides to launch a rocket to the moon, a basic requirement is that there be enough fuel to propel the rocket past the earth's escape velocity, which is about

11 kilometers per second (25,000 miles per hour) for something launched from the earth's surface, independent of its mass. This value is calculated just as I have described it here, using the energy equation.

The analogy between the alternatives for the motion of objects thrown from the earth's surface with positive, negative, or zero total energy and the three different labels that can characterize a matter-dominated universe is more than suggestive; it is exact. Because of its "negative" curvature, an open universe will continue to expand forever. This simply means that every galaxy will escape from the gravitational pull of every other galaxy. To put this another way, think of a spherical region of the universe which is small on the scale of the universe, but large on the scale of any clusters or superclusters, and consider the motion of the galaxies located on the surface of this region. Because of the uniform Hubble expansion velocity, these galaxies will on average be moving radially outward away from the center of the sphere. Are these galaxies moving fast enough to escape the gravitational pull of the matter contained inside the sphere? The answer is given by the same energy equation just described. If their total energy is positive, they will continue to move outward forever. If not, they will eventually stop, and collapse inward.

Each such sufficiently large region of galaxies is a microcosm of the expanding universe. Determining its behavior determines the behavior of the universe as a whole, since the universe is assumed to be isotropic and homogeneous. Thus, it should perhaps not be so surprising that the simple energy equation just described gives directly every term of Einstein's equation which governs the motion of a matter-dominated universe. Even the factors of 2 and π come out right. At no point do we need to discuss "curvature," or geometry. In fact, when Einstein's equations are derived this way, the term representing the curvature turns out to be exactly the negative of the total gravitational energy. In an open matter-dominated universe with negative curvature, therefore, systems on a large enough scale have positive total gravitational energy and will expand indefinitely. In a closed matter-dominated universe, the total gravitational energy of objects in these systems would be negative, so they will eventually recollapse.

The energy equation tells us that to determine which condition prevails, and hence to determine the geometry of the universe, we must measure both the mean kinetic energy and the mean gravitational potential energy and compare the two. We measure mean kinetic energy, which gives the energy of motion, by measuring the Hubble constant, which determines how fast galaxies at a given distance are moving apart. The mean gravitational potential energy, which is related to the gravitational attraction tending to slow the expansion, depends upon the mean mass density of the universe. For a given Hubble constant, there exists a precise value of the mass density, called the "critical" or "closure" density, for which the two terms balance, yielding zero total energy and a flat universe. Larger mass densities would result in a closed universe, and smaller ones in an open universe.

How does our tally of matter thus far compare with this critical density? We can write the actual mass density of the universe today as some constant, traditionally labeled Ω (the Greek letter *omega*) times the critical density. In this way we define the very important parameter in cosmology, Ω, as *the ratio of the actual mass density in our universe to the critical density that would result in an exactly flat universe.* If Ω is less than 1, our universe is open. If it is equal to 1, our universe is flat. If it is greater than 1, the universe is closed. One of the central goals in cosmology, to attempt to determine the future evolution of the universe, simply reduces to determining the value of Ω today.

Since the kinetic energy piece of the equation which determines the critical density depends upon the uncertain Hubble constant, the critical density, too, will depend upon the "fudge factor" h. We can express this density in terms of the measured luminosity and (inferred) mass density of the universe. The luminosity density, which we discussed in the previous chapter, can be written in terms of the luminosity of the sun. Similarly, I described there how the mass density can be written in terms of the mass of the sun. In these units of solar mass versus solar luminosity, the ratio of the mean mass density in the universe to its mean luminosity equals about 1,500 Ωh. In other words, *if the universe is flat, this ratio should be measured to between about 750 to 1,500 today.*

Now recall that this ratio, measured for the *visible* matter in the universe, was determined to be about 20–25 h. Thus, the sum total of all visible matter in the universe results in an inferred value of Ω about equal to 0.01–0.02. In other words, *visible matter provides only about 1–2 percent of the mass that would be required to close the universe today.* All the virial estimates I have displayed for the abundance of dark matter in the universe suggest that within a few million light-years of visible galaxies, dark matter is roughly ten to twenty times more abundant than visible matter. If we include this dark matter, we find that the ratio between mass and light is about 100 to 500, allowing for uncertainties in h.

This is the culmination of all our labors in this chapter. *Dark matter, while it dominates visible matter by at least a factor of 10, seems to result in a value for Ω of just about 0.2–0.3 today.* Only if the dynamical estimates I have described in this chapter are off by a factor of 3–5 would there be enough matter around to close our universe. If this were the case, dark matter would be roughly fifty to one hundred times more abundant than everything we can see through our telescopes today.

Now, note that throughout this discussion I have included the cautionary phrase "matter dominated" whenever I have discussed the dynamical arguments relating the density of the universe, the Hubble constant, and the geometry of the universe. If matter provides the dominant energy in the universe, then this relation is exact, and the ratio of the actual density of matter in the universe to the critical density completely determines both the geometry of the universe and its future.

However, there is an additional possible complication that, until about a decade ago, it seemed safe to ignore. What if empty space carries energy? Einstein's equations of General Relativity tell us that all forms of energy affect the curvature of space and thus the expansion of the universe. Thus, if empty space carries energy, all of the above arguments will be changed.

One's first reaction to this possibility is incredulity. After all, how can empty space carry energy? By definition, empty space is "empty"! However, recall our earlier discussions about virtual particles. We have learned that the laws of quantum mechanics, when combined with the laws of relativity, tell us that empty space is not so simple. It is full of virtual particles. What if these particles carry energy?

Remarkably, in 1916, long before anyone had any idea that virtual particles might exist, Albert Einstein proposed an idea that presaged this very notion. When he first developed his theory of General Relativity, he recognized that by tying together space, time, and matter, his theory presented the very first attempt not merely to describe the dynamical evolution of objects moving within the universe, but the dynamical evolution of the universe itself.

There was a problem, however. In 1916, the expansion of the universe had not yet been discovered. The common wisdom held that the universe was static. After all, our galaxy seemed remarkable stable and constant, at least on a human timescale. But Einstein realized that his theory of General Relativity shared a common characteristic with its Newtonian counterpart: Gravity sucks. The gravitational attraction of matter is universally attractive. Gravity only pulls, it never pushes.

A universe full of matter cannot therefore remain static. The unrelenting mutual gravitational attraction of matter on matter will eventually cause objects to start moving toward each other, causing massive collapse. This was a huge problem for Einstein's theory, but Einstein recognized that he could add an extra term to his equations, one that was consistent with all of the principles that had led him to develop General Relativity in the first place. This extra term, which he labeled the "Cosmological term," would stabilize the universe by producing a new long-range force throughout space, even empty space. If this term were positive, it would produce a repulsive force—a kind of cosmic antigravity—that could hold the universe up under its own weight!

Alas, within five years Einstein abandoned this kludge. The stability offered by the term proved to be illusory, but more important, evidence had already begun to mount that the universe is expanding. In an expanding universe, there is no need for an additional repulsive force. Gravity can be universally attractive and can work merely to slow or ultimately stop the expansion. Like the aether before it, this new term appeared to be headed for the dustbin of history.

Physicists were happy to do without such an intrusion. As I have described, in General Relativity, the source of gravitational forces is energy. Matter is merely one form of energy. Radiation too is both subject to gravitational forces (witness the bending of light) and a source for a gravitational field. However, Einstein's cosmological term was different. The energy associated with it did not depend upon either position or time. The force it produced operated in the complete absence of matter or radiation. Its source, therefore, must be a strange energy that resides in empty space. By dispensing with the cosmological "constant," as Einstein's term became known, empty space was truly empty, and nature once again appeared to be reasonable.

However, like putting the toothpaste back in the tube after you have squeezed it, getting rid of cosmological constant was not so easy. In fact, we now understand that had Einstein not first proposed it, it would have cropped up anyway. This is because the virtual particles resulting from quantum mechanics not only can endow the vacuum of space with strange new properties, but they can provide it with energy! I earlier described how empty space contains a sea of particles of differing energies, but that we usually describe the vacuum state as the state with zero energy when it comes to nongravitational interactions, because the absolute value of the energy of empty space is irrelevant in all interactions but gravity. However, when it comes to the expansion of the universe, it is gravity that dominates, and one must consider the energy of empty space more carefully. Moreover, it turns out that the energy associated with the vacuum has precisely the form that results in a cosmological constant of the type Einstein invented ad hoc in 1916!

We are now faced with an entirely different problem on a whole new scale. We now have to ask not why the energy of empty space might be nonzero, but rather why it isn't much larger than is allowed by current observations. For if we attempt to calculate, using quantum mechanics and relativity, combined with our knowledge of the nature of the fundamental forces, the a priori magnitude expected for the energy density associated with empty space, we come up with an estimate that is over 120 orders of magnitude too large.

The Cosmological Constant Problem, as the above conundrum has become known, is the most severe fine-tuning problem in physics. Somehow we have to explain how the cosmological constant could be at least 120 orders of magnitude smaller than we would naively estimate it should be. To date, no one has the slightest clue how this could result.

In the absence of an explicit solution for this problem, elementary particle physicists have assumed that the ultimate solution would involve a theory that reconciles quantum mechanics and gravity. Moreover, it has been assumed that this ultimate solution would predict that the energy of empty space should be precisely zero. Only this value seems sensible, since one could imagine, at least in principle, some symmetry principles that might cause the cosmological constant to vanish, but it is very difficult to imagine a mechanism that would yield nonzero but be so very small.

This has been the preferred theoretical bias since I was in graduate school. However, one of the most interesting aspects of the universe is that it is the way it is, whether we like it or not. So, it is certainly possible that the energy of empty space might play a role in governing the expansion of the universe. Indeed, as early as 1984, I and my colleagues Michael Turner of the University of Chicago and Gary Steigman of Ohio State, and independently Jim Peebles at Princeton, argued that perhaps there was enough energy in empty space to result in an exactly flat universe today. However, I have to say that while we argued this might assist in making the observed universe conform more closely to the predictions from early universe cosmology discussed in the next chapter, I don't think any of us really believed that this possibility might actually reflect the real universe. Nevertheless, Turner and I, a little over a decade later, in 1995, argued that the fundamental independent observations in cosmology could be combined to compellingly argue that the cosmological constant is nonzero. I shall return to present our arguments at a later point, when the proper context has been laid down.

I think it is fair to suggest that when we first presented our claim, the standard response was a polite smile, sometimes followed by a chuckle. This situation completely changed, however, in 1998, when two observational groups working independently to use a certain type of exploding star, a type 1a supernova, as a standard candle to measure distance versus redshift presented evidence that the Hubble expansion has been accelerating over time!

Recall that the Hubble constant measures the relation between velocity and distance in an expanding universe. As long as the dominant source of gravity produces an attractive force, such as would be the case if the universe were dominated by matter or radiation, then the expansion of the universe will slow down with time. Thus, the Hubble constant is not really constant over cosmic time. If we measure the expansion rate today, it should be slower than it was at earlier times.

At first it may seem that we have no choice but to measure the Hubble constant at the present epoch, and thus we cannot explore directly the temporal evolution of this quantity. However, recall that as we explore to farther and farther distances, we are looking back into time. If we measure the recession velocities of objects that are a sizable fraction of the currently observable universe away from us, then we are looking back a sizable fraction of the age of the universe. If the universe has been decelerating, even slowly, there should be an observable change in the rate of expansion at that time, compared to the present, *if* we are able to resolve the distance-redshift relation accurately enough at these distances.

This is precisely what these two groups have been trying to do. In order to see objects at very large distances, they must be very bright. Among the brightest objects in the universe are supernovae, exploding stars. These can be observed in galaxies that are literally billions of light years away. While supernovae occur very rarely in each galaxy, say once every 30–100 years, if one observes enough galaxies on a given night, say 5,000, one is guaranteed to see at least one supernova. This recognition spurred these two ambitious groups to systematically explore large regions of the sky on several nights each year, scanning their plates to see if one of the host galaxies contained a supernova.

However, any given supernova will not work for this purpose. After all, supernovae can occur in stars of very different mass and as a result have very different intrinsic brightnesses. If these objects are to be used as standard candles to infer distances to distant galaxies, one must have a way of recognizing a supernova of a given brightness.

These are the quandaries that faced a group at Lawrence Berkeley Laboratory headed by Saul Perlmutter, and independently an international collaboration led by Brian Schmidt at Mount Stromlo Observatory in Australia and Robert Kirshner at Harvard University. They had to develop not only a method of discovering supernovae at high distances by careful observations and scanning of plates, but they also had to determine if there were certain types of supernovae that could be used as standard candles. For some time astronomers had focused on one specific type of supernova for this purpose.

When the most massive stars complete their nuclear burning, they undergo a sudden gravitationally-induced collapse, blowing off their outer shells in the process. These become supernovae. Stars less than about twice the mass of our sun will simply cool down as they complete their nuclear burning, ending this phase of their lives as white dwarf stars.

So-called type 1a supernovae occur when a white dwarf slowly gains mass due to the presence of a companion star or other nearby mass. Slowly, by gravitational accretion, the mass of the white dwarf increases, and it becomes denser and hotter. When the star reaches a critical mass, it becomes dense enough for thermonuclear reactions to begin, and these create a runaway explosion, such as occurs in a hydrogen bomb. This literally blows the star to pieces.

Because such explosions generally occur at a specific mass and density, they are all very similar. Moreover, since the same basic physics governs the explosion, one has a good underlying theoretical reason to suspect that these may be good standard candles.

Of course, the detailed dynamics behind such explosions have not yet been worked out, so observers must still rely on empirical rather than theoretical results. A technique refined by Robert Kirshner's group at Harvard and Perlmutter's group at Berkeley suggested that there was a good empirical relation between the amount of time that a type 1a supernova burns brightly and its intrinsic luminosity. With this relationship in hand, the two groups began to search for high redshift supernovae, in order to probe the distance-redshift relation, and from that, the possibility of measuring the Hubble constant as a function of time.

In early 1998, both groups reported what to many was an astonishing result (see figure 4.12). The universe indeed appeared to be accelerating. As I mentioned earlier, there are many independent reasons to suspect that this is the case, but this was the first direct observation supporting this idea. If true, it is one of the most astonishing results to emerge from cosmology in the past several decades. However, it is worthwhile being cautious. All other efforts to use distant objects to explore the global features of our universe have in the end succumbed to the great uncertainty: evolution. If, for one of many possible reasons, distant type 1a supernovae behave differently than the nearby objects that are used to calibrate their luminosity, than objects that appear to be more distant may instead merely be dimmer than expected.

Thankfully, we will not have to rely solely on type 1a supernova measurements to determine if the cosmological constant is nonzero. As described earlier, CMB measurements should definitively help us determine the geometry of the universe. When combined with better determinations of the matter content and matter density, we will know for certain if there is a need for, or room for, a vast quantity of

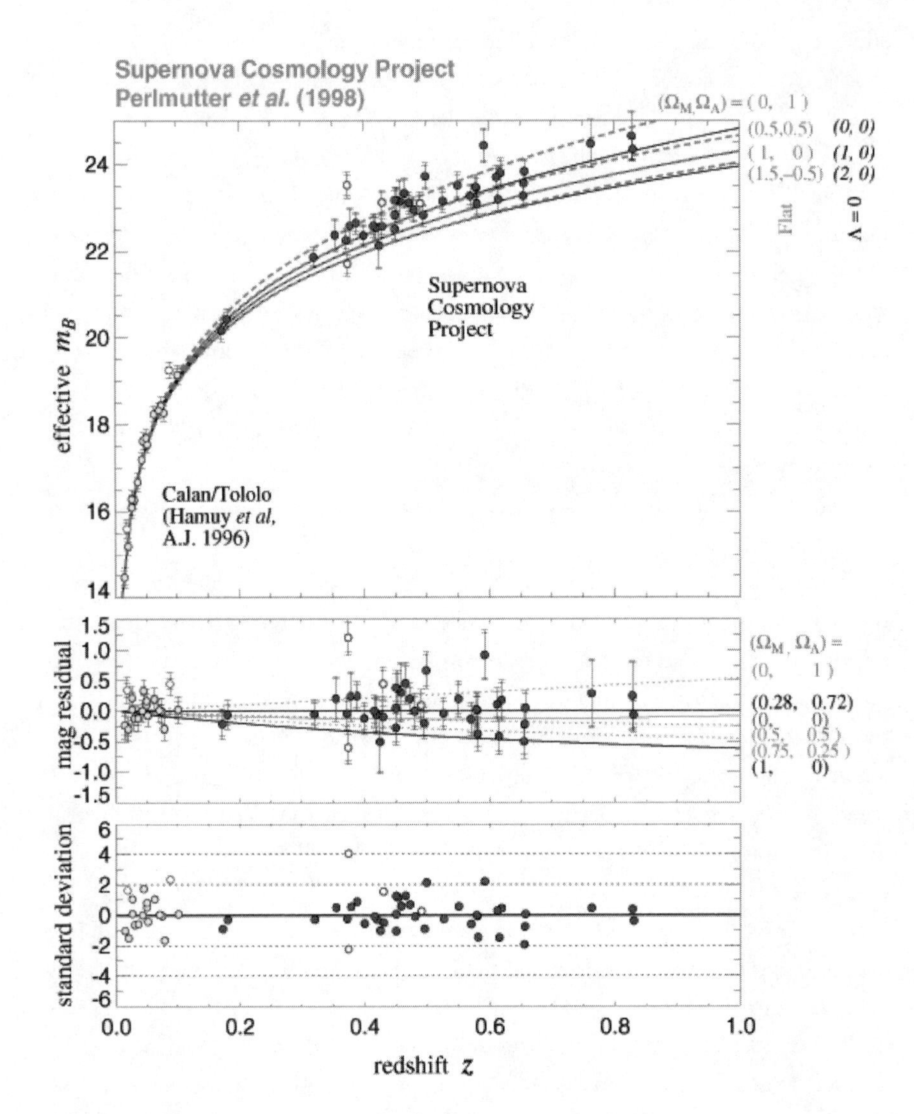

FIGURE 4.12

Data from the Supernova Cosmology Project of the claimed distance versus redshift for a collection of distant supernovae compared to more nearby supernovae observed by another group. The distance-redshift relation provides a sensitive probe of the evolution of the expansion rate of the universe, which depends on the form of the dominant cosmic energy density. The data shown here are compared to a flat universe with energy in matter, and a cosmological constant. The best fit has approximately 25 percent of the energy in matter, and 75 percent in a cosmological constant.

dark energy associated with empty space. All recent observations point in the same direction, suggesting the cosmological constant is in fact nonzero.

If the cosmological constant actually is nonzero, however, it forces us to change completely our notions of how geometry reflects destiny. In a universe dominated by a cosmological constant, which can be dominated at large scales by a gravitational repulsion rather than an attraction, the connection between open, closed, and flat universes and the ultimate expansion or collapse of the universe is removed. A closed universe can expand forever, and an open universe can collapse (if the sign of the cosmological constant is negative!). In fact, as Michael Turner and I have recently claimed, once the possibility of a nonzero cosmological constant arises, it becomes impossible to make any series of observations that will be guaranteed to unambiguously determine the ultimate fate of the universe. What we lose in predictive power, however, we may gain in insight about the early universe. If the cosmological constant is nonzero, it tells us something is very strange indeed about the physics of the early universe, something that may be intimately related to the nature of both ordinary matter and the dark matter that seems to dominate gravitational dynamics on the scale of galaxies and clusters.

A great many questions remain unanswered, yet we have still made great strides since the late 1960s. Dark matter is now an accepted part of astrophysics and astronomy. The evidence for its existence is indisputable. Many and varied observations all seem to converge on about the same value for the amount of dark matter associated with galaxies: approximately a factor of 10 for dark versus visible material. Based on independent measurements of the expansion rate of our universe, this result would imply, in the absence of either a cosmological constant, or some diffusely distributed matter on large scales, that the universe is "open"—it will go on expanding forever, never slowing to a complete stop, never recollapsing—a whimper rather than a bang. The value for the amount of dark compared to luminous matter also coincides neatly with the upper limit on the total amount of "normal" matter—made from protons and neutrons—which could be in the universe today. This value is also about a factor of 10, at least according to arguments I shall describe later. It thus seems quite possible that an open universe need contain nothing *too* unexpected and exotic, unless we consider unobserved Jupiter-sized objects, dead stars, or black holes exotic.

A number of theoretical thunder clouds have appeared on the horizon, however, threatening to destroy this cozy picture. You may have wondered why in the preceding pages I kept pounding away at the possibility that there may be something awry in our estimates of the amount of matter in the universe today—that there might be much more dark matter or energy than our observations on the scale of galaxies suggest. It is because we believe that this is the case. Our current ideas about how the large-scale structure that we observe today is formed seem to imply that dark matter cannot be made of conventional stuff. At the same time, elegant and powerful ideas emerging from research in particle physics have provided apparent solutions to three of the most nagging problems in cosmology. However, they do this at a cost. If they are correct, the universe must be flat today; Ω must equal 1 almost exactly. And this would mean that all the matter, whether dark or light, that we have so far uncovered with our brilliant efforts merely represents the tip of an iceberg in a vast cosmic sea of dark matter or energy.

Whether dark matter is made of exotic material, whether it permeates the earth or resides in dead stars or black holes far out in the heavens, whether our ideas about the formation of galaxies and the evolution of the universe itself have to be changed, whether the dominant energy of the universe might reside in empty space, . . . it all depends crucially on the confrontation of these ideas with current observations. It all depends on a factor of 5. I have tried to present the cosmological observers' direct case for dark matter here. In the next two chapters I will present the theoretical challenge.

PART THREE

Why the Universe Is Flat: The Big Bang, Large-Scale Structure, and the Need for Something New

CHAPTER FIVE

COOKING WITH GAS

There are too many stars in some places and not enough in others, but that can be remedied presently, no doubt.

—MARK TWAIN

As usual, Mark Twain was accurate and to the point. And while filling in the details requires a little more discussion, it is true that our efforts to explain large-scale structure—how stars are organized in galaxies, galaxies in clusters, clusters in superclusters, and so on to ever larger scales—sometimes boil down to simply an attempt to resolve this problem of having too many stars in some places and not enough in others. The present "remedy," which I shall describe, is achieved by a route very different from the direct empirical one described in the preceding chapter. This solution requires the existence of dark matter and its dominance of the gravitational dynamics of the universe. The dark matter required in this case has an important qualifier: it must be material in form, yet made of something completely different from anything we have yet discovered in nature.

Requiring an entirely new form of matter in the universe is a radical concept that must not be taken lightly. Yet theories about the nature of large-scale structure are not the only arguments that signal the need for something exotic "out there." Rather, a series of independent arguments taken together bolsters the case that some exotic material dominates the expansion of the present-day universe. There are two caveats, however. First, these arguments utilize some of the deepest and most subtle ideas in cosmology and particle physics—ideas that can strain the limits of credibility until one gets used to them. Second, many of these arguments arise from a consideration of the early universe, a time very far removed from our direct observations.

It is inevitable, however, once one begins to consider the emergence of structure in the universe, for our imaginations to be drawn ever further back in time: back to the earliest moments of creation, when any local density enhancements of matter which would eventually collapse to form stars and galaxies were still imperceptible amid the uniformly expanding cosmic background of matter and radiation. Each step backward has brought us closer and closer to an inescapable truth. At some stage, the formation of everything we see around us, perhaps the entire visible universe itself, was governed by the very microphysical laws that govern matter today on its smallest scales. As our previous discussion of the cosmological constant made clear, we have thus been forced to come to grips with the necessity of relating macrophysics and microphysics.

In accepting this challenge we have opened the possibility of explaining how the initial conditions, which made possible the evolution of our universe, themselves resulted dynamically according to the laws of physics. Once on this road, there is no turning back. Comfortable notions may have to be tossed aside. We may find that the universe today is a very different place than we had imagined, more mysterious perhaps than even the observations described in the previous chapters might suggest. We are also likely to find our own cosmic significance once again diminished.

Considerations of large-scale structure have played a central role in motivating the idea that dark matter is "different," and when they are placed in the context of these fundamental attempts to understand the early universe they take on much greater significance. Because the theoretical prejudices that have arisen from these considerations are so important in guiding much of the debate about the nature of dark matter, I want to proceed step by step through the ideas that have molded the way we now think about the universe. You may then judge for yourself whether you think they warrant the belief that the material of the stars cannot be all there is.

I will begin in familiar territory—with the origin of matter that we can see. One of the greatest successes of the standard Big Bang theory of cosmology—that model of the universe that is obtained by the most straightforward possible extrapolation backward of the presently observed expansion of matter amid a cosmic microwave radiation bath—has been the development of a consistent model of the origin and evolution of all the light elements that we observe today, such as hydrogen and helium. The general agreement between theory and observation has emboldened us to believe in the possibility of consistently making such cosmic leaps backward. More important for our purpose, this scenario of sequentially building up light nuclei, called "Big Bang nucleosynthesis," also provides probably the strongest theoretical handle we have on the absolute abundance of normal matter in the universe today—seen or unseen. Thus, Big Bang nucleosynthesis predictions, when compared with the dynamical inferences of the total mass around galaxies, in principle could settle the question of the need for exotic matter in the universe. It is against this background that we shall have to weigh the implications for dark matter of the somewhat more esoteric problems associated with forming galaxies, and then those associated with forming the universe itself.

Essentially all of the light elements (whose nuclei contain fewer than seven protons and neutrons) we observe in the universe today first formed during a period that lasted only a few minutes, about 10–13 billion years ago. At this time the temperature of matter and radiation in the universe was in excess of 1 billion degrees Kelvin—one hundred times hotter than the fiery interior of the sun is today. I described in chapter 2 how the temperature of the universe drops inversely as its size expands. Since the temperature of the microwave background today is about 2.7 degrees K, the time to which I now refer occurred when the portion of the universe which is now visible was nearly a billion times smaller than it is now—a few light-years across. The Big Bang expansion had then been under way for less than one minute. The temperature of the universe ranged from about 10 billion to about 1 billion degrees K, as the lifetime of the universe varied between about one and one hundred seconds.* Recall that this is some 10 billion times earlier than the earliest time for which we have knowledge through direct observations of the cosmic microwave background.

The temperatures up to the time when the universe was about ten seconds old were so high that the individual nuclear constituents—protons and neutrons, also referred to as *baryons*—had not yet cooled sufficiently to coalesce to form atomic nuclei, that is, to begin the process of nucleosynthesis. These nuclei, in turn, would, some 100,000 years later, capture free electrons, which by then would have cooled sufficiently to bind with the nuclei to form atoms. These would then over the course of cosmic time bind together to form molecules, crys-

*Appendix B gives a short graphical history of the universe. Interested readers may wish to consult it to gain some perspective on the detailed Big Bang model predictions for how the temperature and size of the universe varied with time.

tals, and all of the materials that make up the world we see around us. At this early time, however, the universe consisted of a hot gas of particles, primarily protons, neutrons, electrons, and of course photons. These particles in thermal equilibrium could scatter and interact, momentarily binding together to form nuclei before further collisions with energetic particles would once again disassociate them. Typical scattering processes inducing "nuclear reactions" would include the following four reactions:

proton (p) + neutron (n) $\leftrightarrow$ deuterium (pn) + photon (γ)

n + n $\leftrightarrow$ pn + electron (e) + antineutrino ($\bar{\upsilon}$)

n + $\bar{\upsilon}$ $\leftrightarrow$ p + e

n $\leftrightarrow$ p + e + $\bar{\upsilon}$

(THE $\leftrightarrow$ SYMBOL INDICATES THAT THE REACTION MIGHT PROCEED IN BOTH DIRECTIONS.)

In each case the sum of the masses of the objects on the right-hand side is less than those on the left-hand side. Thus, if the right-hand objects were at rest, energy conservation would ensure that the reactions could only proceed one way, from left to right. However, if the temperature was sufficiently high so that on average each particle had enough thermal energy to make up for this energy deficit due to the mass differences, the reactions could have proceeded in both directions. Below temperatures of about 10 billion degrees Kelvin, this would no longer have been the case. The mass difference between a neutron and a proton is great enough so that at this temperature and below, the mean thermal energy of particles was too small for the last three reactions to have continued to proceed at any significant rate from right to left. Up to this time the reaction network just outlined would have ensured that thermal equilibrium was maintained so that, because neutrons are heavier than protons, a uniform partition of the available energy would have caused there to be fewer neutrons around than protons.

Now, the binding energy holding together the proton and neutron in deuterium is much smaller than the mass difference between a proton and a neutron, so that deuterium weighs almost as much as a free proton plus a free neutron. Because this implies that the mass difference between the left-hand and right-hand sides in the first reaction are much smaller, this reaction would have continued to proceed in both directions until the temperature was about a factor of 5 lower than that for which the other reactions would have begun proceeding primarily from left to right. In particular, during the time it took for the universe to cool to this lower temperature, the last reaction would have been proceeding preferentially from left to right, causing free neutrons to decay into protons, plus electrons and the antiparticles of neutrinos. Thus the number of neutrons compared to the number of protons would have kept on decreasing. Once the universe became so "cool" that the first reaction too proceeded only from left to right, no more free neutrons would have been created. Those remaining would either decay or undergo a nuclear reaction such as the first reaction to form deuterium.

These reactions are not the end of the line. Once deuterium is formed, more nuclear reactions can take place to form yet heavier nuclei, such as helium, by the following reaction:

$$\text{deuterium (pn)} + \text{deuterium (pn)} \leftrightarrow \text{helium 3 (ppn)} + \text{n} \leftrightarrow \text{tritium (pnn)} + \text{p}$$

$$\text{pnn} + \text{pn} \leftrightarrow \text{helium (ppnn)} + \text{n}$$

The binding energy holding together these nuclei is larger than that which holds together a proton and neutron in deuterium. This means that the mass difference between these nuclei and the particles that collide to form them is greater than the mass difference between deuterium and its constituents. For this reason, these latter reactions proceed immediately from left to right once the universe has cooled sufficiently for deuterium to form and start the whole process. The end result is that almost all the free neutrons left in the universe quickly bind to form helium, a stable nucleus containing two protons and two neutrons. The process does not continue efficiently beyond the formation of helium for two reasons. There are no stable nuclei with a total of either 5 or 8 nucleons (protons or neutrons). Thus, collisions of protons or neutrons with helium, or helium with helium, do not produce stable nuclei. Collisions of deuterium and tritium with helium can, however, produce nuclei with nucleon number 7. But both deuterium and tritium are largely used up in the process of producing helium, so very little is left that could react with helium to form these yet heavier nuclei. Therefore, the abundance of nuclei beyond helium produced in the primordial Big Bang expansion is expected to be smaller by many orders of magnitude.

The number of free neutrons available to form helium via the second set of reactions turns out to be only about 12 percent of the total number of free baryons—that is, protons and neutrons. This is (1) because at the time that the first reaction network began to proceed primarily from right to left, the number of neutrons was already smaller than the number of protons, and (2) because neutrons continued to decay during the extra time that it took the universe to cool sufficiently for enough deuterium to form for these latter helium production reactions to proceed. Since it takes two neutrons to form helium, the fraction by number of helium atoms compared to free protons left in the universe was half the ratio of free neutrons to protons at that time, or about 6 percent. But since a helium nucleus is about four times as heavy as a free proton, the fraction by weight of helium nuclei compared to protons is expected to be roughly 24 percent. This is the key result.

This dynamical process of Big Bang nucleosynthesis has been calculated in great detail since the 1960s. Each of the rates for the reactions described earlier, determined by experiments in the laboratory, is compared to the calculated expansion rate of the universe at early times; the final equilibrium abundances of the light elements produced can then be determined. Before the microwave background was discovered, one might not have taken such a calculation seriously. But when the microwave background is interpreted as an afterglow of the Big Bang explosion, then a natural extrapolation backward implies that these microphysical processes would have had to occur in the early universe.

The original estimate of helium production in the Big Bang described here was performed by Peebles in 1965, within a year of the discovery of the microwave background. In 1967, Robert Wagoner, Willy Fowler, and Fred Hoyle wrote a definitive paper detailing the calculation of primordial production of light element abundances: their calculation has served as the basis for all further efforts. Later, David Schramn and his colleagues demonstrated that deuterium, the intermediate product made on the road to producing helium, could not be produced in stars. Thus any deuterium observed today would have to be primordial. The possibility that one could attempt to explain the observed abundance of light elements in the universe today by such simple dynamical arguments using physics that can be measured in the laboratory is by itself exciting. If measurements of the remnant abundance of these elements could be made and found to agree with predictions, then one could test the Big Bang theory and the interpretation of the observed microwave background as primordial.

Indeed, the theory has been a resounding success. The range of predicted values fit the range of "observed" values remarkably well for nuclei whose abundance varies by 9 orders of magnitude. Of course, some uncertainty exists for the nuclear reaction rates that enter into the calculation, but a more important uncertainty rests in extrapolating to the primordial abundance of various light elements from the observed stellar abundances today. This imprecision is particularly evident in the case of the rare element lithium (nucleon number 7). Here even small amounts of stellar production or destruction could significantly alter the abundance produced during the epoch of Big Bang nucleosynthesis. Suffice it to note here that the primordial abundance of this element could freely vary by perhaps 1 order of magnitude and not be completely inconsistent with some stellar models. Even allowing for this uncertainty, the general agreement between theory and observation is striking.

Of the light elements other than hydrogen, helium is by far the most abundant. It is here that Big Bang nucleosynthesis theory had its clearest smell of success. In the early 1960s, before the processes I have described were investigated and before the microwave background had been discovered, astrophysicists had noticed that the helium abundance in the galaxy is not only large but fairly uniform throughout. Both these facts suggested that this helium had been produced primordially. And one need not follow in detail the dynamics of all of the nucleosynthesis reactions to predict this primordial helium abundance. If Big Bang nucleosynthesis theory is correct, helium must have been produced primordially with an abundance by weight of somewhere around 24 percent. This percentage is so large that it is difficult if not impossible to imagine how stellar processes could alter it significantly. The fact that the observed abundance by weight of helium in interstellar material falls at this same value (with an uncertainty of a few percentage points) gives us a great deal of confidence that the Big Bang model can safely be extrapolated backward some 15 billion years.

Although predictions for helium abundance agree roughly with the observed value, independent of most of the fine details of Big Bang nucleosynthesis, the agreement I earlier touted for the other less-abundant elements depends upon understanding those very details. In particular, these abundances depend strongly on one of the most fundamental cosmological parameters: the ratio of the total number of protons and neutrons (baryons) to photons in the universe. Now, the number density of photons in the microwave background is known once we determine the temperature of this background. Then, if we know the ratio of total baryons to photons, we can determine the density in the universe today of "baryonic," or normal, matter. We could then compare this density to the total density of matter that we dynamically infer to be located around galaxies to find how much of the dark matter might be made up of baryons. One of the most exhilarating results of Big Bang nucleosynthesis calculations is that they not only put cosmological theory on a firmer basis, but they also can give us an independent way of constraining this all-important ratio of baryons to photons, and hence constraining the upper limit on the present density of normal matter in the universe.

How do the predictions of Big Bang nucleosynthesis vary as we change this ratio? As we increase the number of protons and neutrons in the universe, nucleosynthesis takes place at progressively earlier times. This is because the reactions that form deuterium happen more quickly if the participant particles that produce it are more abundant. Once enough deuterium is produced, the other light elements, particularly helium, can be produced via the second set of reactions. What does this imply? The predicted helium abundance will rise slightly if nucleosynthesis occurs earlier because fewer neutrons will have decayed and thus more will be left around to bind with protons when helium forms. However, much more sensitive to the ratio of baryons to photons during nucleosynthesis are the abundances of the rarer light elements such as deuterium and lithium. In the first place, if the density of neutrons and protons is increased, the reactions that turn deuterium into helium not only begin earlier, they act more efficiently as well. This means less deuterium will remain after nucleosynthesis is complete. If we decrease the ratio of baryons to photons by a factor of 10, from 1 part in 10 billion to 1 part in 1 billion, we find that the remnant deuterium abundance decreases by almost 2 orders of magnitude. If we increase this ratio by still another factor of 10, the amount of remnant deuterium decreases by another 3–4 orders of magnitude. This large leverage in the ratio suggests that if we could accurately measure the remnant deuterium abundance today we could pin down the baryon to photon ratio with great precision. Similar arguments apply to another rare light element, helium 3 (a rare version of helium containing one neutron, instead of two, in its nucleus). Unfortunately, because these light elements are so rare, one can imagine scenarios by which the remnant abundance might be significantly altered between those early times and today, via the process of nuclear burning in stars.

Nevertheless, by utilizing measurements of the deuterium abundance in combination with other rare nuclei, such as helium 3 and lithium, one can attempt to obtain firmer estimates for the primordial abundances of these elements in a way that is less sensitive to the vicissitudes of stellar evolution. In addition, recent developments in stellar evolution theory—governing the processing of these elements in stars—allow us to extrapolate through the period of star formation with a little more confidence. At various times, controversies have abounded. By 1995, however accumulated data on the light elements, however, combined with refined calculations of Big Bang nucleosynthesis, pointed unambiguously to a fixed range for the baryon to photon ratio for which theoretical predictions might agree with observations. From this, and from measurements of the microwave background temperature today, the density of normal matter in the universe could be constrained to lie in some range, independent of whether this material is now associated with luminous stars or hidden Jupiter-sized planets.

It is remarkable that the data for different elements, deuterium, helium, lithium, and so on, which vary by more than 8 orders of magnitude in their abundance today, yield predictions that are more or less consistent with observations, only for a fixed and universal range in the baryon to photon ratio. Table 5.1 displays the theoretical predictions compared to the inferred primordial abundances as they were known in 1990.

Even with the uncertainties, the agreement between the predictions and the inferred primordial abundances from observations was impressive. The detailed predictions of Big Bang nucleosynthesis were consistent with observations when the number of baryons compared to photons fell in the narrow range between about 0.2 and 0.8 parts per billion. Translated into a mass density today, this implied that the fraction of the critical density in baryons is constrained to lie in the range $(0.03–0.005)/h^2$. Even if we take the Hubble fudge factor h to be equal to 0.5, this result implied that baryonic or "normal" matter accounts for less than about 12 percent of the critical density today, or at most slightly less than ten times the inferred mass density of luminous material.

This was a very important result, if it is true. Not only does it limit the baryon density to be less than or equal to the density inferred for dark matter on galactic

Nucleus	Predicted Abundance	Actual Primordial Abundance Limits as Inferred from Observation
Helium	.22–.26 (by weight)	.23–.25 (by weight)
Deuterium	10^{-3}–10^{-5} (by number)	greater than 10^{-5}
Helium 3	1–4 $\times 10^{-5}$ (by number)	less than about 2 $\times 10^{-5}$
Deuterium plus helium 3	10^{-3}-2 $\times 10^{-5}$ (by number)	less than 10^{-4}
Lithium	.08–10 $\times 10^{-10}$ (by number)	1–2.5 $\times 10^{-10}$ (controversial)

TABLE 5.1 Big Bang Nucleosynthesis: A Comparison of Theory and Observation, Circa 1990

scales, it suggests that some of this dark matter is likely to be baryonic, made up of protons and neutrons. This is because the lower limit from nucleosynthesis is, for most values of the Hubble factor h, slightly higher than the inferred fraction of the critical density in visible material (about 0.01).

I hope that by now, however, you have become slightly skeptical about conclusions drawn on data of cosmological interest. The aforementioned result is so severe that we should step back and examine exactly the assumptions that went into its making.

The first of these assumptions, that the microwave background is truly primordial, that it dates back to the recombination time, is the foundation for all of our theories of the Big Bang expansion. The general agreement between the results of Big Bang nucleosynthesis calculations and our observations does add further credibility to this scenario. Still, we should recognize that the assumed uniform expansion of the universe since the primordial nucleosynthesis epoch, based on the observed temperature of the microwave background, is an essential ingredient of our analysis. The circumstantial evidence for the self-consistency of this scenario is very strong, yet it is not etched in stone.

What would be the impact had the temperature of the universe *not* varied exactly as I have assumed it did between early nucleosynthesis and the present time? In the first place, the time at which nucleosynthesis occurred could have been somewhat earlier or later. This would imply that the expansion rate during the epoch of nucleosynthesis would differ from the rate that went into the dynamical calculations I have described. A faster or slower expansion rate would alter the equilibrium rates of the nuclear reactions that produce light elements, and thus would alter the remnant abundances that are predicted. Of course, some new physics input would be required for this to be the case. Exotic possibilities have been suggested, but they are just that: exotic. Without some rather extreme events occurring, the time-temperature relationship assumed in the standard Big Bang model cannot be tampered with significantly.

On the other hand, what if the ratio of baryons to photons at the time of nucleosynthesis was not the same as that ratio today? If some cataclysmic release of energy had occurred at a *later time,* heating up the photon background, the number density of photons could increase compared to that of baryons. In this case, constraints on the baryon to photon ratio at nucleosynthesis may be toothless, as they would not tell us anything about that ratio today. Fortunately, there are strong independent limitations on this possibility. As the universe expands, the density of material becomes more dilute, collisions happen less frequently, and it becomes harder to establish equilibrium between matter and radiation, if that equilibrium is disrupted at any time. Thus it can be shown that any cataclysmic reheating well after the epoch of nucleosynthesis would have been likely to alter the thermal nature of the microwave background in a measurable way. The launch of the COBE satellite laid this issue to rest by 1990. The CMB was measured to agree with a thermal black body to better than 1 part in 10,000. It is, in fact, the best measured black body in nature. This convincingly demonstrates that no significant nonequilibrium generation of radiation occurred between the Big Bang and recombination.

Less outlandish than any of these nonstandard scenarios is the possibility that one or several of the detailed dynamical assumptions that went into Big Bang nucleosynthesis might be too naive. For example, the physical conditions during nucleosynthesis may have differed in some way from the simple state of uniform hot gas in equilibrium, which we generally assume. Indeed, this very issue has been raised in at least two different contexts.

While we understand physics very well up to the energy scales relevant at nucleosynthesis, it is quite possible that something that happened earlier might have upset the simple uniformity of the gas of matter in equilibrium with radiation which is normally assumed to have existed during the nucleosynthesis epoch. One does not even need to resort to very exotic new physics input to imagine how this might happen. For example, we now recognize that the fundamental constituents of matter—protons and neutrons—are themselves composite particles made up of elementary objects called *quarks*. These quarks are bound together by the most powerful of the four known forces in nature, called "the strong interaction," into the familiar particles—protons and neutrons—which we observe (the other forces are gravity, electromagnetism, and the weak interaction). One of the great triumphs in the field of physics during the 1960s occurred when Murray Gell-Mann and collaborators showed that the "zoo" of hundreds of strange types of new particles that had been observed in collisions in particle accelerators could be understood in terms of simple combinations of three different types of fundamental constituents, the quarks.

Now, at sufficiently high temperatures, when the universe was about one-millionth of a second old and the density of matter was so great that the mean separation of protons and neutrons was less than the mean separation of quarks inside a proton, our traditional description of matter in terms of protons and neutrons becomes inadequate. Instead, a description of matter in terms of a quark "gas" seems much more appropriate. At very high temperatures, the interactions of the quarks in this gas are weak enough for us to predict its properties. Unfortunately, just at the borderline in density where a quark gas turns into baryonic gas, the interactions of quarks are very intense; at present we really have only inklings—obtained with supercomputers and measurements of the collision of heavy nuclei in accelerators—of how the transition between these two configurations of matter might have taken place in the early universe.

Many possible phenomena could have occurred then. As quarks begin to bind together to form baryons, their interactions could effect other changes. For example, we believe that it becomes energetically favorable for bound states of quarks to "condense" in the vacuum, a phenomenon I discussed in the first chapter. This background condensate can, in turn, affect the energetics of quarks as they bind together into baryons. In any case, it seems reasonable to say that a number of complicated processes may have occurred as the universe cooled from a state consisting of a hot gas, made up of primarily quarks and electrons in equilibrium with photons, to a hot gas consisting primarily of protons, neutrons, and electrons in equilibrium with photons.

It has been proposed that, during these early transitions, protons and neutrons would not be formed uniformly in space, but rather they would form preferentially in "lumpy" agglomerations—just as water vapor does not condense uniformly into water, but rather into separate drops. If this did occur, one might expect that by the time of nucleosynthesis, regions might have developed in which protons far exceeded neutrons in number, and other regions might have developed in which neutrons far exceeded protons. As we might imagine, the synthesis of light elements could take place very differently in these kinds of regions compared to the synthesis that would occur in the standard uniform scenario.

Several groups of scientists have attempted to discover what one might expect in such a "lumpy" nucleosynthesis scenario. For certain ranges of the still-undetermined phase transition parameters it appeared that an acceptable abundance of most light elements might be produced if the universe has perhaps as much as ten times the baryon density that is allowed in the standard nucleosynthesis scenario. But one abundance prediction appears to differ sharply from the standard model. This prediction involves the remnant abundance of one form of lithium (containing three protons and four neutrons). In the new scenarios, almost ten times as much lithium may have been produced primordially. This value appears to conflict with our observations. It is now clear that in order to agree with all observations, no significant "lumpiness" was possible during the epoch of BBN.

Another more exotic scenario could alter the nucleosynthesis constraints on baryon densities observed today. This theory posits the existence of new massive but unstable elementary particles. If these could live long enough to decay very shortly after the standard nucleosynthesis processes in the universe had been completed, their decay remnants could still contain enough energy to initiate a new phase of nucleosynthesis at later times. As in the earlier scenario, general agreement with observed abundances is possible for much higher initial baryon densities, but lithium—this time also a rarer form (containing three neutrons and three protons)—is produced at potentially unacceptable levels.

Whether or not these scenarios remain viable depends upon our ability to extrapolate our present-day observations back through the period of star formation to confront the predictions of Big Bang nucleosynthesis. Whereas there can be very little doubt that most of the helium (containing two neutrons and two protons) now present in the universe was created in those first instants of the Big Bang expansion, the abundance of deuterium, lithium, and so on is so small that stellar processes could significantly alter the abundance between the BBN epoch and today. Since it is these nuclei that most tightly restrict nucleo-synthesis parameters—notably the baryon to photon ratio—any constraints that we derive depend in the end on our ability to separate primordial and later stellar contributions.

It is here that perhaps the greatest breakthrough has taken place, all within the past five years. While many different measurements of the local deuterium abundance in the interstellar medium have been performed, the ability to extrapolate back from the present day abundance to the primordial abundance is complicated by our lack of detailed understanding of galactic chemical evolution. However, using new telescopes and high precision instruments, astronomers have now been able to directly measure the primordial deuterium abundance at several different remote locations in the universe.

Among the most distant objects known today are quasars. These compact objects shine with a brilliance of an entire galaxy, although their luminosity varies on a timescale implying that they must be much smaller. The only mechanism proposed for generating such luminosities involves accretion onto a large black hole at the center of a protogalactic mass. During the infall process, light would be radiated across many wavelengths, from the visible to X-rays and beyond.

We can now take detailed spectra from distant quasars and find that these spectra have a characteristic broad continuum of emission. This has been very useful in trying to untangle their detailed nature. However, more relevant for our purposes is the fact that since these objects are located in remote corners of our universe, the light they emit traverses much of that universe before arriving at our telescopes.

By looking for absorption of the quasar light, one can attempt to limit the amount of matter in intergalactic space. While no such universal absorption troughs exist at a level that might allow for vast, otherwise undetected amounts of neutral hydrogen gas, there are, along certain lines of sight, enough such pregalactic hydrogen clouds that they do produce detectable absorption of quasar light.

The predicted primordial deuterium-to-hydrogen ratio is less than 1 part in 10,000. Thus, one might imagine that it would be extremely difficult to directly observe deuterium in such systems. It is difficult. However, in the past several years this feat has been unambiguously achieved. Deuterium is slightly heavier than hydrogen, and therefore the wavelength of light it absorbs is slightly different. This difference is remarkably small, however. The shift in wavelength is less than 1 part in 1,000. Nevertheless, for a few systems in which the hydrogen absorption is particularly strong, David Tytler at the University of California at San Diego and his student, Scott Burles, now at the University of Chicago, have convincingly demonstrated the existence of a small deuterium absorption sidelobe in the hydrogen absorption trough associated with absorption of light from a distant quasar (see figure 5.1).

When these observations were first made, many people were skeptical that deuterium was truly being seen. Since these hydrogen clouds are moving away from us due to the expansion of the universe, and farther clouds are moving away from us faster, several clouds absorbing light at different intervening distances between the quasar and us due to the Doppler effect will absorb light at different observed frequencies here on earth. A relative velocity of about 82 kilometers/sec between clouds would be sufficient to cause hydrogen absorption from one cloud to mimic hydrogen absorption from the other.

For this reason, several early claims have now been shown to be incorrect. However, the Tytler-Burles data and analysis is convincing. Most cosmologists now believe that the primordial abundance of deuterium has been essentially pinned down!

Why is the deuterium observed in distant hydrogen clouds thought to be primordial? Because these systems are at cosmological distances, we are looking back a significant fraction of the age of the universe. Moreover, these systems are diffuse and appear to be protogalactic. That is, they had not yet undergone significant star formation at these early times. Thus, there is no mechanism to significantly alter the deuterium fraction.

When all the dust is settled (theoretical as well as observational), when the comparisons of standard BBN models are compared to the observed deuterium

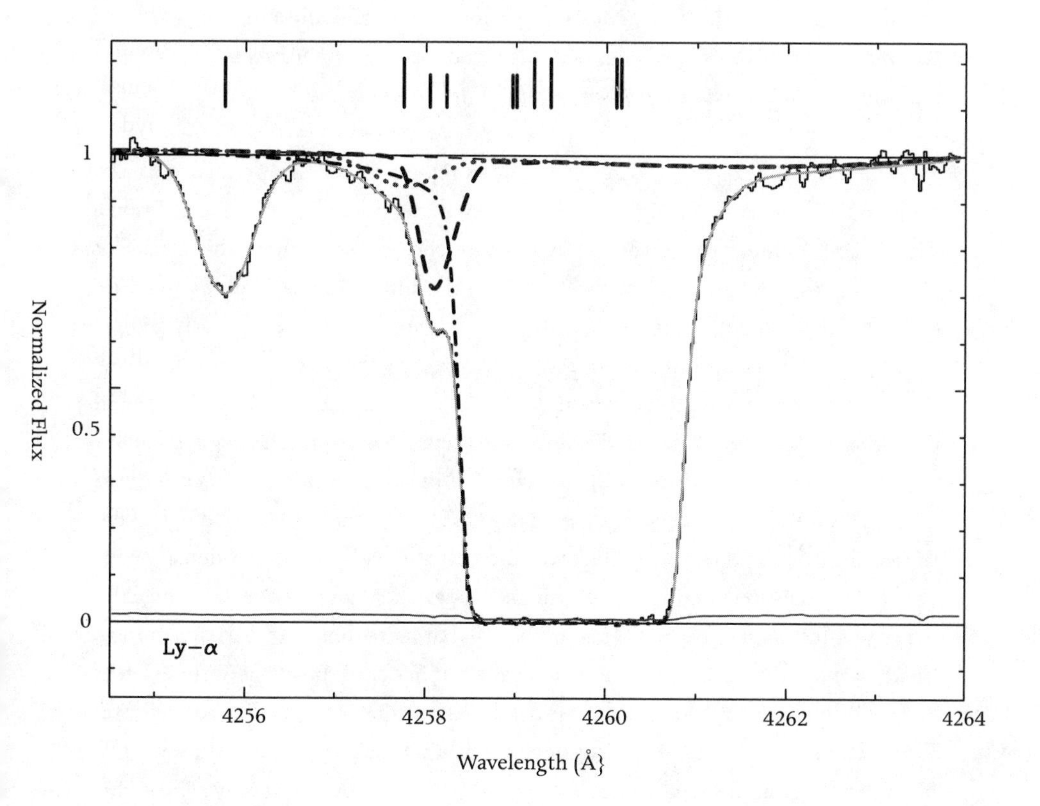

FIGURE 5.1

Deuterium absorption from distant quasars gives the first direct probe of the primordial deuterium abundance. Combined with calculations from Big Bang Nucleosynthesis, this puts a direct limit on the baryon density of the Universe. Shown here is a Keck+HIRES spectrum (resolution of 0.12 Ang) of the quasar 1009+2956. Shown (courtesy of S. Burles) is the data (solid histogram) of the Lya feature at redshift 2.504 overlaid with the best model fit (solid gray). The deuterium absorption (long slashes, short tick marks), lies 1.15 Ang blueward of the associated hydrogen absorption (dash-dotted, and tall tick marks). A weak hydrogen line (dotted) lies 0.25 Ang blueward of the deuterium profile. (Courtesy NASA.)

fractions, we can derive a precise measure of the actual fraction of baryonic matter compared to the critical density today. The results tell us that the fraction of the critical density in baryons is $.019/h^2$, with an uncertainty of plus or minus .002. This is, happily, within the range earlier predicted on the basis of the information available at the time. More important, given the allowed range for the Hubble parameter *h*, this implies that *the actual density of baryons in the universe today must be less than about 8 percent of the closure density!*

The standard nucleosynthesis scenario works remarkably well. It explains all the observed abundances of light elements using very simple initial conditions. There is no evidence whatsoever that it is incorrect. Of course, this does not allow us to rule out all other possibilities, but the simple Big Bang predictions are still the most consistent; moreover, they require no new exotic or poorly understood physical mechanisms. If simplicity and accuracy are the hallmarks of correct physical theory, we must conclude that in the absence of new experimental data, the standard Big Bang nucleosynthesis scenario is at present the best bet.

If we accept the predictions of the standard model at face value, then baryons alone cannot yield a critical density of matter in the universe. However, nucleosynthesis constraints go further. The lower limit on the baryon number density suggested by measured deuterium abundance is already well above the observed density of luminous material. This implies that at least some and perhaps most of the dark matter in galaxies is made up of normal material, possibly in the form of hidden Jupiter-sized planets or long-dead stars. More interesting still, the preferred range determined from the nucleosynthesis results suggests a density in baryonic material of up to almost 10 percent of closure density today. This value is very close, within a factor of 2, to the virial estimates for the total density of matter in the universe today, both dark and light.

Such a numerical coincidence should be heartening. The simplest conclusion to infer from this near agreement seems to be that the nucleosynthesis constraints signal the need for baryonic, or normal, dark matter; and the virial estimates, allowing for the maximum uncertainties, allow this to be all that there is—*there seems no need for anything else.* Of course, it is also important to note that the median virial estimates taken at face value yield systematically more dark matter than can be accounted for by baryons. Whether one therefore believes in nonbaryonic dark matter on this basis alone depends upon the degree to which one is willing to restrict the accuracy of virial estimates.

To be sure, it is daunting to imagine how to make ten times as much normal matter dark compared to that which is visible, but it is by no means impossible to do. If dark matter is made from baryons, the dark matter problem remains interesting, but primarily of interest to astrophysicists. This is because the dark material would likely consist of astrophysically sized objects, and even in the most exotic possibility—when dark matter is in the form of massive black holes scattered throughout the galaxies—the implications for learning something new about the stuff of matter here on earth would seem to be reduced.

Yet, *in spite of* the possible concordance of prediction and observation, our picture is neither complete nor consistent as it stands. Something else *should* be "out there," or our whole notion of the evolution of large-scale structure, and perhaps of the whole universe itself, must be altered. Moreover, whatever "it" is, it should have an abundance far greater than that which Big Bang nucleosynthesis constrains baryons to have today. To explore the rather deep theoretical reasons for these beliefs requires, first, a great leap forward: we need to travel forward as much as several billion years after the era of nucleosynthesis. Then, in the next chapter, we need to leap backward to the first few billionths of a second of creation. Hang on to your hats.

Because gravity is universally attractive for matter, systems acting only under gravity's influence are inherently unstable, tending to collapse under their own gravitational attraction. This holds for systems ranging from a house of cards you may construct in your living room to the universe as a whole. Somewhere in between lies the scale of galaxies. Because gravity is attractive, we can be reasonably sure that the clumped objects we see in the universe today were less clumped at earlier times, as long as only gravity has been directing the action. The current theory of galaxy formation is largely just a quantitative reflection of our understanding of the action of gravity. For example, take an expanding region of the universe and assign it a little higher mass density than the surrounding regions. Then the extra gravitational attraction due to this extra mass will cause this region to expand less quickly than its surroundings. As expansion proceeds, the density difference between the region and its surroundings will continue to widen. This increased density will further suppress the relative expansion rates, and so on. Eventually the density contrast between the region and its surroundings may become so great that the gravitational attraction at its surface will overcome the Hubble expansion velocity and the region will decouple from the background expansion and begin to collapse. Depending upon the dimensions of such a region, we then will witness the formation of a star, a small cluster of stars, a galaxy, or a cluster of galaxies.

When some region of space has mass density a little higher than the average, this is termed a "fluctuation" or "perturbation" from the mean value. The mathematical technique for handling such small fluctuations is termed perturbation theory. Though perturbation theory is rarely mentioned in news accounts of physics, in fact 90 percent of what most working physicists undertake on a day-to-day basis involves just this. In essentially all areas of physics we can arrive at exact solutions only for problems that are extremely simple. For systems that are not extremely simple, we can analytically explore them with good accuracy if they deviate only marginally from the simplest case. These deviations are what perturbation theory is designed to handle.

In terms of gravitational collapse, as described in the previous example, as long as perturbations from the mean background expansion are small, we can analytically describe how these perturbations will grow with time. Once they reach magnitudes large enough to approach unity—that is, when the deviations from the background density grow to become the same order of magnitude as the background density itself—the system begins to decouple from the background expansion under its *own* gravitational attraction. From this time onward, we need computers to numerically determine the details of what will happen. One thing is certain, however. In order to be large enough to reach this latter stage, any density fluctuation must pass through the earlier stages, stages when these fluctuations are small enough to permit us to know analytically what is happening. Given an initial density perturbation in a given region, we can calculate exactly how long it will take before the system is *ready to begin its collapse*.

These considerations lay the groundwork for hoping to understand why we see structures on some scales in the universe, say galactic sizes, and why we do not see structures on certain other size scales. In other words, we hope that cosmological theory can provide us with an initial set of fluctuations for structures on different size scales, and that simple classical mechanics and gravity will then tell us why these systems collapse to form the structures we see today.

So far so good. But to understand when and how things collapse, we must look at the history of the universe a little more carefully than we have so far in the preceding chapters.

If the universe has existed for 10 billion or so years since the initial Big Bang, and no information can be transferred faster than the speed of light, the farthest distances we can learn about—even in principle—are now about 10 billion light-years away from us. If the universe is infinite, then there can exist objects much farther away from us than even this distance, but we cannot hope to know of their existence at this time. It does not matter whether our telescopes would be powerful enough to see such objects. The light from them simply *has not had time to reach us*. The region of the universe with which we could have had contact, that is, could have exchanged information via the transmission of light, is finite if the universe has had a finite lifetime up to the present time. As the lifetime of the universe increases, the size of this region increases, because we can see progressively farther as the light from farther distances can reach us.

The farthest distance out to which we can see has a special name: it is called the "horizon," in analogy to the horizon on earth, which is as far as we can see on the earth's surface. The region inside our "horizon" in space is the largest region with which we could have communicated since the beginning of the universe (had we existed since that time). Put another way, it is the region with which we have had "causal" contact. Something that happens here, within the horizon, cannot be the cause of anything that has happened outside the horizon because light cannot have transmitted the information that something was happening here in the time available. It becomes extremely difficult to formulate physical questions for scales larger than the horizon, because such questions are usually framed in terms of observable quantities, and we cannot make observations on scales larger than those defined by our horizon.

The existence of a horizon dramatically alters the way we understand the formation of structure in the universe. Microscopic physical processes can affect the nature of collapse only for volumes whose size is smaller than the horizon volume at any time. Thus, if we are to explore how structure forms in the universe, we must differentiate between regions that are smaller or larger than a horizon volume at any given time. Consider what happens in a region containing some density fluctuation—a region in which the density is initially slightly higher than its surroundings.* If this region has a uniform density inside, then as long as it is itself larger than the horizon, no physical process inside the region will be able to respond to the inhomogeneity on yet larger scales because no physical process can probe its existence. If the horizon encompasses the region and its surroundings, on the other hand, then microscopic physical processes can "sense" the inhomogeneity and may play a role in governing its evolution. In particular, vari-

---

*The term *density fluctuation* itself has no direct physical meaning unless the region encompassed is smaller than that encompassed within a horizon-sized volume, since there is no physical way to measure relative densities on larger scales. For this reason, although one can discuss the evolution of systems larger than the horizon in the general theory of relativity, one must be very careful in so doing. I should note here that regions larger than the horizon can collapse. Such regions then essentially behave as isolated universes. If the density is high enough inside such a region relative to its expansion rate, it will eventually collapse, just as our whole universe, if it is closed, will eventually recollapse. No microscopic physical processes can act to stop this collapse because they cannot act over distances larger than the horizon. Accordingly, a theorem, initially formulated in 1970 by Stephen Hawking and Roger Penrose, says that if a system larger than the horizon begins to collapse, it must eventually form a black hole. Here, however, we are interested in how such structures as galaxies emerge, so I shall restrict my discussion to evolution on scales smaller than the horizon at any time.

ous processes I shall describe can work against the natural attraction of gravity in order to suppress or halt collapse.

One particularly effective way to halt collapse is by the pressure of matter and radiation. The sun remains very stable, for example, and does not collapse inward in spite of the huge gravitational pull at its surface. This is because the heat generated inside produces a strong enough gas pressure to withstand gravitational collapse. When the pressure that balances the gravitational forces in a star is removed, the results can be spectacular. We may witness a supernova, in which the entire inner core of a star collapses, from a region larger than the size of Earth to a region the size of Manhattan, in a fraction of a second.

Consider early times in the universe, when matter and radiation were coupled in thermal equilibrium. As the matter in a region with a density enhancement tries to collapse inward, it must do so against a background of photons, moving at the speed of light, which can escape the gravitational pull of any object except a black hole. This photon bath provides a pressure support that can halt the collapse of matter. The matter-radiation mixture is best imagined as a single entity as long as the two media are in equilibrium. This combined medium can have a significant pressure—calculated as the response to an external force. If the pressure is high enough, then when a force is exerted inward, the response can be springlike, setting up vibrations that produce waves that travel through the medium. If the speed of these waves is great enough to traverse a region before the time it would take for the region to collapse under its own gravitational attraction, then the medium will bounce back and oscillate, and the material will not collapse. If, on the other hand, the time it takes for these density waves (which are in fact just like sound waves inside conventional materials) to cause the medium to bounce back is longer than the time-scale on which it would take the object to collapse, then the gravitational attraction wins out and collapse is inevitable.

In a medium composed primarily of radiation coupled to matter, the "springiness" is such that speed of density waves approaches the speed of light. This means that pressure will in general beat out gravity in regions that can have been traversed by a wave traveling at near the speed of light since the Big Bang. In other words, the pressure can equalize or overcome gravity for regions up to the size of the horizon at any time. Any small density fluctuation that is encompassed in a region smaller than the horizon will not increase. In fact, the situation is actually more extreme. What begins as an overall density fluctuation will eventually be dissipated as a series of density waves moves through the medium and as the pressure responds to the initial excess gravitational force inward. Ultimately, any initial density excess will be smoothed out.

Therefore, in a universe containing materials such as baryons coupled to radiation, any small density fluctuation that is located inside the horizon at some time—that is, inside the ever-growing volume over which light (and other information) can have traversed since the beginning of the Big Bang—will eventually be washed out! This means that structure cannot begin to form gravitationally on such scales as long as this is the case. Once the matter becomes neutral, however, namely, at the recombination time, any region with a slight density excess in normal matter can collapse freely without contending against the pressure of radiation—since matter and radiation will then be decoupled.

This process gives us a characteristic signature of structure formation by gravitational collapse in a universe dominated by baryons alone. Until the epoch of recombination, any density fluctuations on scales smaller than the horizon would have been dissipated. After recombination, fluctuations on scales smaller than the horizon will be free to begin to collapse inward. After some thought, we can see that this implies that the smallest scale on which fluctuations would be free to start growing is a scale on the order of the horizon volume at the time of recombination. Fluctuations on smaller scales will have been dissipated due to pressure effects before recombination, but these pressure effects cannot have operated on this scale or larger because the horizon size at earlier times was always smaller. The distance traversable by light *grows* over time.

One might hope that the eventual gravitational collapse of initial density fluctuations on this scale and larger would have led to the development of the structures that we observe in the universe. But with a baryon-dominated universe, it appears, both theoretically and observationally, that this is unlikely.

The first problem is one of timing. After initial density fluctuations are free to increase under gravity, it takes time for such local density enhancements to become extreme enough so that the regions they encompass can begin to collapse inward, instead of continuing to flow outward with the background expansion. Recall, however, that there has been only a limited amount of time since the epoch of recombination up to the present for such growth to have taken place. This implies that fluctuations in matter on scales of the size of the horizon or greater at recombination must have had a certain minimum amplitude at that time if they were to have grown sufficiently by the present for the matter inside of them to have collapsed inward under gravity, to form self-bound systems such as galaxies or clusters.

Between the time of recombination and today, the radius of the region that would evolve to be the presently observable universe expanded by a factor of about 1,000. One can calculate that as soon as small density fluctuations are free to start the route to collapse under gravity, their amplitude can increase in direct proportion to the increase in this size of the universe. As the background universe expands, these regions expand by a little less, so that the density difference between the matter inside and that outside the fluctuation grows in proportion to the background expansion rate. Thus, the amplitude of such initial density fluctuations could have increased by, at best, a factor of 1,000 since the time of recombination, as long as they remained small enough for this kind of simple analysis based on perturbation theory to remain valid. Once the magnitude of such density fluctuations approaches unity (and perturbation theory then breaks down), collapse rapidly proceeds.

Hence, *in order for the material in some region to have collapsed to form galaxies and clusters by the present time, the initial density in such a region—of a size equal to or larger than the horizon scale at recombination—must have exceeded the mean background density by at least 1 part in 1,000 at that time.*

This does not seem to be much of a constraint, but it is quite forceful. It leads to a *potentially* observable effect that we do not observe. Remember that the cosmic microwave background was produced at the recombination time and has not interacted with matter since then. This cosmic microwave background gives us a picture of what the distribution of radiation in the universe looked like at that time. Recall that baryonic matter and radiation were strongly coupled together up to the recombination time. In general, one finds that any initial fluctuations in the matter (baryon) density will therefore leave an imprint of comparable magnitude on the radiation density at that time. So, if we scan the microwave background in different directions, searching for slight alterations in its density—or equivalently in its effective temperature—at different places, we may hope to see a signature of the primeval matter fluctuations at the recombination time, which eventually formed the galaxies we see today.

Over the years since its discovery, the microwave background has been scoured by balloon-launched and satellite-launched detectors that can compare the radiation density in different directions, seeking differences. On the size scales that would have corresponded to or exceeded the horizon size at recombination, no anisotropies in temperature had been detected even at a level of almost 1 part in 100,000. This suggests that any density fluctuations in baryons at that recombination time on such scales would not have been likely to exceed this level.

In 1992, the COBE satellite reported one of the most important discoveries in the history of cosmology. This detector, sensitive to fluctuations in temperature of only a few millionths of a degree on separations of 7 degrees or more on the sky, finally observed what cosmologists had been seeking for over twenty-five years. Small temperature fluctuations, on the order of 1 part in 100,000 in the

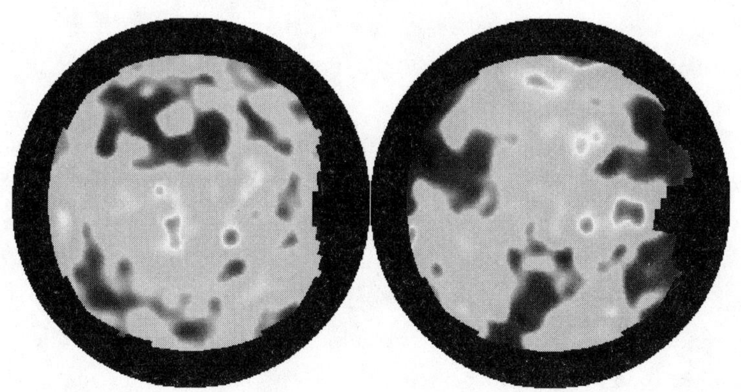

FIGURE 5.2
This image shows tiny variations in the intensity of the cosmic microwave background measured in four years of observations by the Differential Microwave Radiometers on NASA's Cosmic Background Explorer (COBE). the dark and light spots correspond to regions of greater or lesser density in the early Universe. Corresponding to temperature variations in the cosmic background radiation, these record the distribution of matter and energy in the early Universe before the matter became organized into stars and galaxies. The largest features seen by optical telescopes today would fit neatly within the smallest feature in this map.

CMB temperature, were unambiguously distinguished. Since the angular scale corresponding to 7 degrees is larger than the size of the horizon at the time the CMB was created, these anisotropies *could not have been created by causal processes since the earliest moments of the Big Bang!* Thus, COBE had finally discovered primordial fluctuations of the sort that we believe eventually produced all of the structures we observe in the universe today (see Figure 5.2).

While the scales observed by COBE are larger than those corresponding to the scale on which the first growth of fluctuations inside the horizon would have occurred, they are not that far off. The fact that the magnitude of these fluctuations is *almost 100 times* smaller than is required of observable structure formed by the present time due to gravitational collapse in a universe with only baryons and radiation is quite telling. Moreover, since COBE, a host of ground-based CMB experiments have probed the magnitude of fluctuations on even smaller scales. It is now clear that fluctuations in the CMB do not, by well over an order of magnitude, reflect the necessary size or shape as a function of scale that one would expect in a universe dominated by baryons.

This is a strong statement. *It suggests that galaxy formation by gravitational collapse is not compatible with the assumption that protons and neutrons dominated the density of the universe in the time since recombination.*

This statement depends on the assumption that the microwave background today truly reflects the radiation distribution at recombination and on the assumption that fluctuations in the baryons at that time were reflected in the radiation as well. Neither is indisputable, although it is getting increasingly difficult to bypass them. Catastrophic processes affecting the microwave background since recombination might have washed out small-scale fluctuations, although such processes are not a part of the simplest cosmological models and current CMB measurements show no tell tale evidence for such processes. Alternatively, perhaps baryon fluctuations are not reflected in the photon background. It is very difficult to imagine cosmological scenarios where this is the case, but it is not impossible.

Even though this conclusion is not irrefutable, there is another related result that perhaps depends on fewer assumptions. This argument also points to the need for "something else" to dominate the density of the universe during the period of galaxy formation. As I have described, density fluctuations in baryons in regions smaller than the horizon size are dissipated by pressure waves before recombination. Hence, only fluctuations on scales larger than this scale will remain to continue to grow after recombination. Thus, the horizon size at recombination sets a *minimum* size on the scale of regions that will be the first on which structures will emerge due to gravitational collapse.

Because of the expansion of the universe, one can calculate that a distance scale comparable to the horizon size at recombination would today have expanded in size to encompass a distance of about 10^{26} centimeters or about 100 mega-parsecs (for a graphical display see appendix B). This distance is about ten times larger than the largest clusters of galaxies we observe today. If this scale were characteristic of the first scale on which structure could begin to form—as one would imagine in a baryon-dominated universe—then even if these structures later fragmented to form galaxies and clusters we should expect to see significant remnants of this earlier clumping on these large scales.

But this is not the case. On scales of 100 mega-parsecs, the average density of matter does not vary with any significance as far as we can tell. The largest collapsed structures seem to be clusters of galaxies, on sizes of less than 20 mega-parsecs in size. Thus, even if the microwave background's constraints on primordial density fluctuations in a baryon-dominated universe were somehow circumvented, so that structure would have had time to form, the qualitative features of such structure would be expected to be quite different from what we see.

No matter what, large-scale structure arguments seem to suggest the need for *something else*.

The preceding arguments are compelling, but they are not yet airtight. We are not yet certain that the process of galaxy formation due to gravitational collapse is the only acceptable possibility. Until we are sure that we know how galaxies formed, we cannot be certain that no loopholes exist. Alternatives have been suggested, including large-scale explosive processes that could push matter together much more quickly than the slower gravitational process I discussed here. Of course, these scenarios themselves require new physics, but we should not rule them out completely until we have firmer evidence in favor of our standard assumptions. In any case, it is obvious that a baryon-dominated universe would require an overhaul of much of present cosmological thought if it is to remain a viable theory.

While the above caveat remains valid, the observations over the past decade of the CMB have made it far more difficult to imagine how nongravitational processes could have resulted in the formation of structure. As we shall see, the primordial fluctuations discovered by COBE are precisely of the magnitude and form predicted if the dominant matter in the universe is something quite different than anything that makes up the visible universe.

Combining observation with astrophysical theory, I have argued in this chapter that while in principle baryons could be abundant enough to make up a significant fraction of the inferred dark matter on galactic scales, some form of exotic dark matter seems necessary to explain the general features of large-scale structure in the universe. It may surprise you to learn that many physicists find at least as convincing an argument based almost purely on theory. Moreover, this theoretical argument requires more than just some exotic substance; it requires almost ten times more of it than any dynamical arguments based on observation suggest. In a science that is supposed to be empirically based, this might seem absurd. Yet sometimes the logic of a theoretical argument is so strong that faced with considering an illogical alternative or the possibility that experimental results are incomplete or even incorrect, physicists choose the latter. The history of twentieth-century physics is replete with noble examples of the acceptance of theory in a period of experimental uncertainty. Einstein's theory of relativity, the development of quantum theory, the hypothesized existence of the neutrino, the unification of weak and electromagnetic interactions are but a few. I do not suggest that the argument I shall outline in the next chapter is equally noble, nor is it necessarily correct. Nevertheless, I, and many of my colleagues, find it almost irresistible.

CHAPTER SIX

THE TIP OF THE ICEBERG

In the 1970s Robert Dicke and Jim Peebles of Princeton pointed out an apparent problem with the internal consistency of Big Bang cosmology. This paradox, which has since become known as the "flatness" problem, inspired a young particle physicist named Alan Guth to propose in 1981 a solution that has revolutionized cosmology. I will discuss both the problem and Guth's solution here. First, I want to clear up a possible misconception. Although the flatness problem and Guth's solution are of course related, one should remember that they are logically distinct. Whether Guth's cosmology is correct, the flatness problem poses a fundamental challenge for theorists and observers alike.

The flatness problem can be stated in many different ways; it will be useful to outline each of them in progression here. It is my experience that viewing a deep problem from several different vantage points allows its significance and ramifications to sink in more easily.

Most succinctly, one can phrase the flatness problem by asking, Why is the universe so old? To understand why this might present a problem, recall our earlier analogy between the expansion of the universe and the trajectory of a ball thrown in the air from the earth's surface. There I pointed out that if one examines the trajectories of balls thrown with different initial velocities, these trajectories all converge at very early times. These early trajectories are relatively independent of the details of the later trajectories over time. In a similar way I argued that the early evolution of the universe can be traced with some accuracy without knowing exactly all of the parameters governing the present-day expansion.

However, if we turn this analogy around, it forewarns of a potential problem. If we manipulate ever so slightly the initial conditions of the Big Bang expansion, we can drastically alter the late time evolution of the universe.

Consider the trajectories shown in figure 6.1. One curve corresponds to throwing a ball up in the air and having it fall back, and the second curve corresponds to throwing a ball up with the escape velocity, so that it does not return. It is a general rule of thumb that the time it takes the two paths to deviate by a significant amount corresponds roughly to the time it will take for the first ball to begin to fall back to earth. Now consider the two trajectories shown in figure 6.2. This time, imagine that they represent the size of some region of the universe in the Big Bang expansion. One trajectory corresponds to a matter-dominated closed universe that will recollapse; the other corresponds to a matter-dominated flat universe that will continue to expand forever. Once again, the time it takes before a closed universe begins to deviate significantly from a flat one is roughly the time it will take for such a universe to begin to recollapse.

In this context we may frame the flatness problem as follows: If the universe is not exactly flat, then why has it evolved for so long without deviating significantly from being flat? If the universe is closed and matter-dominated, it most likely should have collapsed a long time ago. If the universe is open, it should have expanded fast enough to reduce its density to a small fraction of what we observe today. Let us quantify this problem.

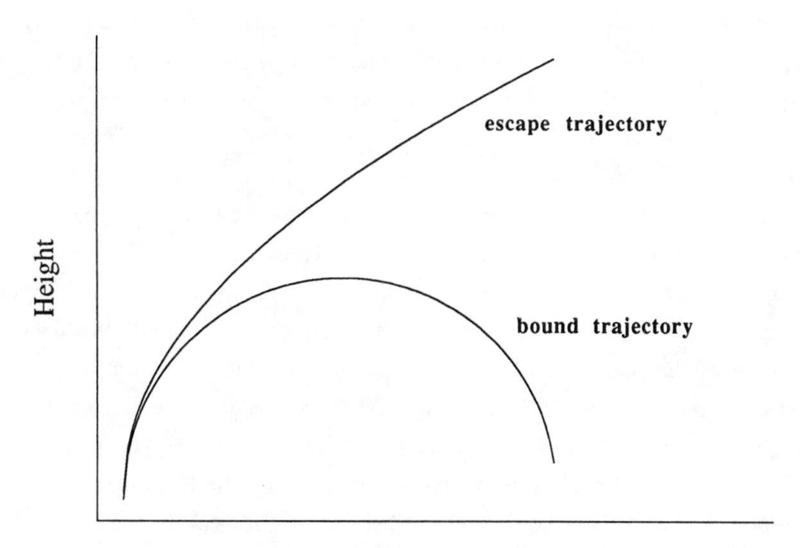

FIGURE 6.1

Shown here are two trajectories giving the position as a function of time of two objects thrown up from the earth's surface with different initial velocities.

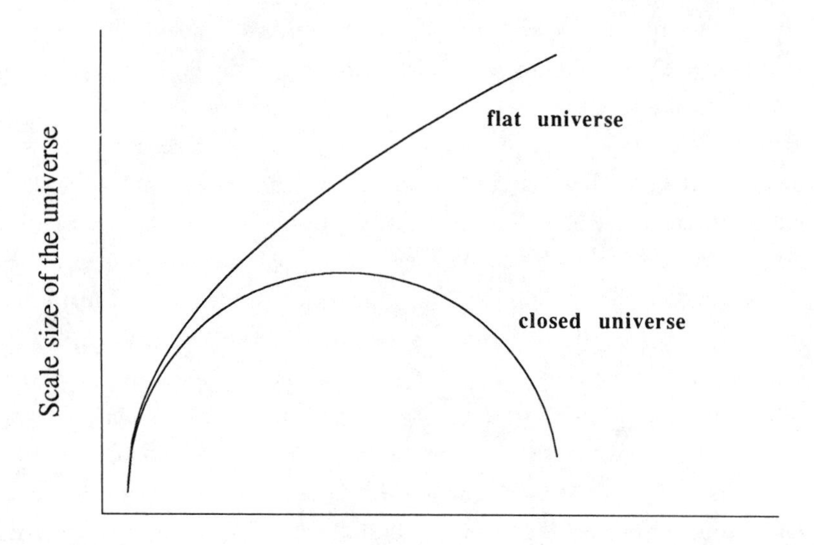

FIGURE 6.2

In this figure, the curves represent the size of some reference region during expansions that correspond to either a flat or closed and matter-dominated geometry for the universe.

In an exactly flat universe, the density parameter Ω (equal to the ratio of the actual density to the critical density; see chapter 4) is exactly equal to 1 today; in other words, the density of the universe is equal to the critical density. However, it is important to recognize that Ω is not in general a constant, independent of time. If Ω is slightly less than or slightly greater than 1 at any time and if the vacuum energy is zero, it tends to evolve farther away from one as time progresses. This is just a reflection of the phenomenon pictured in figure 6.2. Classical trajectories that differ at one instant tend in this case to diverge further as time goes on. Physicists would call the special value of 1 for Ω an *unstable equilibrium point,* just as a ball resting on top of a very sharp mountain peak is at an unstable equilibrium. The slightest force from any direction will cause it to come toppling down. Indeed, for a matter- or radiation-dominated universe one can show that the ratio of the departure of Ω from 1 divided by the magnitude of Ω itself increases by at least as fast a rate as the rate of expansion of the universe. If the universe becomes twice as large at some later time, then this ratio of the charge in Ω divided by the initial value of Ω will also double. The flatness problem can now be phrased in the following way. Observations indicate that the density of the universe is today somewhere between one-tenth and perhaps twice the critical density. Why is it, if the universe has had some 10 billion years to reach this point, that Ω remains within a factor of 10 of being exactly equal to 1 today? I can put this question more strongly. If Ω deviates at all from the value 1 today, then this is essentially the first epoch in the history of the universe where this is the case. At all earlier times, Ω was much closer to 1, and for most of the evolutionary stages it must have been almost exactly equal to 1. Why should the universe have waited about 1–10 billion years before beginning this departure?

The flatness problem is a prime example of what has become known in particle physics as a "naturalness" or "fine-tuning" problem. The Cosmological Constant Problem, previously mentioned, is another. Naturalness is a subtle concept to define. Indeed, "unnatural acts" undoubtedly seem quite natural while they are being enacted; particle physicists often hear outside observers exclaim that no matter how unpalatable theorists may find it, and no matter how unnatural it may seem, Ω is what Ω is—and it may not be equal to 1 today. Indeed, I know from experience that the foregoing arguments in favor of an $\Omega = 1$ universe today strike some as less than convincing.

However, "naturalness" has a quite well defined meaning in particle physics. If we confine our studies to the evolution of astrophysical structures with lifetimes of billions of years, and witness the relative insensitivity of many of these evolutionary processes to the exact value of Ω today, in turn, Ω may seem a somewhat arbitrary parameter that just happens to determine the future evolution of the universe. Yet if we accept that we must eventually understand the universe in terms of the microscopic fundamental laws that govern its activity on the smallest scales and at the earliest times, only then does the utter *implausibility* of a nonflat universe today make itself apparent.

A modern particle theory is classified "unnatural" if it possesses two specific characteristics. First, the theory may become suspect if in order to explain observed phenomenon extremely large or small *dimensionless* numbers are required. A dimensionless number is a value expressed without units of measure. The size of an atom, for example, has the dimension of length; it may be very small when measured in centimeters, but we can always find another unit of measure so that it will not be very small when expressed in this new set of units. A dimensionless number, on the other hand, such as the ratio of energy levels in an atom, or the ratio of masses that emerge in a given family of elementary particles, cannot be so altered. The emergence of a large or small number in such a ratio is not itself fatal in the eyes of theorists. There may be a physical reason why some process happens very infrequently compared to similar processes, or why some particle mass is very tiny. It is unacceptable, however, when an otherwise unjustified "fine-tuning" of parameters is required to keep some dimensionless number in agreement with observation. Here is an example. Say that the rates for two independently calculated physical processes must cancel out with an accuracy of 1 part in 10^{20}, leaving a residue that is 20 orders of magnitude smaller than either of these rates taken separately, in order to agree with some observation. In this case one is tempted to say that a theory in which such a cancellation is required without some physical rationale has about as much probability of being right as the probability of two numbers chosen at random have of being identical up to the twentieth decimal place. In this context, then, the flatness problem is the second worst fine-tuning problem we know of in physics.* To understand the dimensions of this dilemma, we have to take one more giant leap backward in the history of the universe.

We have as of yet no complete microscopic theory that allows us to calculate what the universe was like at time zero of the Big Bang, or why. At early times in the Big Bang expansion, the temperature of the universe was so high that energy densities were greatly in excess of anything we can probe directly in the laboratory today. Our present direct laboratory data can take us up to energies corresponding to those available when the temperature of the universe was almost 10^{16} degrees Kelvin, which occurred when the universe was about 10^{-12} (= .000000000001; see appendix A) seconds old in the standard Big Bang picture. The most massive particle that we have produced in the laboratory is the top quark, about 175 times the mass of the proton, observed at the Fermi National Accelerator Laboratory near Chicago. The next heaviest are the so-called W and Z particles, which are about one hundred times the mass of the proton; these particles were observed in the accelerator at the European Center for Nuclear Research in Geneva in 1984. At the energies available when the universe was at 10^{16} degrees K, the top quark and the W and Z particles existed in plentiful abundance in thermal equilibrium. We do have some indirect empirical knowledge about much higher energy scales from certain ultrasensitive nonaccelerator-based experiments, but this information provides only bits and pieces of the picture. Thus we cannot make definitive statements about the nature of matter and energy at higher temperatures. We can only make such statements for conditions at and

---

*In fact, the worst fine-tuning problem in physics relates to the cosmological constant. If that quantity is nonzero, but small, then a fine-tuning of about 125 decimal places seems called for. One might suspect that these two fine-tuning problems, being the worst in physics, are related, but this does not seem to be the case. As we shall see in this chapter, the flatness problem has a natural solution in terms of calculable physics. To date, no one even understands how to address the cosmological constant problem.

below the scale of the W and Z particles, where we understand in some detail the nature of the forces that govern the behavior of elementary particles observed in the laboratory.

If we want to apply microscopic physical theory to describe the behavior of matter as the universe expands, we might want to begin when the temperature of the universe was, say, 10^{16} degrees K and work forward, since presumably we comprehend most of the relevant physics after this time. We can then ask ourselves how carefully we would have to fix the parameters of the expansion at that time so that they would match those we observe today. We find that in order to make the density parameter Ω fall within a factor of 10 of unity today without being exactly equal to 1, Ω would have had to differ from exactly equaling 1 by only 1 part in about 10^{27} at that time. In other words, the density parameter Ω would have to be fixed to an accuracy of *twenty-seven decimal places*.

There is another way of putting this. If we fix initial conditions for the subsequent expansion of hypothetical universes at this era when W and Z particles were plentiful, when the universe was 10^{-12} seconds old, and choose values for Ω to lie randomly between zero and 2, roughly half the universes we create in this way will collapse again on a timescale of 10^{-12} seconds. And roughly half will quickly expand and dilute Ω to almost zero on a time-scale of 10^{-12} seconds. Only one in 10^{27} universes with Ω initially set different from the value of 1 will last 10 billion years before Ω begins to deviate significantly in either direction from 1!

We might want to be more conservative and begin applying microscopic laws to the evolution of the universe at, say, the nucleosynthesis epoch, when the universe was approximately one second old. Here, based on the success of Big Bang nucleosynthesis, we have some confidence that our picture of the expansion is accurate. In this case, we still must fix Ω to 1 part in 10^{15} at nucleosynthesis in order to end up within a factor of 10 of unity today.

Of course, all of these speculations are unsatisfactory. At some point, we must stop being content to fix random initial conditions before letting the universe expand under the laws of classical gravity supplemented by the equations governing the state of matter. Eventually we must come to grips with the fact that even these initial conditions for the classical Big Bang expansion too must result from physical processes governed by a microscopic physical theory, a theory that covers energy scales well beyond those we presently can probe (or, perhaps, that the initial conditions are indeed random, and that our observable universe is just one of a potential infinite series of universes in which the fundamental parameters may vary).

When one asks physicists at what energy scale these initial conditions are likely to be determined, the scale they are most likely to select is the "Planck scale," associated with a temperature of about 10^{32} degrees Kelvin, when gravity becomes so strong that quantum mechanical effects become important. It is at this point that the two great theoretical revolutions of twentieth-century physics, quantum mechanics and the general theory of relativity, come head to head.

Nearly a century after Einstein first attempted to unify gravity and the other forces in nature, we still have no explicit quantum theory of gravity. Some particle theorists hold out a great deal of hope that a theoretical framework, referred to as "superstring" theory or, more recently "M" theory, in which elementary particles are replaced by microscopic one-dimensional "strings" as the fundamental objects created out of the vacuum and whose oscillations are quantized to produce the observed spectrum of particles in the universe, may someday provide a true theory of quantum gravity. It is not a complete theory at present, however. There exist fundamental interpretational, if not computational, problems in allowing space itself, which is the fundamental degree of freedom in the general theory of relativity, to undergo the weird quantum behavior that we associate with elementary particles and atoms. Perhaps we have become accustomed to particles popping spontaneously in and out of the vacuum, but any theory of "quantum gravity" may force us to assimilate the notion of *universes* popping in and out of the vacuum.

Not having a complete theory of "quantum gravity" presents no real problem for dealing with physics at ordinary energy scales. This is because the effects of gravity at the scale of protons or atoms are negligible. Although gravity is basically the only force we regularly "feel" on a day-to-day basis, it is actually the weakest force we know of in nature. It is about 28 orders of magnitude feebler than the next weakest force, called the "weak interaction," and almost 40 orders of magnitude weaker than electromagnetism. The only reason gravity seems so omnipresent in our daily lives is that the gravitational attraction from every atom in the earth acts coherently to pull us down when we jump. Luckily for us, the earth is almost exactly electrically neutral, so no large-scale electric attraction can build up. (Very small charge abundances do build up locally; they are responsible for the lightning storms we witness on earth.) For a good example, first proposed by Richard Feynman, of the relative strengths of electromagnetism and gravity, let us imagine that while you are engrossed in reading this book, you step into an empty elevator shaft on the thirteenth floor of a building. It takes gravity several seconds, and several hundred feet, to accelerate you to your final velocity just before you rather abruptly stop. However, it only takes electromagnetism an infinitesimal fraction of an inch to do the stopping. It is the electronic forces between the atoms in your body and those in the floor which stop you from falling through the floor, and, in this case, end your fall.

Because gravity is so weak at ordinary scales, it usually does not worry us that we have not incorporated gravity properly in our quantum theory governing the behavior of elementary particles. The classical theory—general relativity—provides a perfectly good approximation for any foreseeable purpose. But as we pack matter together more and more closely, the effects of gravity eventually become relevant. For example, when two elementary particles approach each other and are unimaginably close together, at a distance of less than 10^{-33} centimeters, some 19 orders of magnitude smaller than the size of a proton, then gravity can no longer be ignored. So too, in the earliest moments of the Big Bang, when the temperature attained the unimaginably high value of 10^{32} degrees Kelvin, matter and radiation were compacted together so tightly that classical general relativity no longer provides us with an adequate description of the effects of gravity. At this temperature and higher, when in the standard Big Bang scenario the universe was less than 10^{-44} seconds old, our traditional picture must break down. To understand the physics that sets up the expansion which proceeds to a point where our classical laws of physics take hold, we must come up with a theory of quantum gravity.

Because we do not yet have such a theory, we generally use the "Planck" time (when classical gravity no longer applies) as the starting point for our models of the expanding universe. A quantum theory that can incorporate gravity beyond this scale, perhaps based on superstrings or the like, would provide the initial value data with which we may begin. For now, for particle theorists, the age of the universe is best expressed not in billions of years, but rather in multiples of the Planck time. In this scale the universe is 10^{62} Planck times old. And here the flatness problem really rears its ugly head. If quantum gravity is to fix the initial conditions at the Planck time, it must in the standard Big Bang scenario fix the value of Ω to differ from 1 by only 1 part in about 10^{59} at this time, if Ω is to differ from 1 by less than a factor of 10 today.

It is for this reason that many physicists believe that whatever physics laws determine the initial conditions for the observed Big Bang expansion, this physics must determine that Ω is set to be essentially exactly equal to 1, presumably somewhere near the Planck time.* In this case, the most natural result would

*For a similar reason, most physicists used to assume that some new physical laws must set the cosmological constant equal to zero. However, the data, in this case, suggest that there may be something fundamentally different about the cosmological constant.

be that the mass density of the universe remains essentially exactly equal to the critical value today. We must live in a flat universe!

This kind of theoretical leap of faith may be unsettling to anyone who is not accustomed to seeing such leaps regularly pay off when in certain rare cases experiment is led by theory. It may be especially unsettling without a specific theory to examine. Happily there exists such a theory which predicts that $\Omega = 1$ today . . . and a lot more. This is the "inflationary universe" theory, the brainchild of a particle physicist-turned-cosmologist named Alan Guth. I cannot do complete justice to this theory in the scope of this chapter, but we can briefly discuss the ideas that led to Guth's discovery. While a host of others have refined Guth's theories, some of his predictions remain as much a part of modern cosmology as is dark matter.

Inflation, as the theory has become known, has at its source one of the basic building blocks of modern physical theory: symmetry. We now understand that the fundamental dynamical quantities that govern the universe appear to result from symmetries of nature. We can prove, for example, that the existence of quantities such as energy and momentum, which are conserved throughout time, is a direct consequence of the invariance of the laws of physics under time and space translations: in other words, the laws of physics will be the same tomorrow as they are today, and they are the same at your house as they are at mine. As long as these two simple properties remain true, there must exist quantities that we recognize as energy and momentum and that govern the laws of motion.

Associated with each of the four known forces in nature is a certain symmetry that governs the form of the dynamical equations describing the interactions of matter. Fundamental symmetries are also associated with the behavior of bulk materials. Symmetries determine crystal lattices which, in turn, determine the properties of materials. Symmetries also control the dynamics of phase transitions, such as the transformation of water into snowflakes, or the transition that turns a piece of metal into a magnet, mentioned earlier.

The importance of symmetry for particle physicists was established by the work of Einstein on special relativity. Special relativity completes the unification of electricity and magnetism into a single theory first described by James Clerk Maxwell in his set of four equations. It was Einstein who then demonstrated why these relations between electricity and magnetism follow physically from a simple symmetry principle, associated with an invariance of nature first described by the Dutch theoretical physicist Hendrik Lorentz in the nineteenth century.

The connection between electricity and magnetism, which resulted in Maxwell's unification, was manifested in easily observable physical phenomena (for example, move a magnet near any wire and a current will begin to flow) and thus did not require any explicit symmetry arguments to elucidate the connections. But symmetry played a much more explicit role in the next unification of forces in nature: the unification of electromagnetism and the weak interaction.

At terrestrial energy scales, electromagnetism and the weak interaction (that interaction responsible for many nuclear reactions, including the decay of the neutron) appear to be quite unconnected. Electromagnetism is a long-range interaction; the weak interaction is short range. Electromagnetism is relatively strong; the weak interaction is at least a million times weaker. Electromagnetism involves charged particles; the weak interaction takes place between neutral as well as charged objects. For me, the story of their unification presents theoretical physics at its best. It demonstrates the power of deductive reasoning, the application of beautiful mathematics to understanding fundamental processes, and the boldness to follow a good idea into the realms of the unexplored.

The story begins with the union of quantum mechanics and relativity, which I discussed in the first chapter, which led to the development of the quantum theory of electromagnetism, or *quantum electrodynamics*. The formulation of the theory by Feynman in terms of the physical diagrams discussed in the first chapter led both to the ability to do calculations with the theory and to the development of some intuition about elementary particle interactions. Most important, the electromagnetic force could be seen as due to the exchange of virtual photons, as I described.

As a result of these developments, it was soon suggested that perhaps all forces in nature could be so viewed. Perhaps the nature and strength of a force could be interpreted in terms of the properties of the virtual quanta that mediated the force through their exchange. As I described earlier, one can argue that only massless quanta can transmit long-range forces. Massive particles cost more energy to create, and therefore the uncertainty principle ensures that they can propagate for only a short time without measurably violating energy-momentum conservation. Since the other two forces known to exist between elementary particles, the strong interaction and the weak interaction, were apparently short range, it was proposed that some new massive particles might exist to mediate these forces.

Now we return to symmetry. The fundamental symmetry of electrodynamics is related to the conservation of electric charge. This symmetry is called *gauge invariance*. It turns out that this symmetry is completely responsible for the form of the equations governing electrodynamics. This means that if one requires the dynamical equations governing the theory to be "gauge invariant" (that is, they keep exactly the same form when the fields in them are all multiplied by a specific mathematical function, whose form can vary from point to point), this of necessity forces us to write down a unique theory describing the interactions of electrons and light. That theory is quantum electrodynamics. One of the most important consequences of gauge invariance is that it restricts the form of the equations governing electromagnetism so that the quantum of the electromagnetic field, the photon, is exactly massless, and the electromagnetic force is therefore long range.

Electrodynamics was understood as a classical theory in the nineteenth century, long before quantum mechanics was developed. Not until the 1930s did the brilliant Italian physicist Enrico Fermi develop the first operationally successful theory of the weak interaction, which mediates neutron decay as well as the other processes of primordial nucleosynthesis described earlier, and which also powers the sun and stars. Fermi emigrated to the United States shortly before the Second World War and led the team that developed the first working nuclear reactor as part of the Manhattan Project. He was adept at experiment and theory alike, one of the few great physicists of this century to be so diverse.

Fermi's weak-interaction theory was not "fundamental," but rather "phenomenological." It was designed to give the correct rates of interactions but was not based on any underlying physical theory. Using a modern Feynman diagram language, the fundamental weak interaction process of the decay of a neutron into a proton, an electron, and an antineutrino would, in Fermi's theory, be pictured as in figure 6.3.

By analogy to the Feynman diagrams of quantum electrodynamics, it became clear that both the short range and the weak strength of the interaction might be explained if the central interaction point of the diagram really masked the

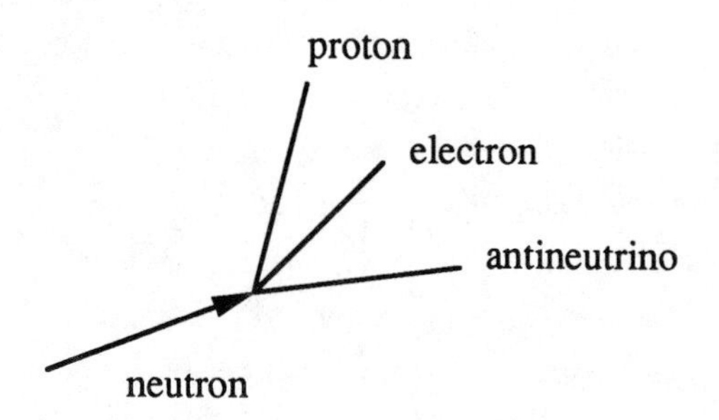

FIGURE 6.3

A Feynman diagram showing the decay of a neutron into a proton, an electron, and an antineutrino.

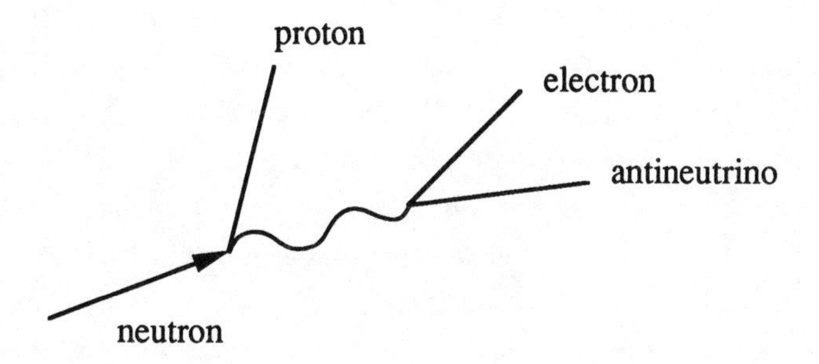

FIGURE 6.4
The same decay as displayed in figure 6.3, but this time "magnifying" the interaction region to display explicitly the exchange of a virtual particle, in analogy with the processes described in quantum electrodynamics.

exchange of a very massive virtual particle whose range is so short that the neutron decay reaction, shown in figure 6.3, actually can be represented by the emission and subsequent absorption of this virtual particle, as shown in figure 6.4.

From our knowledge of the strength of the weak interaction, the mass of the virtual particle being exchanged could be estimated; that mass was huge, about one hundred times the proton mass. Since such a massive particle could not be produced in the laboratory, and since no one had devised a fundamental theory that predicted the existence of such particles, their possible existence as real objects remained little more than an interesting possibility throughout the 1950s.

Once the weak interaction was depicted as in the diagram of figure 6.4, the analogies with quantum electrodynamics became tantalizing. Largely through the work of two preeminent particle physicists, Richard Feynman and Murray Gell-Mann, the possible couplings of these intermediaries were elucidated. It was found that the intermediaries could be of the same type as the photon—that is, particles with the same angular momentum quantum numbers. These hypothetical quanta of the weak interaction were called "intermediate vector bosons."* A group of brave individuals, many of them closely associated with the U.S. physicist Julian Schwinger, began to dream of a unification of electromagnetism and

*Because of the way the mathematics describing such particles behaves when we want to make a rotation in space, they are called vector particles, and because their angular momentum quantum number is an integral rather than a half-integral number, ensembles of such particles obey certain kinds of statistical laws, called Bose-Einstein statistics, after Einstein and Satendra Nath Bose. Such particles are termed "bosons."

the weak interaction. One of Schwinger's students, Sheldon Glashow, was assigned the task of investigating how the gauge symmetry associated with quantum electrodynamics might be related to the weak interaction.

Some years earlier, in about 1954, two theorists, Chen Ning Yang and Robert Mills, came to a profound realization. They recognized how the gauge symmetry of electromagnetism could be generalized into a much richer structure. Once again, the type of theories that could be written down to manifest this symmetry were very limited. At the time they had no definite proposals for a physical application, but the structure was so attractive that they published their results, along with some speculations about future applications. As is often the case, elegant mathematics eventually finds a place in physical theory; Yang and Mills uncovered a gold mine. We now recognize that every one of the four known forces in nature has a kind of gauge invariance; these theories can be described by some type of Yang-Mills gauge theory. Generalizing and extrapolating what had worked so well for electromagnetism once again proved worthwhile.

Schwinger and Glashow were interested in whether Yang-Mills gauge invariance might be useful for describing the form of the weak-interaction theory. Glashow made a fundamental breakthrough. He discovered that if he envisaged the existence of three very heavy particles, now called W^+, W^-, and Z particles, and combined them with the photon, together with Yang-Mills-type gauge symmetry, he could create a theory in which the photon coupled to electric charge, and the other particles coupled in just the proper way to mediate the observed weak interaction.

Yet such a theory made no sense as it stood. If the gauge symmetry were manifest, then the W and Z particles would be massless just like the photon. Recall, however, that the arguments described earlier required such particles to be extremely heavy, if they existed at all. Glashow could find no way around this dilemma, except to suggest that perhaps somehow the symmetry could be "spoiled" in such a way as to give the W and Z particles a heavy mass without changing the form of the Yang-Mills theory. Since gauge symmetry is crucial in determining this form, Glashow's "out" seemed remote at best.

Glashow's theory lay dormant until a theoretical ingredient from the physics of materials found its way into particle physics. This new ingredient was the theory of phase transitions, and an associated notion called *spontaneous symmetry breaking*. We encountered the concept of a phase transition earlier, in chapter 5, when I discussed the quark-baryon phase transition, and also in chapter 1, where I discussed the spontaneous magnetization of a ferromagnet. Both of these examples demonstrate the ideas that helped complete the theory of the weak interaction.

"Spontaneous symmetry breaking" sounds pretty intimidating, but it is a physical phenomenon that is fairly common in daily life. The laws of physics have many symmetries that are not manifest as we look about us in the everyday world. For example, when I first drafted this chapter I happened to be flying in a jet plane north along the California coastline; to my right I saw mountains and to my left ocean. There is no manifestation whatsoever of any left-right symmetry in the observed topography, yet most people would find it very difficult to believe that the fundamental physical processes that formed the coastline distinguished between right and left. Indeed, neither gravity nor electromagnetism, the two forces at work, do make such a distinction. No quantity appears in the equations describing the theories of these two forces which selects any preferred direction in space. That the underlying left-right symmetry of these fundamental theories is hidden because of the characteristics of the particular location I happen to find myself in is an example of spontaneous symmetry breaking.

The existence of California's coast may not seem very spontaneous, so let me give you another example, which I think is attributable to physicist Abdus Salam and which may be more apt and perhaps more familiar. Imagine that you are the first person seated at a round dining table set out for eight people. If you are like me, you might forget which wine glass is yours, the one on the left or the one on the right. Because of the circular symmetry of the place-setting situation, it is impossible to tell. Only standard "table manners," in which my wife claims I have grave deficiencies, can provide the correct answer. Now before you do anything, the table is perfectly symmetric. However, the moment you choose a wine glass—say the one on the right—you break this symmetry "spontaneously." The table setting is still symmetric, but once you drink from the glass on the right, everyone else at the table must do the same, or, heaven forbid, someone would end up with no wine.

Nature, too, is often forced to choose some particular configuration that does not maintain a symmetry of the underlying physical laws. I discussed such a situation in the first chapter, although I did not stress it then. Recall that the favored configuration of the little atomic magnets in a bulk piece of iron either can involve an alignment of all the magnets, in which case the iron will be a bulk magnet, or else all the magnets can point in random directions, in which case there will be no bulk magnetic field. At high temperatures the tendency to randomize wins out, and at low temperatures the tendency to align dominates.

Now, I have stated that the theory of electromagnetism does not pick out as special any particular direction in space. Nowhere do Maxwell's equations suggest that if you point a magnet in one direction something fundamentally different will happen than if you point it in another (ignoring for the moment the earth's magnetic field). As long as the individual magnets point in random directions, the favored "ground state" (that is, lowest energy) configuration respects this symmetry, because no specific direction is selected. As I have described earlier, however, once the material is cooled, a spontaneous thermal fluctuation can cause the magnets to line up in some direction, and this direction is determined as serendipitously as the choice of wine glass at the table. After this happens, there is certainly some preferred direction inside the material. If you put a little magnet inside the piece of iron, it will line up along with the bulk magnetic field created by the other aligned microscopic magnets. In this case, the underlying rotational symmetry of electrodynamics is no longer manifest in the ground state of the material, although the fundamental equations that govern the dynamics still possess this symmetry. This is a classic case of spontaneous symmetry breaking.

This might be no more than a mere curiosity were it not for an important phenomenon that it implies. When the symmetry of the ground state of the theory changes, the properties of the material change—iron gets magnetized, coal changes to diamonds, raindrops become snowflakes, and so on. Even the properties of propagating particles can change, as in the example of the phase transition that affects the way quarks bind together into baryons, which I discussed in chapter 5.

Another example of this phenomenon is more famous. In 1911, the Dutch physicist H. Kamerlingh Onnes discovered that when he cooled mercury down to a dozen or so degrees above absolute zero (more than 400 degrees below the freezing point of water on the Fahrenheit scale), it suddenly lost *all* electrical resistance. This was the discovery of a phenomenon known as *superconductivity.* In superconducting materials, currents can flow resistance-free for years.

It took about sixty years before the observation by Onnes was fully explained theoretically. The phase transition that resulted in superconductivity was understood to be due to a very subtle remnant attraction between electrons when they move in a solid. Electrons normally repel one another, but they interact with the crystal lattice in metal in such a way that a small residual attraction results. When the temperature is low enough, this attraction causes the electrons to bind into pairs in a very special new ground state. In this state, currents can flow freely because the dynamics of these electron pairs turns out to be governed by quantum mechanics and not classical mechanics. Just as electrons in energy levels in atoms can remain in "stationary" states forever with no alterations in energy, coherent motion of these electron pairs in a pure quantum state can yield a net current with no dissipation of energy.

A related property results when pairs of electrons "condense" into this new superconducting ground state configuration. The gauge invariance of electromagnetism is spontaneously broken. Recall that gauge invariance is what ensures that the photon remains massless. Thus, when a photon propagates inside a superconductor, where this symmetry is no longer present, the photon behaves as if it were a massive particle. This is because the photon's interaction with the condensed pairs of electrons present in the material in which it is propagating results in an extra contribution to the photon's energy and momentum. Since it is the masslessness of the photon which in turn ensures that electromagnetism is long range, this means that inside the superconducting material the electromagnetic forces are *no longer long range*. We can actually test for this experimentally. When we bring a magnet near a superconductor we find that the magnetic field cannot penetrate the material as long as it is superconducting. If we heat up the material so that superconductivity disappears, the magnetic field can once again permeate the material.

This phenomenon in condensed matter physics was first examined in the context of particle physics by the U.S. physicist Yoichiro Nambu as early as 1959. (Nambu is a remarkable physicist who is not as well known among nonphysicists as perhaps he should be. Aside from his work on symmetry breaking, he has been ahead of his time on a number of other occasions, as when, for example, he established the fundamental quantum mechanics of string theory—now all the rage—in the 1960s.) As noted in chapter 1, an important property of the vacuum in quantum theory is that if the interactions between particles are just right, the vacuum need not contain only virtual particles; instead, interactions can cause a finite density of real particles to condense together into a zero momentum configuration in a new vacuum state. If the vacuum state initially respects the symmetries of a particular theory, just as the ground state configuration of randomly aligned atomic scale magnets in an unmagnetized metal respects rotational symmetry, the new ground state condensate can alter the vacuum in such a way that the symmetry is spontaneously broken—just as the aligned microscopic magnets in a ferromagnet break rotational symmetry.

After Nambu's work, other physicists in the early 1960s established a number of related results. After it was shown that phase transitions and spontaneous symmetry breaking could occur in the "vacuum" of particle physics, it was demonstrated that when a Yang-Mills-type gauge symmetry is spontaneously broken, something remarkable happens. Just as the photon becomes massive in a superconductor when the gauge symmetry of electromagnetism is broken, the "vector bosons" that transmit the force associated with any gauge symmetry also acquire a mass when the symmetry is broken.

Once this phenomenon was established, it did not take long for two and two to be put together. In 1967, Steven Weinberg, an active participant in the above developments, reinvented the Glashow theory, but this time he posited a spontaneous symmetry-breaking mechanism to explain consistently why the W and Z particles were massive and the photon massless. Independently, Abdus Salam, through his earlier related work with J. C. Ward, hit upon the same idea. The mass of the W and Z particles could be related to the scale of energy associated with the symmetry-breaking phenomenon. In phase transition language, this scale could be interpreted as approximately the temperature at which the symmetry-breaking ground state of the theory becomes favored. For the weak interaction, this scale would have to be almost 10^{16} degrees Kelvin, where the mean energies of particles would be about 100–1,000 times the energy associated with the masses of familiar particles such as the proton and neutron. This theory was so compelling that in 1979, nearly twenty years after Glashow's initial paper, Glashow, Salam, and Weinberg were awarded the Nobel Prize in physics. This was still five years before the W and Z particles were experimentally observed (with the predicted masses) at the European Center for Nuclear Research in Geneva.

I have dwelled on the details of this story in part to demonstrate how central the notions of phase transitions and spontaneous symmetry breaking are in modern particle theory. Together they play an integral role in our understanding of the fundamental forces in nature. As it turns out, they also play a fundamental role in modern cosmological theory, although they were not developed with that purpose in mind. Later on we will discuss possible phase transitions that can result in many seemingly fantastic phenomena in cosmology. It is important to recognize that regardless of their effect on the evolution of the universe, we know with certainty that some phase transitions, such as that involved with weak interaction symmetry breaking, must have occurred at early times during the Big Bang expansion.

In any case, with the successful unification of the weak and electromagnetic interactions completed, it was not long before physicists began to speculate about further unification of the one other known force in nature besides gravity—the strong interaction between quarks. Glashow, this time in collaboration with Howard Georgi at Harvard in 1975, first raised the possibility that the strong, weak, and electromagnetic interactions could all be unified into a single, simple Yang-Mills-type gauge theory, whose symmetries would be spontaneously broken at presently observable scales. Investigating the implications of such a "Grand Unified Theory" (GUT) became the hottest topic of interest in particle physics. Among other things, a theory of Grand Unification might explain such hitherto unexplained but fundamental observations as why all elementary particles have electric charges that are multiples of the charge of the electron.

In collaboration with Weinberg and Helen Quinn, Georgi demonstrated that laboratory measurements of the different relative strengths of the three known interactions were consistent with such a unification. But they also found that the energy or temperature scale at which symmetry breaking should occur in the simplest model would be about 13–14 orders of magnitude larger than the scale at which electromagnetic-weak symmetry breaking occurs.

Never before had physicists seriously discussed physics at such large energy scales. No terrestrial particle physics accelerator will ever reach such energies. The uncertainty principle tells us that for two particles to be close enough to exchange virtual particles with masses in this range, and thus to undergo any of the phenomena described in the new theories, the particles would have to be less than about 10^{-29} centimeters apart. This is 15 orders of magnitude smaller than the size of a single proton. Nevertheless, even though terrestrial accelerators cannot make particles crash with enough energy for this to happen, the probabilistic nature of quantum mechanics implies that the new Grand Unified Theories might be tested.

Quantum mechanics says that we must make probabilistic statements in advance of measurement for such quantities as the position or energy of a particle. Because of this need, there always exists a small probability that the particle may have energy or position different from the mean value that we determine after a series of measurements on identical systems. Indeed, one can show that, just as there is a small probability that two quarks inside a neutron will come close enough together to exchange a W particle, causing the transition between a neutron and a proton, which we recognize will implement neutron decay, so too there is an infinitesimal possibility that the two quarks might get close enough together to exchange a virtual particle of mass characteristic of the Grand Unified scale.

One of the most outrageous predictions of GUTs was that a particle such as the proton, which was earlier presumed to be absolutely stable (we *exist*, after all), could in fact decay. In Glashow's words, "Diamonds aren't forever!" However, because of the extremely small probabilities involved, the average lifetime of a proton could be calculated to be about 10^{30} years, about 20 orders of magnitude longer than the present age of the universe.

One might imagine that such a decay could never be measured, but probability once again comes to the rescue. A given proton may take, on average, 10^{30} years to decay, but if we can amass 10^{30} protons in an experiment, chances are that one of them will be observed to decay if we monitor them continuously over the course of a year. As fantastic as this sounds, Grand Unified Theories seemed so compelling—following as they did on the resounding successes of particle physics in the previous decade during which all the forces in nature were explained as gauge theories—that several experimental groups planned extensive experiments to look for proton decay. Some of these involved enormous underground tanks that can contain more than 10,000 tons of water. The largest proton decay detector now contains more than 40,000 tons of water (see figure 6.5). These vessels, up to 60 meters on a side, are placed deep underground and monitored with detectors that wait in the dark for a signal that might indicate that a single proton in the water may have decayed. Proton decay detectors have been installed in the United States, Europe, Japan, and Africa. We shall have cause to return to them in the latter half of this book, for beyond being advanced particle physics devices, these proton decay detectors may become the astrophysical observatories of the future.

Proton decay emerged as the chief experimental probe of a theory that promised an understanding of elementary particle interactions spanning 20 orders of magnitude. Moreover, the possibility that protons might decay was quickly parlayed into the first great potential cosmological success of the theory. One of the most important parameters in cosmology, as we have seen in our earlier discussion of primordial nucleosynthesis, is the ratio of the number of baryons—protons and neutrons—in the universe to the number of photons in the microwave background. This number determines everything from the nature of the present and past universal expansion and the total amount of observable matter in the universe today, to the abundance of light elements produced in the Big Bang. Unfortunately, until the advent of GUTs, no one had the slightest idea how this all-important ratio evolved. One had to postulate it by fiat.

If the reactions described in GUTs can mediate proton decay, then they can change the number of baryons in the universe. Each time a proton decays, there is one less baryon. Once we know that the baryon number in the universe can be changed by dynamical processes, the possibility arises that the observed baryon number might similarly result from dynamical processes. This idea dawned on several different groups of theorists in the late 1970s. By doing dynamical calculations similar to those appropriate to primordial nucleosynthesis—but this time involving the reactions described in GUTs, and occurring not during the first minute of the expansion, but rather during the first 10^{-35} seconds—they showed that the observed baryon to photon ratio of the universe might naturally emerge dynamically. For the first time, physicists and astrophysicists had a microphysical

FIGURE 6.5

A photo of the Super-Kamiokande Underground Detector being filled up with water. Thousands of phototubes surround this tank containing over 40 thousand tons of water, ready to record light emitted when a proton in the water decays, or when neutrinos from the sun or supernovae interact with atoms in the water. (Photo courtesy of ICRR [Institute for Cosmic Ray Research], The University of Tokyo.)

theory that they might hope could explain the origin of one or all of the previously theoretical undetermined global parameters governing the expansion of the universe.

But GUTs were not free of cosmological problems. During any phase transition, such as that which must occur when the symmetries of the interactions described by a Grand Unified Theory break down and the strong and electroweak interactions become distinct, significant inhomogeneities may develop in the universe. When ice freezes it need not do so uniformly. Anyone who has noticed the patterns that ice crystals make on a window on a cold winter morning can see that the crystals form in different places, and point in different directions. As the ice crystal "domains" grow and merge, beautiful points or lines of discontinuity form to reflect the morning sun. These are localized regions at the points or lines of contact between domains where the crystal structure must change orientation to match that of the domain on *either* side. Such discontinuities are called "defects" in condensed matter physics. Their formation is due not to dynamics but rather to topology. These defects are forced into existence by the existence of the different domains that merge.

In the early universe, a similar phenomenon might have occurred. Because the distance traversed by light increases in comparison to the size of what is now the presently observable universe, the region within our horizon was at earlier times smaller than it is now. Indeed, in our presently observable horizon, there exist regions which *earlier* could not have been in contact with one another, because light would not have been able to traverse the distance between them in the time available since the Big Bang. If we extrapolate further and further back in the standard Big Bang scenario, we find that this trend continues. The number of different regions one could fit into the region that would eventually become our presently observable universe, and which were not in causal contact with each other, increases as one considers earlier and earlier times.

If we extrapolate backward to the time when the GUT symmetry-breaking phase transition was supposed to have taken place, we find that there were roughly 10^{90} different regions that could fit into that volume which would correspond to the observable horizon today, and which were not in thermal contact with each other at that time. If a phase transition occurred at this time, these different regions in the universe would have had no chance to "communicate" with each other by the exchange of anything moving at the speed of light or less. In this case, the symmetry-breaking ground state in each different region could have been different, since the physical processes in each region would have been occurring independently.

Eventually, as time went on, these regions would come into contact. At the boundary between the regions, topological "defects" could result, due to a mismatch of the nature of the symmetry-breaking ground states in each region. I shall describe later in more graphic detail how this could happen. For now, accept that one can show that in the GUT phase transition such defects are inevitable. No dynamics that we know of would stop their formation at the interface of these different regions. Such defects could store a significant amount of energy—an amount comparable to the energy difference between the symmetric state favored at very high temperatures and the symmetry-breaking state favored at low temperatures.

As the universe expands, the temperature, and hence the energy density of normal matter and radiation, decreases. Topological discontinuities or defects may trap huge amounts of energy, however; and they may also have no way of decaying or dissipating. These defects may eventually come to dominate the energy of the universe.

This situation would seem to be one way of generating dark matter. It might even be phenomenologically desirable. But for GUTs, it was a case of too much of a good thing. The *density* of such defects would have been prohibitively high, if on average such a defect formed at the intersection of causally disconnected regions at the time of the GUT symmetry-breaking transition. If these defects had no way of dissipating their energy, or otherwise disappearing, they would have caused the universe to recollapse eons before I had this chance to write about it.

All this talk about symmetry-breaking defects and GUTs may seem to be just a way of inventing problems that we can then occupy our time by solving. But even without GUTs, the "horizon problem" poses some mystifying questions. Since the region of our universe which we see today was at earlier times composed of many smaller regions that, in the standard Big Bang model, had not yet had time to establish thermal contact with each other, there is really no reason for us to believe that these regions should all be alike today. Why, then, is the cosmic microwave background temperature so uniform as to be identical in different directions to better than 1 part in 10,000? What made the expansion of the universe so uniform, if the different regions that were expanding had no way to communicate with each other, to "smooth" out any differences?

Enter Alan Guth. As a particle physicist at Stanford interested in gauge theory, Guth had been collaborating with Henry Tye at Cornell to investigate the formation of the topological defects associated with Grand Unified Theories. A public lecture by astrophysicist Robert Dicke in 1978 on the Big Bang theory had made him aware of the flatness problem. After learning enough general relativity to understand the basic dynamics of the Big Bang expansion, Guth proceeded to investigate not just the implications for cosmology of a GUT phase transition that might have taken place during the Big Bang expansion, but also the more fundamental question of how such a phase transition might affect the expansion itself. He discovered something remarkable.

Phase transitions can take place either smoothly or discontinuously as external conditions slowly vary. The freezing of water is an example of the discontinuous type, which is called a first-order transition. Guth decided to examine the possible consequences if a GUT transition were first order. As we have seen, in a first-order transition, separate "domains" of the new preferred ground state configuration form. Eventually, these domains grow and merge, so that all the material settles into the new phase. If the formation of domains is slow, the phase transition may take some time to complete. Supercooled water is an example. If one cools water with very few impurities while mixing it with a clean stirrer, as its temperature falls far below 32 degrees Fahrenheit it may still remain liquid. Eventually, however, ice crystals will form at various locations and suddenly spread out to freeze the whole sample. Because the frozen configuration of water at these low temperatures is energetically favored, as the ice crystals form, the energy that was stored in the liquid state is released. Even though the water sample continues to be cooled, it may maintain the same temperature until the transition to the ice state is completed. The energy difference between the water's liquid and frozen states, which is released as the transition is completed, is called *latent heat*.

If this kind of phenomenon could be shown to occur in a GUT transition, Guth discovered that it might solve not only the problem of topological defects but also all of the classic paradoxes of the Big Bang theory, including the flatness and horizon problems. If the universe had "supercooled" through a GUT transition, then those regions in which the phase transition had not yet taken place would remain for some time in a "false" vacuum state—the configuration with no particles condensed in the vacuum, which was no longer energetically favored. This state therefore had associated with it some fixed energy density, a latent heat if you wish, which was stored to be released upon transition to the favored phase. As the universe expanded during this time, regions of false vacuum that had not yet made any transition would continue to store this constant energy density. Unlike the energy of real particles, there was no way to dilute this kind of energy stored in the vacuum.

Einstein's equations relate the expansion rate of the universe to its energy density at any time. Since the energy density of normal matter is diluted as the universe expands, the expansion rate, characterized by the Hubble parameter, gets smaller as time progresses. However, as in the case of a cosmological constant-dominated universe for a region of false vacuum where the energy density remains constant, the expansion rate remains constant. This makes a drastic difference in the net expansion. In the first case, where the energy density of normal matter evolves, the scale factor of the universe increases with time but the rate of growth decreases with time. But in the case of a false vacuum, the expansion rate of the universe remains constant. If the GUT phase transition took place at some characteristic time, say 10^{-35} seconds into the Big Bang, then any region that managed to maintain its false vacuum for some time longer than this could have expanded by huge factors before completing a transition. If the region lived sixty times longer—still only about 10^{-33} seconds—then during this very short time, each such region could have expanded by *more than a factor of* 10^{25}.

After the transition is completed in any such region, all the stored vacuum energy (latent heat) is released in the form of matter and radiation; the universe "reheats" to almost the same temperature it had before the rapid expansion began, and we return to the standard Big Bang picture. The only thing that has changed is the size of the universe.

Guth appropriately named this amazing phenomenon "inflation." It would change our entire image of the early development of the observable universe. If each region of the universe expanded by some huge factor during this short time, then the region we observe today need not have encompassed many horizon volumes before the GUT transition. Instead, it may have originally existed inside a single such volume. Thus, for times before the brief inflationary epoch, the region we observe today was much smaller than we would have imagined in the absence of inflation. It then could have had sufficient time, given its smaller size, for the matter and radiation throughout the region to have come into equilibrium, distributing the energy quite uniformly. Then such a region would have quickly expanded, so that at the end of the inflationary epoch it was so large that it would then take light about 10 billion years—until today—before we could once again "see" all of it. Because the universe might quickly reheat back to its original temperature after this very brief but violent inflation, the standard Big Bang relationships between temperature, time, and so on would take on again, after inflation, exactly the form they had in our earlier discussions.

I have been a little cavalier in talking about "the universe" expanding, reheating, and so forth. In reality only those regions in which a false vacuum persisted for some time would have expanded in such a fashion. The point of inflation, however, is that the whole universe we see today could have initially existed inside one such region. In this regard, I want to point out that Guth's original inflationary model contained a serious flaw. Guth envisaged domains, or "bubbles," of "true vacuum" forming amid the rapidly expanding false vacuum phase. But in this case it becomes virtually impossible for the transition to complete itself, because the bubbles cannot fill up the intervening space fast enough. This problem in Guth's model was solved independently by Andrei Linde in the Soviet Union and Paul Steinhardt and his student Andreas Albrecht in the United States. In what they called "new inflation," they showed that a single "bubble," inside of which a domain of true vacuum forms, can itself "inflate" during the period of its formation to be so large as to encompass easily the observed universe today. In their model, what is the entire visible universe today could have initially been encompassed inside a single domain. What happened outside such a region becomes merely a matter of metaphysical interest since we will never, in any practical time, be able to observe out to such large distances.

Since this invention of "new inflation," a plethora of other inflationary models have been proposed, either in attempting to make closer contact with realistic particle models or in resolving various ugly features of the preexisting models. Among these models, several explored by Andrei Linde—now at Stanford—and colleagues, seem to be the most appealing. They go by the exotic names of chaotic inflation and hybrid inflation. Both models suggest the existence of meta-regions, well outside our observable universe, in which many other inflationary epochs may occur, eternally.

Now at this point most people's minds begin to reel. To try to imagine that the whole visible region of the universe once lay inside some primordial bubble produced during a phase transition is quite a challenge. I really cannot do justice to the whole mechanics of inflation theory here.[1] What is important to keep in mind, even if the details seem mysterious, is that during a phase transition in the early universe—an occurrence that we believe was very likely, *given our current ideas of particle physics*—the evolution of the universe could have briefly but significantly altered from that predicted in the standard, hot, Big Bang model.

The idea of rapid inflation of the universe associated with such a phase transition can solve in one fell swoop almost all of the outstanding problems of our classical Big Bang model. In the first place, if what is now the observable universe originally occupied a region smaller than one horizon volume before the GUT transition and inflation, then the symmetry-breaking ground state that resulted from the transition would be expected to be uniform, because dynamical physical processes could have operated over such a small volume. Topological defects tend to form at the interface of horizon-sized regions, where different symmetry-breaking ground state configurations would come in contact. Hence, due to inflation, we would expect to have no such remnant defects in the region that we identify with the observable universe today, instead of about 10^{90} such vestigial defects, as would be expected in the absence of inflation. What is now the observable universe was much smaller before the inflationary epoch compared to the size we would have predicted had inflation not taken place. Moreover, since the observable universe today *once existed inside a single horizon volume* and presumably all parts of it were at one time in thermal contact, the fact that it is apparently so uniform today is no longer a mystery. The "horizon problem" has disappeared.

More important for our purpose is what happens to the universe's mass density parameter Ω during the inflationary period. It is possible to show that as the universe rapidly expands it is driven quickly toward being flat: think of the way the surface of a balloon gets less curved as it is blown up. With no extreme fine-tuning it is easy to construct models where W is driven toward 1 to such high accuracy that it ends up being equal to unity not merely within 60 decimal places, but perhaps 600. Inflation therefore *predicts* a flat universe today!*

Another amazing aspect of inflation is that, during the rapid expansion, any matter and radiation that happened to be around beforehand would become diluted to essentially zero. Were it not for the latent heat of the vacuum that is released after inflation comes to a close, the universe would end up cold, dark, and empty. Instead, all the matter and radiation we see today in the universe would have derived from the energy that was stored in literally empty space during the inflationary expansion. Alan Guth calls this possibility, that everything we see may have come from the energy of the vacuum itself, the ultimate "free lunch." We are reminded once more of the "indefinite" of Anaximander, out of which all matter emerged, and into which it is annihilated.

Inflation not only generically predicts that the universe is flat today, and explains why the universe is so uniform on large scales, it also has the potential to explain even the small deviations from uniformity. When I spoke earlier of primordial fluctuations that might collapse into galaxies, I treated those fluctuations as being established by fiat at some early time. This is exactly how they were viewed before the inflationary model. Nowhere in the standard Big Bang model is there an explanation of *initial* conditions. The theory of inflation allows the possibility

---

*It has been pointed out, by those who like to hedge their bets, that it is possible for inflation to result in an open universe today. However, while this is possible, it is not generic at all in inflationary models.

that causal processes before or during inflation established the conditions that governed the expansion afterward. Indeed, for the first time, the idea of an inflationary epoch allows us to undertake a dynamical derivation to describe the spectrum of primordial fluctuations. In this scenario, the fluctuations, which on extremely small scales are due to quantum mechanical effects, determine just when each different region would begin its transition out of the inflationary phase. Regions that begin the transition at slightly different times end up with slightly different energies. This leaves remnant energy density fluctuations in the universe afterward.

The spectrum of fluctuations predicted by inflationary models exactly matches the spectrum of fluctuations that had been argued for earlier on phenomenological grounds by astrophysicists. We know today that primordial fluctuations cannot have been too large on very small or very big scales. If they had been large on small scales, we should see too many remnant black holes in the universe today. We have no evidence for any such primordial black holes, however. On the other hand, if fluctuations on big scales had been large, they would have left an imprint on the microwave background today which would be observable. Thus, a spectrum of fluctuations that was roughly uniform on all scales seems required. And this is exactly what results from inflation.

This situation may have seemed largely academic in 1982, because there was then no direct empirical evidence constraining the form of primordial density fluctuations. A decade later, everything changed. Two years into its mission, the Cosmic Background Explorer Satellite science working group stunned the scientific community and the world by announcing the discovery of minute temperature fluctuations in the CMB that could not be attributed to any causal processes associated with foreground contamination of the signal. After twenty years of searching, density fluctuations at the level of 1 part in 100,000 had been discovered. Moreover, the angular size of the fluctuations across the sky was so large—indeed, larger than the horizon at the time the CMB was formed—that no causal process since the Big Bang could have produced them—that is, in the absence of inflation.

While the COBE detectors had limited angular range, the scale dependence of the primordial fluctuations in the CMB was completely consistent with the rough scale invariance predicted in inflationary models. Of course, the data at hand could not be used to make definitive statements about the viability of specific inflationary models, but for the first time a direct probe of physics associated with the very early universe seemed at hand.

The COBE anisotropy signal energized the CMB research community. Within months of the COBE discovery, land-based and balloon-based detectors started to report positive detections, on angular comparable to COBE, and on much smaller scales. While none of these experiments could probe the whole sky as COBE had, they have provided important confirmation both of the magnitude of the CMB primordial anisotropy and of the scale-dependence of the primordial density fluctuations that the anisotropy presumably reflects.

The magnitude of the primordial fluctuations confirmed that it is virtually impossible for galaxies to form by gravitational collapse if baryons are all there is in the universe. Many exotic models have now been developed in order to try and reconcile a baryon-dominated universe with the data, and none has succeeded. Structure formation by gravitational collapse can apparently only be reconciled with the data if something else dominates the gravitational dynamics of the expanding universe.

More important for our immediate purpose, however, are tentative results that have emerged over the past year on the basis of new high resolution CMB studies. While the primordial fluctuation spectrum predicted by inflation is essentially scale-invariant, it does not remain that way as the universe evolves. After all, galaxies seem to have characteristic sizes, and as scales get larger and larger, clustering of matter in the universe gets less and less significant.

Once the scale of fluctuations is smaller than the horizon, causal physics can affect their evolution. And as I discussed earlier, there is a certain special angular scale on which one would expect to see a characteristic signature of the physics associated with the growth of primordial density fluctuations.

The size of the horizon at the time of recombination—when matter and radiation first decoupled—is the smallest scale on which primordial fluctuations in matter will not have been damped due to the presence of the background radiation pressure. On this scale and larger, light rays cannot have propagated across the full size of the fluctuation. Thus, the horizon scale is the first scale on which one would expect to see gravitational attraction begin to produce clumping after radiation decoupled with matter. One expects, when examining the magnitude of temperature fluctuations in a statistical sense across the CMB sky, to see a peak corresponding to the angular scale associated with the size of the horizon at the time the CMB decoupled from matter.

This is an important test of our ideas about structure formation, but as I described earlier, it provides a test of much greater significance for our understanding of the global geometry of our universe. The horizon size at the time of recombination is a fixed physical distance, determined by the time when recombination occurred, and thus the distance light could have traveled in this time.

The angular scale spanned by this distance reflects the geometry of the universe (see figure 4.10). While a definitive measurement of the expected bump in the CMBR spectrum of fluctuations at this scale will probably await one of the next space-based CMBR experiments, to be launched in the first decade of the twenty-first century, ground-based and balloon-born experiments have already begun to frame out the general features of such a peak. Moreover, the peak seems to occur at an angular scale precisely where one would expect it to be if the universe was flat! If this result is confirmed, it will provide further support for the basic predictions of inflationary cosmology, and moreover, it will definitively establish that the gravitational dynamics of the universe must be dominated by something other than normal matter!

After having sung the praises of inflationary models, and for a flat universe, let us return to earth. In the two decades since they were established, proton decay detectors have found no evidence for a single proton decay. The simplest Grand Unified Theory proposed by Georgi and Glashow can now be ruled out. No other direct evidence for observable processes mediated by interactions predicted by GUTs has been amassed. On the cosmological side, when examined in detail the simplest GUT resulted in inflationary epochs that were in conflict with our observations. In particular, although the spectrum of primordial fluctuations resulting from inflation has the right shape, in most models, the magnitude of the fluctuations is far too large. As I alluded to earlier, models have been developed to alleviate this problem, but none are very appealing.

Should we therefore give up on GUTs, or inflation? I think not. Extrapolating by 15 orders of magnitude from observed physical scales to the scale at which the strong, electromagnetic, and weak forces might be unified was a very bold move. It was also likely that the first guesses would be wrong in detail. But these ideas solve too many problems to be totally wrong. In a theorist's jargon, they "smell" right. Inflation probably takes place by a very different mechanism than was initially envisaged, or it may take a very different form. But it is hard to imagine why the universe looks anything like it does if something like an inflationary phase did not occur. My guess is that the kind of solutions offered by the notion of inflation will persist, even if none of the specific models remain. In fact, I would be very surprised if any of these initial models turned out to be true. After all, no mention of quantum gravity entered into their derivation. It is hard to imagine that physics at the Planck scale, where quantum gravitational effects become important, does not play some role in determining the subsequent evolution of the universe.

The inflation theory leaves several legacies. By solving the flatness problem, it made an obscure curiosity a central issue in cosmology. By providing the first microphysical theory of initial conditions, it raised the challenge that must be met by any such theory. Recall again, however, that while inflation solves the flatness problem, this problem is not *tied* to inflation. In other words, we cannot say that if the inflation theory is incorrect, the flatness problem and the suggestion that Ω should equal 1 today disappear. Nothing could be further from the truth. Inflation provides only one explanation of why Ω should equal unity. Without the inflation theory, the puzzle of why Ω should be so close to 1 today looms even larger. Inflation also provided the first theoretical basis for any discussion of primordial fluctuations, and confirmed a "flat" spectrum as one of the few acceptable primordial fluctuation spectra; indeed it suggests that the flat spectrum is the most likely one. Basically, all astrophysical simulations of galaxy formation from random primordial fluctuations now assume this initial condition. In the past decade, the CMB observations have established both the empirical consistency of a flat primordial spectrum and most recently of the flat universe predicted by inflation. And finally, by bestowing legitimacy on particle physics models of the early universe, the inflation idea created a new era of "particle astrophysics." The dark matter question is a chief beneficiary of these new activities. The inflation theory demonstrates how new ideas in particle theory can have a dramatic impact on the large-scale structure of the universe. Phase transitions and spontaneous symmetry breaking are not far-out notions invented merely to give cosmologists something to talk about at cocktail parties. These microphysical concepts play a paramount role in modern particle theory, and inflation has made them a key part of modern cosmology.

The theoretical discussions contained herein are among the deepest and most abstract in cosmology, and I apologize if it has been rough going. The salient point is that independent arguments about large-scale structure, galaxy formation, and the evolution of the visible universe most recently combined with CMBR observations all suggest the need for "something else" that dominates the density of the universe today. Galaxy formation in a universe dominated by baryons (normal particles) now appears to be problematic. Something that was not coupled to radiation for so long a time seems to be called for. Such a material, which does not have electromagnetic couplings, would naturally be dark today. If the universe is flat today and Ω equals unity, then Big Bang nucleosynthesis implies that at most only 10 percent or so of the mass in the universe can be baryonic.

Significantly, however, this last-mentioned argument implies consequences of prodigious magnitude. If we accept the argument and the emerging evidence that the universe is flat, then we must confront the fact that there must exist, not ten to twenty times, as is inferred from observation, but almost *one hundred times as much dark matter or energy in the universe as visible matter.* The universe visible through our telescopes would then be *only about 1 percent of all there is!* And not only must most of this other stuff be made of "something else," it must be distributed in such a way that all the virial mass measurements discussed in chapter 5 have *missed most of it*. This is a very large pill for many to swallow. Yet I have dwelled on the flatness problem, and its inflationary solution, because many of my colleagues and I believe that the alternative is at least equally, if not more, difficult to accept.

Whether or not the universe is flat, whether one accepts the theoretical need for Ω to be equal to 1 today, or the tentative CMBR evidence that it is, the other arguments about large-scale structure provide ample theoretical reason to suppose that baryonic matter is not all there is. Thus, whether dark matter or dark energy provides merely 90 percent of the mass in the universe today, as virial estimates indicate, or 99 percent, as the flatness argument suggests, there is now overwhelming reason to believe that all, or most dark matter is made from something else. The rest of this book is devoted to discussing what else this may be, and how we may experimentally probe to find out what it is.

PART FOUR

The Neutrino Saga and the Birth of Cold Dark Matter

CHAPTER SEVEN

THE OBVIOUS CHOICE?

Many are called, but few are chosen.

—Matthew 21.14

If ever a particle was born to be dark matter, it was the neutrino. Among all the exotic particles I shall discuss in this book, the neutrino has one distinct advantage: we know that it exists. Moreover, we know that neutrinos are naturally abundant in the universe, as I shall soon describe. Finally, the neutrino is so weakly interacting that it is the "darkest" particle we have yet observed. What we do not know is whether it has any mass, and thus whether a cosmic neutrino background would make any measurable contribution to the overall mass density of the universe.

The neutrino is the most elusive particle yet discovered in nature. While the average garden variety nuclear reactor produces about a billion billion neutrinos every second, you would not know it if you were standing right next to one. Unlike other radiation from reactors, we do not have to shield ourselves against neutrinos. In fact we cannot. Neutrinos coming from a reactor can travel on average about a million billion miles through rock without interacting once! Thus, trying to stop all or even a significant fraction of the neutrinos produced by a reactor would require a formidable amount of material indeed—much greater than the mass of the earth. This fact leads to another question: If neutrinos are so invisible, how do we even know that they exist?

Like the W and the Z particles, the neutrino was postulated to exist as an elementary particle some twenty years or so before it was actually observed. The spontaneous decay of a neutron—called beta decay—which I have discussed several times in chapters 5 and 6, was first observed by Antoine Henri Becquerel when he discovered nuclear radioactivity in 1896. Beta decay was later analyzed in detail by Sir James Chadwick in 1930, when he discovered the existence of the neutron as an isolated particle. As I have described, it is the beta decay process that revealed the existence of the weak interaction, which mediates this process.

When beta decay was first discovered, however, a serious problem threatened to undermine the very fabric of physical theory. Two particles could be observed emanating from the site where a neutron had decayed: a proton and an electron. The neutron is neutral; the proton is positively charged and the electron is negatively charged. Hence the total charge was conserved during decay. But if the momenta and energy of the outgoing proton and electron were measured, something very unusual was observed. If one started from a neutron at rest with zero momentum, and measured the final momenta of the outgoing proton and electron, their total momentum did not add up to zero as it should have if momentum were conserved and the range of observed energies of the outgoing particles are not consistent with strict energy conservation. It is much easier to visualize this problem than to write about it. Because electrons and protons are charged parti-

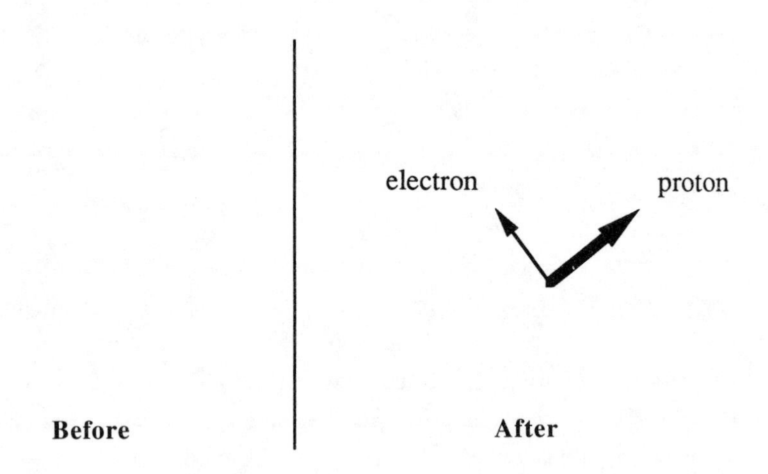

FIGURE 7.1

The process of beta decay, that is, the decay of a neutron, as it might be observed in a particle detector in which the "tracks" left by charged particles are observable.

cles, we can follow their tracks in matter after a neutron decays. Figure 7.1 shows what the process might look like (I have not shown the neutron, because if it decays at rest, it would not produce a track).

It is obvious from figure 7.1 that something else is needed to balance momentum. When a bomb explodes, not all the particles travel out in the same direction. If the proton and electron shown in the figure both travel upward, then one suspects that something else must be emitted traveling downward to balance things.

Faced with this observation, physicists had to make a choice. Either momentum and energy conservation for elementary particles had to be abandoned, or something was being emitted that could not be observed, but which carried off just the right amount of momentum and energy to make everything work out right. One of the "czars" of theoretical physics in the 1930s, Wolfgang Pauli, declared that the second alternative was the only acceptable one. Later, Fermi coined the name *neutrino*—Italian for "little neutron"—for the unobserved particle that must have been emitted in the reaction. Because it was not observed, this neutrino must first have been electrically neutral like the neutron. Unlike the neutron, however, which can still make its presence known by colliding with nuclei through the strong interaction, the neutrino must not "feel" the strong interaction. For this reason, the neutrino is much more like an electron (which also is not subject to the strong interaction), except without any electric charge. The only force the neutrino is subject to is the weak interaction, and because of this, its interactions with normal matter are incredibly feeble, as we shall see. In addition, because the sum of the proton and electron masses is so close to the neutron's mass, this new particle was required to be very light. Indeed, the momentum and energy analysis of beta decay suggested that, in terms of the experimental accuracy available at the time, this particle was required to be massless, like the photon.

The other alternative, the possibility that momentum was not conserved at the elementary particle level, was so distasteful that Pauli's proposal was easily accepted. In addition, the 1930s had earlier witnessed the discovery of two other new particles, the neutron and the positron (the antiparticle of the electron); the positron also began as a mere theoretical construct. In any case, once Fermi had produced his weak interaction theory, one could estimate how strong the interactions of neutrinos would be with matter. The result was discouraging. A neutrino emitted in a beta decay traveling through matter interacts roughly *20 orders of magnitude more weakly than an electron of the same energy*. Physicists apparently had to be content with accepting this particle sight unseen.

Shortly before Fermi's death in 1954, he was asked whether he thought it would be possible experimentally to detect neutrinos with existing technology. He said no. The young experimentalist who posed the question, Fred Reines, took Fermi's response as a challenge. In a few years, with his collaborator Clyde Cowan, he had assembled a ton-sized detector near a nuclear reactor at Savannah River, in South Carolina. Reines based his hopes on the same kind of statistical argument that later motivated the building of proton decay detectors discussed in the preceding chapter. Normally, a neutrino from the Savannah River reactor would travel through the whole earth unscathed, but if one let enough neutrinos pass through a detector, eventually one should interact inside the detector. The calculations of the day suggested that perhaps five to ten neutrino-induced events per day might occur in a ton-sized detector. Such an event rate was trifling compared to the background of cosmic rays; this is why Fermi had not believed that anyone could ever extract the signal from the background "noise." Reines, however, used a trick. He knew that neutrinos interact with matter by one of several different specific reactions. If one looks for the neutrino-induced reaction that creates both a positron and a neutron, when a neutrino scatters off a proton, one could search for the positron signal, and then a short time later witness the signal that would result when the neutron was captured (that is, became bound) in the nucleus of another atom, via the characteristic emission of an energetic photon. This double signature promised to stand out against the background and make the experiment possible.

In 1956, Reines and his collaborator Clyde Cowan reported a detection. They had witnessed several hundred events in their detector, with exactly the rate predicted by the available theory. The neutrino had been discovered. Later on, it turned out that the good agreement between observation and theory had been somewhat of a fluke. The theoretical predictions they relied upon were actually incorrect by a small factor. Nevertheless, other observations by a number of groups confirmed the correct rate. The neutrino entered the realm of the observable. Reines, who later became one of my predecessors as chair of the Physics department of the then Case Institute of Technology, was awarded the Nobel Prize in 1995 for his discovery of the neutrino.

Since its first detection in 1956, three different types of neutrinos have been discovered—one for each "electron-like" particle that we know of. They are named after their charged partners that are produced in weak decays like the decay of the neutron. The electron neutrino is the neutrino produced with an electron in beta decay, and the *muon* and the *tau* neutrinos are associated with similar weak decays that produce the heavy copies of the electron called the muon and tau particles.* We do not know why there are three types of these "lepton" doublets, as they are called, just as we do not understand why there are two extra heavy copies of the electron. The American physicist I. I. Rabi's remark upon hearing of the discovery of the muon still holds: "Who ordered that?" We do not know whether any other lepton doublets are waiting to be discovered, although as I shall discuss, cosmology—in the form of Big Bang nucleosynthesis estimates—has already put bounds on the number of light neutrinos. These limits are supported by direct terrestrial accelerator-based experiments.

The connection between the number of light neutrinos and nucleosynthesis points to the very significant role that neutrinos have played in astrophysical theory. Indeed, neutrinos have been central in almost every aspect of the developing exchange of ideas between particle physics and astrophysics. Experiments designed to detect neutrinos have given us information on such diverse subjects as the conditions at the interior of the sun, the composition and origin of cosmic rays, and the dynamics that produce supernovae—the brightest fireworks in the universe. Perhaps one of the most exciting events in astrophysics in the 1980s has been the birth of "neutrino astronomy."

The development of modern cosmological theory—and our understanding of dark matter—depends in large part upon considerations of the role neutrinos might have played in the early universe. This "neutrino saga" in cosmology has been something of a roller-coaster ride, however. Once on top—prospective candidates for the dominant stuff in the universe today—neutrinos are now out of fashion as dark matter candidates. From their apparent demise has emerged a new breed of dark matter, and a new understanding of the obstacles that lie ahead for any theory of the evolution of large-scale structure. The following section is devoted to this saga.

Neutrinos first entered cosmological theory via Big Bang nucleosynthesis calculations. In my earlier discussion of nucleosynthesis, among the important reactions leading to the formation of the observed light elements, I listed several that involve the production of neutrinos. This is vital, because it means that the same calculations that predict so well what light elements we should see today can be used with confidence to predict the present abundance of neutrinos in the universe. No exotic physics processes are needed.

---

*The Nobel Prize was awarded in 1988 to the experimentalists L. Lederman, M. Schwartz, and J. Steinberger, who first demonstrated that the electron and muon neutrinos were different particles. We have not yet directly detected the tau neutrino, but all indirect evidence suggests that it should exist along with the tau particle.

At high temperatures, neutrinos are kept in thermal equilibrium with matter via these nucleosynthesis-related and other weak interaction–mediated reactions. Because of this, the laws of thermodynamics imply that at early times neutrinos must have been as abundant as photons. There is no room for quibbling here. If we accept that the reactions that led eventually to light element formation were important at the time of primordial nucleosynthesis, then these same reactions ensured thermal equilibrium before this time.

This has an immediate implication for cosmology. At early times the energy density of radiation dominated the energy density of the universe. It was only much later, after radiation and matter cooled, that energy stored in massive material began to dominate over the energy of radiation. At these early times in the nucleosynthesis epoch, the energy of radiation at these temperatures would have been far in excess of the energy associated with any possible light neutrino mass. Neutrinos in equilibrium with radiation would then have had sufficient total energy to move with relativistic velocities, like photons. Each neutrino type, if it were as abundant as photons, would have contributed roughly an equal amount to the total energy density as the photon bath and could have affected significantly the total energy density at these early times.

Since the expansion rate depends upon the total energy density, this means that the existence of more light neutrino types implies a faster expansion rate at the time of nucleosynthesis. A faster expansion rate, in turn, implies that the weak interactions that keep neutrons and protons in equilibrium drop out of equilibrium at an earlier time, and also that things will cool more quickly. Thus nucleosynthesis would occur earlier. But recall from chapter 5 that the key factor that determines how much helium is produced in the Big Bang is the ratio of neutrons to protons at the time that nucleosynthesis begins. If it begins earlier, then fewer neutrons will have had time to decay into protons and the neutron number density will be larger. In this case, the remnant abundance of helium produced will be higher. Upper limits on the primordial helium abundance then provide upper limits on the expansion rate at the time of nucleosynthesis. These bounds, in turn, place maximum limits on the number of light neutrino types. According to the most recent calculations, this limit is only consistent with fewer than four light neutrino types, or the number we have discovered up to this time. The same experiments at the European Center for Nuclear Research which led to the discovery of the W and Z particles have also provided direct evidence that the number of light neutrino types is precisely three. The concordance between experiment and cosmological arguments once again reinforces our belief that we do understand the physics of the expanding universe at least as far back as the nucleosynthesis era.

Like photons, which later decoupled from matter at the time of recombination, neutrinos also eventually decoupled from matter, but this occurred about the time of nucleosynthesis. Even though this is much earlier than the recombination epoch, when photons decoupled, as long as nothing very peculiar happened in the intervening time, the neutrino background would have continued to redshift during most of this period just as the photon background was redshifting. In this case, the relative number density of neutrinos and photons would have remained fairly close. After the photons decoupled at recombination, the two backgrounds would then have diluted identically and uniformly as the universe expanded. This means that there must exist today a cosmic neutrino background with a density of particles almost as large as the density of photons in the microwave background.

Nothing could be more marvelous for cosmologists and physicists than observing this neutrino background—whether or not it has anything to do with dark matter. Because it would have decoupled from matter at a much earlier time than the photon background, it would give us a "picture" of what the universe looked like at a time almost 12 orders of magnitude earlier than the view provided by the microwave background. Unfortunately, no experiment has been proposed that might be sensitive to this background. I have personally spent a great deal of time trying to devise such an experiment, but nothing seems even to come close.

Nevertheless, I once again stress that as long as we believe in our primordial nucleosynthesis calculations, or, more generally, insofar as we think that we understand the physics of the expansion back to this period, such a neutrino background must exist—whether or not we can observe it directly.

In about 1975, physicists R. Cowsik and J. McLelland made a simple but striking observation. By measuring the temperature of the microwave photon background today, we can, as long as the distribution is actually thermal, determine both the number density of photons in this background and the total energy density contained therein. The number of particles in a thermal bath at temperature T increases as the cube of the temperature. If, on average, each particle has an energy proportional to the temperature, then the total energy density of the bath can be expressed as T times the number of particles, or as T to the fourth power, T^4. When we calculate this energy for the 2.7 degree K microwave background we find that it is about 1/10,000 of the energy density that would be necessary to close the universe today, or at least 1/1,000 of the energy density in baryons today.

If neutrinos are massless like photons, then the neutrino background contributes almost as much to the present energy density as does the photon background. In fact, since the neutrinos decoupled earlier than photons, one can show that their number density is actually about one-tenth of that of photons today, so the neutrino background should contribute about 1/100,000 of the energy density necessary to close the universe. But, as Cowsik and McLelland suggested, if the neutrino is not massless, but instead has a *small* mass, this situation could change drastically. As long as this mass is very small on the scale of energies available during primordial nucleosynthesis, our earlier arguments remain unaltered and the remnant number density of neutrinos existing now would be unchanged. In this case, if one of the neutrino types has a mass, and hence an associated rest energy (according to Einstein) that is larger than the mean energy associated with the present temperature of the photon background, then the foregoing estimate of the neutrino contribution to the energy density of the universe would have to be altered. In this case, the neutrino contribution to the energy density is not proportional to temperature T. Instead it is proportional to its rest energy, which is proportional to the neutrino mass according to Einstein's famous relation, $E = mc^2$. The total energy density in neutrinos today would thus not be T times their number density, but rather mc^2 times their number density. Hence, the neutrino energy density could be larger by the ratio of mc^2 to the temperature T, compared to the earlier estimate.

Since the energy density of a massless neutrino bath would now be about 10^{-5} times the closure density, if the neutrinos have a mass such that the ratio of mc^2 to T is about 10^5, then the present energy density in the background neutrino bath could result in a flat universe. Accordingly, there could be an $\Omega = 1$ worth of mass density in neutrinos.

What kind of mass does this calculation imply? If one makes the appropriate energy conversions, a rest energy equivalent to about 10^5 times 3 degrees Kelvin would result from a mass of about 1/10,000 of the electron mass. Expressed in the units used by particle physicists, this mass would be about 10–50 "electron-volts" (eV). (The mass of the electron expressed in these units is about 500,000 eV.) This is a very small mass indeed—much smaller than any other particle we know of. Nevertheless, this mass would be enough to make neutrinos the dominant matter in the universe today. This provocative speculation spurred the ongoing efforts to measure the mass of the electron neutrino, the particle emitted in beta decay.*

*As I earlier described, it is actually the antiparticle of the electron neutrino that is emitted in beta decay, but since antiparticles must have the same mass as particles, this distinction does not matter, so I will refer to antineutrinos as neutrinos unless it is important for our discussion to make a distinction.

Imagine the shock and excitement that spread through the physics and astronomy communities when in 1980 a research team headed by V. A. Lubimov in Moscow announced that they had just completed such a measurement. Their finding was consistent with a nonzero mass for the neutrino of about 45 eV. By our earlier arguments, this mass could result in a critical, or "closure," density in the universe from the cosmic neutrino background today.

Measuring the mass of the electron neutrino in the laboratory is no easy task, however. The method used by the Soviet group, and several other groups since then, is a variant of the analysis that originally led to the prediction of the neutrino's existence. In the beta decay of a free neutron, or a nucleus containing a neutron which decays, the momenta of the outgoing proton and electron do not in general balance; this requires the neutrino to take up the missing momentum, as I described earlier. Because radioactive beta decay is a quantum mechanical process, however, it must be described probabilistically. Thus, each time a neutron decays, the neutrino, electron, and proton split up the total energy and momentum available in the decay in different amounts. On average, they split these quantities up uniformly. However, in very rare events, one or two of the particles can take up almost all of the energy and momentum, leaving the third with much less. Now, if the neutrino is massless, as its momentum goes to zero, so does its total energy. Contrast this with a case in which a neutrino has a mass. In this case, if its momentum goes to zero, its energy will not go to zero, but will rather approach the energy associated with its rest mass given by Einstein's relation $E = mc^2$. Therefore, if we examine the energy "spectrum" of electrons emitted in beta decay we should see two different distributions, depending upon whether the neutrino is massive or massless. In the massless case, it is possible in rare instances for the electron to take up all the energy associated with the mass difference between the neutron and the proton in the decay. In the massive case, the electron can only take up at most this energy minus the energy associated with the rest mass of the neutrino. This difference is schematically pictured in figure 7.2.

Unfortunately, as is shown in the figure, these spectra only differ near the upper endpoint, where the event rate is very small—down by perhaps more than 5 orders of magnitude compared to the rate when all three particles share more evenly the total available energy. One must have a very strong radioactive source of beta decay electrons, or wait a very long time, before it is possible, even in principle, to distinguish between the two spectra shown in the diagram. Moreover, we must be able to measure very accurately the energy of the outgoing electron in each beta decay event. Remember, the deviation that we are looking for

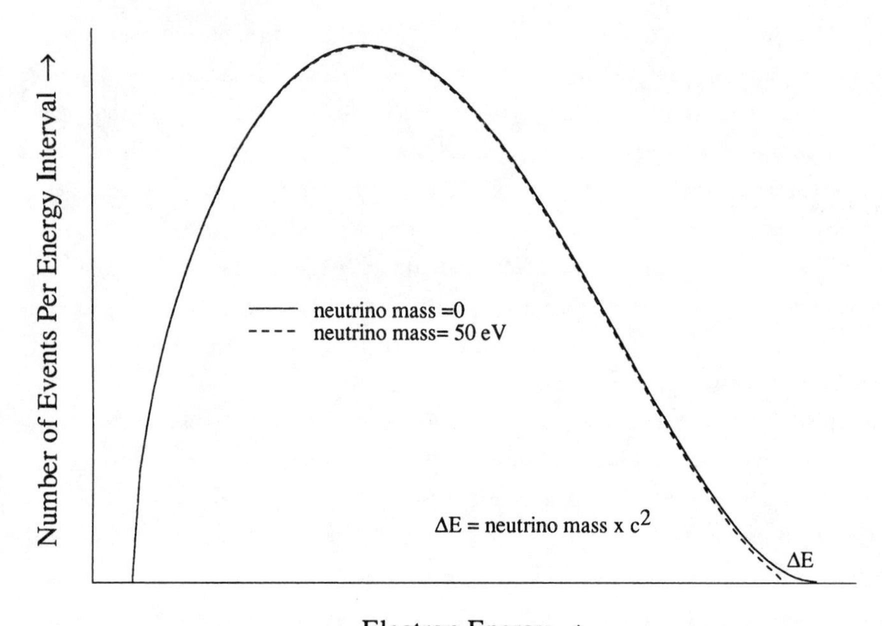

FIGURE 7.2 ELECTRON SPECTRUM FROM NEUTRON BETA DECAY

The predicted number of electrons of energy E emitted in the beta decay of a neutron, as a function of the electron's energy. The solid curve shows the prediction if the neutrino mass is zero, and the dashed curve the prediction if the neutrino mass is 50 eV. In the latter case, the spectrum falls to zero at an energy ΔE smaller than in the case when the neutrino mass is zero.

occurs for energies only tens of eV away from the upper end of the spectrum, whereas the electron may carry away a total energy of millions of eV. Measurements with an accuracy of one part per million or so are necessary.

However, a much more subtle problem is associated with measuring the mass of the neutrino. One does not find free neutrons normally existing in nature, because neutrons decay when free in about eleven minutes. Thus, beta decay must be studied for neutrons inside the nuclei of atoms, which can have radioactive lifetimes that are much longer. In this way large amounts of the radioactive source can be contained throughout the lifetime of an experiment. However, when a neutron decays inside a nucleus, the energetics of the decay can be quite different. One must determine the energy available in the decay by measuring the mass difference, not between the neutron and proton, but rather between the two different nuclei, before and after the neutron decay. If the nuclei are in atoms that are bound together into molecules, then one must also take into account the atomic binding energy which is used up when the nucleus decays and the molecular configuration changes. These binding energies are in the range of tens of eV—just the magnitude that one is trying to discover in a neutrino mass. Moreover, the molecular binding energy changes are notoriously difficult to calculate in advance, if the molecule is sufficiently complex.

These factors make a truly accurate neutrino mass measurement very difficult—and very costly. A number of groups around the world have attempted over the past ten years to verify the Soviet result without having succeeded. Partly for this reason, a great deal of skepticism surrounds the initial Soviet report. The Moscow group looked at the beta decay of tritium inside a very complicated molecule called valene. The atomic physics calculations necessary to address adequately the energetic questions of valene already push the limits of our current computational ability.

To this date at least three other groups have performed endpoint measurements but have been unable to reproduce the Soviet finding. Even the Moscow group reanalyzed their original estimate of 48 eV and lowered their mass estimate to about 27 eV. In about 1986, Swiss-based experimentalists claimed to put an upper limit on the electron neutrino mass of 17 eV, in apparent disagreement with the Moscow limit. The most recent upper limit, from such endpoint experiments, is now about 6 eV, in firm disagreement with the Moscow report.

In the midst of this dispute, Supernova 1987A appeared in the Large Magellanic Cloud. For the first time in history, neutrinos were observed from a stellar collapse, in two different experiments, no less. During the sudden collapse of a star under its own gravitational pressure, energy must be released as the mass of the star collapses inward. One can calculate that during this process, which lasts only a few seconds, the total energy released is about 10^{20} times as much energy as is released by the sun during those same few seconds. Almost all of this energy is released in the form of neutrinos, because they are the only objects that can permeate the hot, dense environs of the collapsing star.

Never before had this theoretical expectation been verified, but in the ten seconds or so during which the two detectors on earth signaled nineteen neutrino events, supernova theory advanced from pure speculation to empirically tested wisdom. Remarkably, in the same ten seconds or so in which the signal was observed, a limit on the electron neutrino mass could be obtained which, if not arguably better, was at least comparable to the limits obtained after a decade of hard work in the laboratory. Had the neutrinos from the supernova been massive, then during their 150,000-year-long voyage those with slightly greater energy would have traveled slightly faster. From the duration of the signal, between the first and the last event of differing energies, one could derive an upper limit on the electron neutrino mass. I think that about half of the high-energy physicists in the world must have made a back-of-an-envelope estimate of this momentous effect within hours after hearing of the data, and some fraction of them then submitted their results for publication. When the dust had settled, and several groups had done as detailed an analysis as possible with just nineteen relevant data points, the conclusions were mixed. Under the assumption that all of the neutrino events were supernova-induced, and making certain assumptions about the initial signal, an upper limit of 12–16 eV could be placed on the mass of the electron neutrino. If cosmic ray background noise is taken into account and if less restrictive assumptions are placed on the signal, the upper limit rises to about 23 eV. In this sense, the supernova supports, but does not further restrict, the preexisting limits obtained in the laboratory.

Nevertheless, this extraordinary occurrence in ten seconds matched anything we could achieve with existing technology on earth. Moreover, the neutrino mass limit obtained from the supernova appears to affirm the apparent disagreement between the recent beta decay upper limits and the initial Soviet mass determination. If we are lucky enough to observe the neutrino signal from a supernova in our own galaxy, we might improve the mass limit by a factor of 3–5. Supernovae are expected only about every ten to fifty years in our galaxy, however; it is not clear whether terrestrial experiments will have by then also achieved this sensitivity.

There are two other neutrino species, either of which could have a mass in the required range to be cosmologically significant. However, the downward slide of the electron neutrino mass limit in the laboratory after 1980 has been followed by equally pessimistic results coming from astrophysics.

After the Cowsik and McLelland results and the Lubimov mass measurement, and the growing evidence for dark matter in galaxies, several groups of astrophysicists decided to examine in detail what a neutrino-dominated universe might be expected to look like. Would it resemble what we actually see?

It is important to realize that it is only in the last twenty years or so that observational technology has advanced to a point where we are beginning to have some inklings of what the universe actually looks like on large scales. The first large sky maps were confined to cataloguing the positions of galaxies as seen on the plane of the sky. The most comprehensive of these, the Shane-Wirtanen catalogue, was completed in the early 1950s and covers the entire sky observable from the Lick Observatory in California. Separated into very small cells covering a small fraction of a degree each, this map lists more than a million galaxies, and it is still useful in delineating structures on very large scales. However, any such map of galaxies as seen on the sky provides only a two-dimensional projection of galaxies. All three-dimensional information—crucial for understanding the details of clustering—is lost. Information on depth, as well as breadth, requires the use of redshift data for each galaxy observed, in order to ascertain its Hubble velocity and hence its distance. Sky surveys with redshift information—so that the three-dimensional distribution of galaxies can be inferred—only began in earnest in the 1980s. The largest survey, done by the Harvard-Smithsonian Center for Astrophysics (CFA), covers all galaxies in the northern sky away from the plane of our galaxy down to some minimum brightness that corresponds to galaxies within a distance of 100 mega-parsecs or so. This survey was completed in 1982; one projection of this view of the nearby universe is shown in figure 7.3.

Other surveys have since probed out to much farther depths, but over smaller solid angles of the sky. Not only has observational technology taken some time to

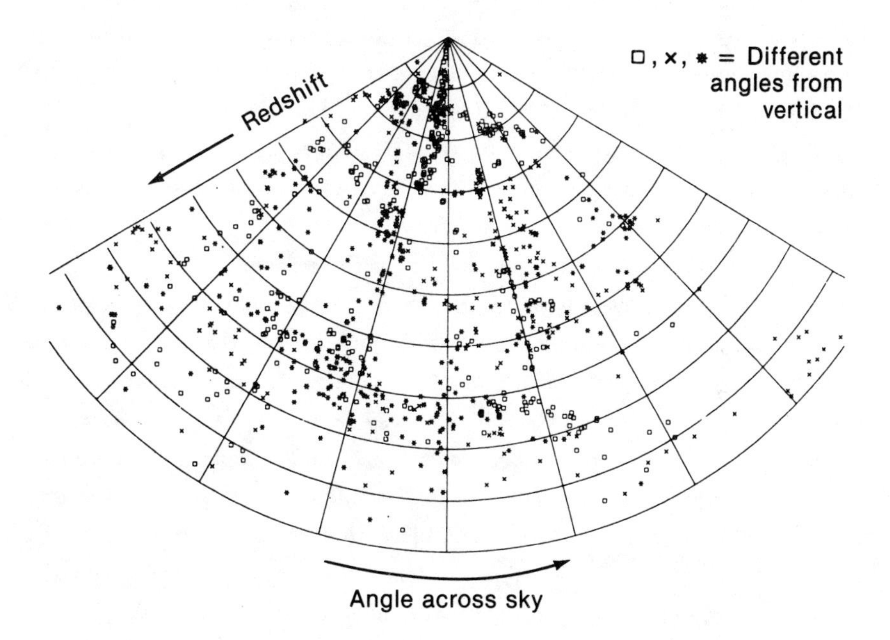

FIGURE 7.3

One projection of the three-dimensional Center for Astrophysics survey of nearby galaxies. In this projection, we would be located at the center point, where all the radial lines converge. The radial direction represents distance away from us, as determined by galactic redshift. The direction along the circular arcs gives the angle of the galaxy as seen across the sky. The third dimension, that is, the angle from the vertical, is not displayed explicitly, but the three different symbols shown represent galaxies located on three different planes across the sky. (From M. Davis et al., *Astrophysical Journal 292* [May 1985].)

advance to the point where redshifts can be systematically taken for many systems, but the computing technology to analyze the data has had to be developed. These facts demonstrate how embryonic observational cosmology really is. My earlier colleague at the CFA, Margaret Geller—an expert on large-scale structure (and the first person to show me a photographic plate of galaxies obtained from a telescope)—has emphasized that we are only now beginning to obtain the data that will give us an accurate picture of the universe on large scales. As if to underscore this, almost every year startling results are reported about the observations of structure on larger and larger scales. As a result, scientists and laypersons alike should be skeptical, or at least should carefully scrutinize, any categorical conclusions drawn on the basis of existing data, because some of our assumptions about the universe on large scales are likely to change. The ambitious Sloan Digital Sky survey has now begun and will measure over 1 million galactic redshifts in the coming decade.

Nevertheless, we already have at our disposal a substantial body of information about large-scale structure. This poses a challenge to both astronomers and physicists. Can any simple dark matter model explain the evolution of this structure? We have already seen that a universe dominated by baryons alone apparently fails to predict qualitatively the structures we see. What about a neutrino-dominated universe?

Initial analytical estimates appeared to be very favorable. Y. B. Zel'dovich, the eminent Soviet astrophysicist, had suggested as early as 1970—perhaps based on the fact that most galaxies are observed within clusters—that structure might "form from the top down." In this way he suggested that the first structures to collapse out of the uniform expanding material were cluster-sized. If these large structures collapsed as coherent entities, then, Zel'dovich argued, "pancake-like" sheets of very high density called "caustics" would form. These could then fragment along filaments into many smaller objects that would later be recognized as the galaxies of today. This "pancake" scenario had many appealing features; notably it would produce a filamentary structure of galaxy clusters that is easily observable on any sky survey. In advance of detailed numerical models, this was considered by many to be a likely scenario for structure formation.

As I have noted, as long as the matter fluctuations that eventually grow due to gravity are small, perturbation theory can be used to calculate their growth. Once they become large enough that nonlinearities can become significant, however, analytical methods break down and detailed studies of evolution require the use of computer simulations. Following the exact behavior of even a few hundred test masses becomes prohibitive without a computer. Since each mass is subject to the gravitational attraction of all the other particles, following the behavior of a few hundred particles requires millions of iterative computations. Since a realistic simulation requires at least thousands, not hundreds, of masses, this kind of analysis was impossible before the advent of superfast computers with large memories.

Before such a task was undertaken, however, simple analytical estimates suggested that neutrinos could produce just the right conditions for a pancake-type clumping of matter in clusters. The arguments used to derive these estimates are very similar to those outlined in chapter 6 for describing structure formation in a baryon-dominated universe. The key is to determine what scale will be the first on which matter fluctuations can grow. Recall that in the case of baryons, fluctuations in regions smaller than the horizon were wiped out or dissipated until the time in which photons decoupled from matter, at recombination. The first scale on which fluctuations could grow was that which just encompassed the horizon at that time, so that fluctuations would not have been smoothed away before that time by any causal process. Unfortunately, such scales would be huge today, encompassing hundreds of mega-parsecs and implying well-defined structures with masses hundreds of times as large as the biggest clusters we have yet seen.

How does a neutrino-dominated scenario differ from this? Neutrinos, because they interact only weakly, do not couple directly to photons. Hence, any initial density fluctuation in neutrinos would not be prevented from collapsing by the radiation pressure of photons. In principle, mass fluctuations in the neutrino distribution could begin to grow *earlier* than baryon fluctuations. How much earlier depends on the size of the neutrino mass.

As far as we can tell, the universe is "matter"-dominated today. In other words, we expect that whatever is dominating the mass density of the universe today is more like matter than like radiation—that is, the dominant stuff is not moving relativistically (near the speed of light). There is no ironclad proof of this; however, a universe dominated by relativistic particles, that is, radiation, would have several important differences from one dominated by nonrelativistic matter. For the same expansion rate, today one can easily show that the universe would have to be 30 percent younger if it is dominated by relativistic particles, a result that now firmly disagrees with observation. More important, the growth of structure is greatly slowed in such a universe. It is already difficult to form galaxies gravitationally by today, and a universe dominated by relativistic particles only compounds the problem. In short, the likelihood that the universe is radiation-dominated today is close to zero.

Yet even if the dominant material in the universe is nonrelativistic today—as almost all scenarios would suggest—it was not always this way. The energy density of relativistic particles decreases in an expanding universe more quickly than does matter. This is easy to understand. Take any gas, made of either relativistic particles such as photons or of nonrelativistic particles such as dust. If the total number of particles does not change as the volume expands, the *density* of particles decreases inversely as the volume increases, because the particles get more dilute. If the energy carried by each particle remained constant, then the energy density of the gas would thus diminish inversely as the volume increased. However, as we have seen, while the universe expands, the wavelength of light increases in direct proportion to the increasing size of the universe. This universal "redshift" of wavelength implies that the frequency of each photon becomes smaller. Since the energy of each photon is proportional to its frequency, this means that energy is reduced as well. Hence, the photon energy density decreases inversely not as the volume of the universe increases, but rather it decreases inversely as the volume *times* an extra power of the scale size of the universe, because of the extra redshift factor. Particles of matter, on the other hand, once they are at rest, maintain a constant energy, proportional to their mass. Thus the energy density of matter does not diminish with this extra redshift factor; instead it simply decreases in the standard way, inversely as the volume of the universe increases.

These relations imply that the ratio of the energy density of radiation (that is, relativistic particles) compared to nonrelativistic matter becomes progressively smaller as time goes on, in inverse proportion to the size of the universe because of the extra redshift factor for radiation. If we take the ratio of the energy density in the microwave background photons today compared to the energy density in visible matter, we find it is about 1 part per 10,000. This means that when the universe was about 10,000 times smaller, this ratio was equal to 1. At earlier times it was greater than 1.

The exact time when the universe was last radiation-dominated depends upon how much other nonrelativistic matter exists now. If there is more massive stuff than we see as visible matter, then the time of matter-radiation energy equality is pushed back slightly. However, if we assume at most a closure, or critical, density in nonrelativistic stuff, we can only push this time back by a factor of 10–100 compared to the value it would have if there were no dark matter at all. Before that time we are assured that the energy density from the photon background alone would have been sufficient to dominate the expansion of the universe. The approximate time when the energy density of matter and radiation were equal, if the universe is flat and matter dominates (that is, $\Omega_{matter} = 1$ today), turns out to be about fifteen times earlier than the recombination time, or when the temperature of radiation was about 20,000 degrees Kelvin.

At this time and earlier, photons need not have been the only relativistic particles to make up the radiation gas that dominated the expansion of the universe. Massive material which is now cool enough so that it is moving slowly enough to be nonrelativistic was once much hotter. For example, if neutrinos have a mass of 30 eV, so that a nonrelativistic neutrino background now dominates the universe, when the temperature of the universe was high enough, so that neutrinos of a mass of 30 eV would be energetic enough to approach the speed of light, they too would have behaved as radiation, at least insofar as the expansion of the universe is concerned. We can determine the time when the neutrinos cooled sufficiently to become nonrelativistic. This would have happened when the temperature of the universe fell to the point where the average energy of motion of neutrinos fell below the energy associated with their mass. If neutrinos have a mass of 30 eV, then this would have happened at a temperature of about 100,000 degrees Kelvin. It is a numerical coincidence that this temperature is close to that for which the universe first becomes matter-dominated in the case of an $\Omega = 1$ universe today.

This time is, in a neutrino-dominated scenario, a turning point for the formation of structure. Before this time, 30 eV neutrinos were relativistic. After this time, they cooled to be nonrelativistic. What would happen to a region containing an excess of neutrinos inside the horizon on either side of this temporal landmark? Before this time the neutrinos would be traveling, on average, near the speed of light. As I have earlier mentioned, particles traveling at or near the speed of light can escape from the gravitational pull of any mass distribution except a black hole. Relativistic neutrinos, therefore, would have been able to travel, or "free stream," away from any initial density excess. In so doing, they would smooth out a density excess. Thus, when the neutrino background was relativistic, any fluctuation in the density of neutrinos inside of any region out of which they would have been able to travel and hence escape in the time available since the Big Bang would have been smoothed away. This distance over which relativistic particles could move is nothing other than the horizon size. Hence, the smallest scale that would contain fluctuations that would *not* have been smoothed out is the scale that corresponds to the horizon size just *after* the neutrino average velocity fell well below the speed of light. In other words, the scale corresponds to the time when the neutrinos became nonrelativistic.

After this point neutrinos would not have been able to escape out of the gravitational "potential wells" created by any existing density fluctuations. Density fluctuations in the neutrinos could then increase due to gravity. Later, after recombination, when the baryons were no longer tied to photons, one would expect that they too could then collapse into the big potential wells created by the background neutrino clumps, forming the structures we observe today. Hence, the smallest scale on which structure might be expected to form in a universe dominated by 30 eV neutrinos is that corresponding to the horizon size at around the time when the neutrinos first became nonrelativistic.

Because this time is about a factor of 30 earlier than the recombination time, the horizon scale was then smaller (light had had less time to travel). So this characteristic scale for the formation of structure in a neutrino-dominated universe is smaller than that in a baryon-dominated universe (see appendix B for a graphical demonstration). This is good news; I have already noted that in a baryon-dominated universe the clumping scale predicted is far too large. Instead the clumping scale for a neutrino-dominated universe turns out to be about 10^{25}–10^{26} centimeters, or about 10 mega-parsecs today. *This is exactly the scale size of the large superclusters of galaxies we now observe.* It is also just what was expected to be the first scale for structure to form in the "pancake" model of galaxy formation.

This estimate, combined with the early positive laboratory findings on the electron neutrino mass, encouraged astrophysicists to try to examine a neutrino-dominated universe scenario in more detail. This meant resorting to computer simulation: a number of groups used computers to see how structure might be expected to evolve in a neutrino-dominated universe well into the range where analytical techniques would break down. The first results showed that the Zel'dovich hypothesis held up under detailed scrutiny. The first objects to condense are pancake-like sheets which link up across the volume under investigation. Filaments form at the intersections of these caustic sheets, and the filaments break into fragments, forming clusters at their intersections.

As in the case of laboratory mass measurements, however, things quickly began to look bad for neutrinos as candidates for dark matter. After laying down an initial spectrum of random fluctuations originally postulated by fiat (independently by Zel'dovich and by Edward Harrison of the United States) and then predicted from first principles in inflationary scenarios, astrophysicists attempted to use computer-simulated projections to examine in more detail what the distribution of large-scale structures in the universe might look like.

Keep in mind that this initial random spectrum was expected to be uniform; that is, primordial density fluctuations are predicted to have roughly the same amplitude on all scales, at the time each successive scale comes inside the horizon—before the causal processes inside a horizon distance can affect their growth one way or another. This is certainly the simplest assumption, and the one that agrees with observation at very large and very small scales. If one arranges a primordial fluctuation spectrum with more structure—a very difficult thing to do without invoking rather exotic physics—one might hope to circumvent some of the problems I will discuss shortly. Using this simplest assumption, however, researchers could compare their numerical results with observations, such as the CFA redshift sky survey.

One of the authors of that survey, Marc Davis, and his collaborators Georges Efstathiou, Carlos Frenk, and Simon White (hereafter DEFW) in about 1983 formed one of several groups to begin numerical simulations aimed at testing the neutrino-as-dark-matter hypothesis for structure formation. Their goal was to follow the growth of fluctuations into the nonlinear regime, where fluctuations are too large to treat with perturbation theory, and then see how well their simulations mimicked the detailed structures observed in the real universe. Of course, even with the aid of a computer, an accurate simulation is easier to talk about than actually perform. If one wants to follow the evolution of a region of, say, 50 or 100 mega-parsecs across—a volume comparable to the Center for Astrophysics survey—such a region would be expected to contain on the order of 10^{80} neutrinos today. Clearly, even a supercomputer cannot follow every single neutrino. Since it is most feasible to follow tens or at best hundreds of thousands of particles, the problem must be made even more mathematically discrete. Davis and collaborators first considered the evolution of about 30,000 "particles" in a cubic volume designed to mimic a region about 65 h^{-2} (the cosmic fudge factor appears everywhere) mega-parsecs on a side. If they were simulating an $\Omega = 1$ universe, each "particle" of their simulation would have then represented about 10^{12} solar masses—a very large galaxy—containing perhaps 10^{78} neutrinos. Because of the fact that the initial distance between the "particles" in their "box" was about 1/32 the size of the box, the range of structure that could be explored was limited—from about 1/32 the size of the box to about half the size of the box, or only about a factor of 16 in scale. One must be very careful that the severe numerical limitations and constrained box size do not introduce computational side effects that overwhelm the actual physics. Many tests were performed until the group was reasonably certain that the numerics were under control.

There is an old saying in computer circles, "garbage in, garbage out." Davis and his colleagues had to make certain that the initial configurations of their simulations represented accurately what one might expect to result from the growth of small fluctuations in the early universe. One of their main improvements over earlier work was to start with initial conditions that arose directly out of an analytical perturbation theory analysis of the growth of small fluctuations in a neutrino-dominated universe performed in 1982 by J. Richard Bond, Georges Efstathiou, and Alex Szaley. Having thus done their "homework," they could begin their simulations confident that the results might be relevant to the real world.

The bottom line is that the agreement between theory and observation was lousy. They could follow an initial configuration of small fluctuations as their model universes expanded by up to a factor of 10. By this point, things went from being relatively smooth to a state where most of the mass was tightly concentrated in clusters. Of course, even in this case, one must be careful to distinguish the "neutrinos," which encompass the particles laid down in the simulations, and actual "galaxy" candidates, where the neutrinos have locally collapsed together to form a bound system in which matter can clump. Tagging systems whose volume had dropped to zero on their scale of resolution as galaxy candidates, they found that, by the time their model universe had expanded by a factor of 2.9, 1 percent of the particles were tagged as galaxies. This presumably signaled the onset of galaxy formation. They could continue to follow the simulations, until, say, 50 percent of the particles had been tagged as galaxies. Three pictures obtained from their simulations at various expansion factors are shown in figure 7.4. On the left-hand side is the distribution of particles, and on the right-hand side is the distribution of tagged "galaxies." The formation of filaments and clusters is clear.

A particularly telling way to compare these results with observations is to generate distributions with the same built-in observational biases that one would obtain from earth-based observations, and compare them with actual galaxy surveys. This is exactly what DEFW did. Figure 7.5 displays the actual Center for Astrophysics survey (*upper left*) compared with three simulated neutrino-dominated universes, normalized to give the correct density of nearby galaxies. One can see that these universes are far too "clumpy" on large scales. The triangles, which represent candidate galaxies from the simulations, may not all produce observable galaxies. And the dots, which do not pass the galaxy candidate cut, cannot represent galaxies. Thus, one can also see that there are much larger empty areas in the neutrino models than in the real data.

In retrospect, these problems of "clumpiness" and empty areas can be expected. In a "from-the-top-down" model of structure formation via neutrino collapse into cluster-sized objects that then fragment, galaxies form after clusters form. By the time galaxies have had time to form, significant agglomeration on larger scales must have first occurred. Such a universe thus tends to be very clumpy on these larger scales—much clumpier than the universe that we live in actually appears. In order for very large-scale structure in these models not to be too pronounced, one must arrange for galaxies to have only just formed. The longer we let them coalesce, the more clumpy will be the cluster-sized regions out of which they formed. As I shall

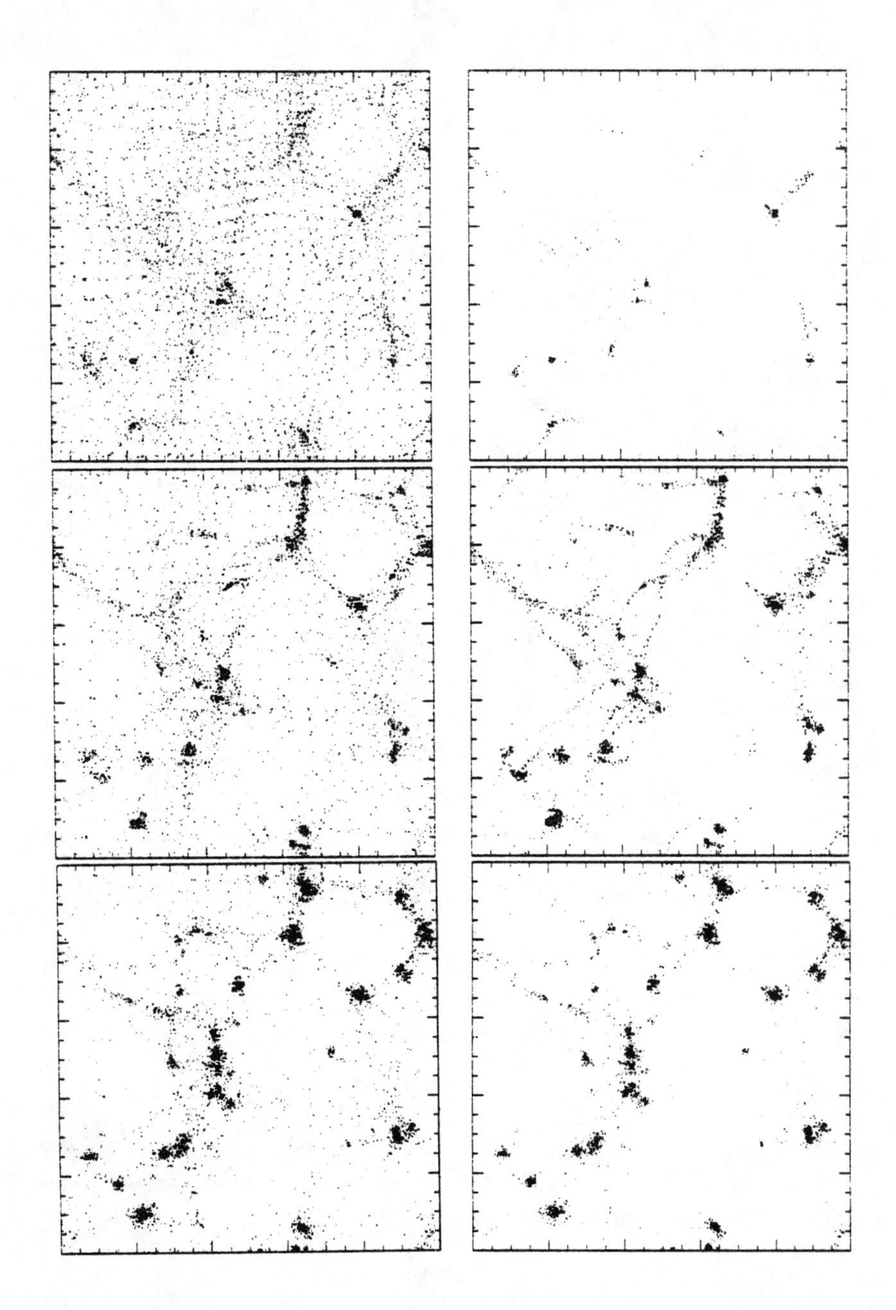

FIGURE 7.4

A numerical simulation of the growth of structure in a neutrino-dominated universe displays on the left all "particles" in the simulation and on the right only the subset tagged as galaxies. The three sets of "snapshots" are taken at different stages in the development of the simulation, as the model universe has expanded by progressively larger amounts and the clustering has advanced even further. (Courtesy of S. White, based on simulations by Davis, Efstathiou, Frenk, and White; from the proceedings of the 7th Annual Grand Unification Workshop, copyright © 1987 World Scientific Press.)

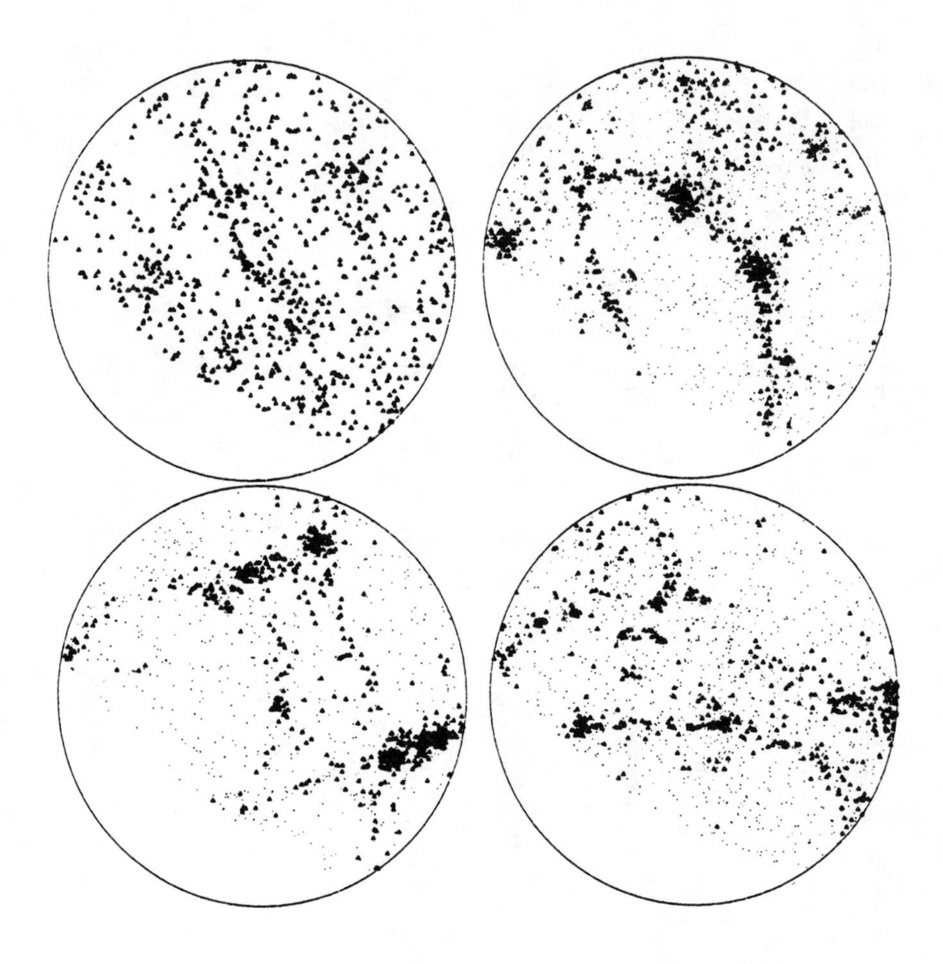

FIGURE 7.5

Three different views of structure in neutrino-dominated simulations as might be seen by hypothetical observers in them, compared with the actual distribution of galaxies as seen from earth displayed in the Center for Astrophysics survey (*upper left*). The triangles represent candidate galaxies in the simulations, while the dots represent regions that do not have a sufficiently high local density to be associated with galaxies. (Courtesy of S. White, based on simulations by Davis, Efstathiou, Frenk, and White.)

describe in more detail later, there is good evidence, however, obtained from observations of distant galaxies, that galaxy formation began in earnest when the universe was less than one-third of its present age.

The subjective visual disagreement between the simulations and the actual universe can be quantified in a number of ways. First, we can compare the numerical correlations in position between different galaxies in both the models and in the real data. The best agreement is obtained if the epoch of first galaxy formation in the simulations as just defined occurred when the universe was at least half as big as it would be at the present time. Light would have redshifted by a factor of 100 percent: its wavelength would have doubled since that time. Hence, this picture would correspond physically to all galaxies that are seen to have a redshift factor of less than 100 percent today. Since galaxies are seen out to redshifts much greater than this amount in the actual universe, this appears in clear conflict with observation. Also, since clumping is more pronounced on large scales in the neutrino models, the potential wells into which galaxies find themselves moving are very large and deep. The virial theorem of classical mechanics states that if the mean gravitational attraction in a system is large, this will be reflected by large mean velocities of the objects in the system. If we compare the velocities predicted in the simulations with the measurements of the average motion of actual galaxies, we discover that the predictions exceed the observations.

In the intervening decade, computer calculations have improved over one-hundred-fold, and the earlier results remain essentially unchanged. Moreover, observations of large scale structure have also been greatly refined, as I shall shortly describe. The disagreement between the predictions of a neutrino-dominated universe and the observed universe have continued to grow.

The idea that light neutrinos could be candidates for dark matter suffers from another flaw, a flaw that also afflicts baryon-dominated cosmologies, although in a less advanced form for a neutrino-dominated universe. While structures can grow somewhat earlier in a neutrino-dominated universe than in a baryon-dominated universe, there is still limited time for initial fluctuations to grow until the present time. Once again, in order for these to be large enough by today to produce structure, the fluctuations in matter at the time of recombination would have left remnant fluctuations in the CMBR larger than those which are observed today. In this case, the disagreement is not a factor of 50, however; it is a factor of a few.

It is clear that a neutrino-dominated universe, without something else, does not resemble the universe in which we live. If tomorrow someone were to perform an experiment demonstrating conclusively that some neutrino had a mass in the range of 30 eV, and so dominated the universe today, we would have to come up with something new to provide consistency between theory and observation. Whatever this new factor is, it would force us to abandon the general qualitative features of the neutrino scenario, which so obviously conflict with our observation. Of course, theorists have prepared several alternative possibilities, some of which may yet resurrect neutrinos as candidates for dark matter. One such possibility involves what are called "cosmic strings."

Cosmic strings (very distant cousins of the microscopic "superstrings" of interest in particle theory today) are huge remnant filament-like configurations of energy density which could be left over as "defects" from some phase transition in the early universe. If they exist, they could alter the arguments I have just presented for the growth of fluctuations. Strings could act as "seeds" for the growth of fluctuations. They could remain largely unaltered as the universe expands to the point where neutrinos or baryons can first collapse. Instead of fluctuations being washed out inside the horizon at early times, strings existing inside the horizon could preserve small-scale energy density fluctuations, so that structure need not begin to grow first on only very large scales. Many physicists have simulated what one might expect structure to look like in a universe that contains remnant cosmic strings. Beauty is in the eye of the beholder, however. At present only an earnest few individuals argue that cosmic string scenarios with or without neutrinos remain viable, but by now I hope readers recognize that early estimates usually do. Actually, even if such cosmic strings do exist, it is not clear that neutrinos would remain viable candidates to dominate the universe today.

Finally, as if it were needed, another very strong and independent argument works against neutrinos as the dark matter on all scales. Elementary particles can have associated with them a quantum mechanical property analogous to the property called angular momentum associated with a spinning top or a gyroscope. Even though these particles may be pointlike, they can behave, at least quantum mechanically, as if they are spinning. In the quantum world, the amount of "spin angular momentum" associated with particles is "quantized," that is, its value is a multiple of some basic number. It turns out that all elementary particles can be divided into two classes: those with a "spin quantum number" that is an integer (that is, 0,1,2,3 . . .); and those with a spin that is a half-integer (that is, 1/2, 3/2 . . .). Neutrinos, like electrons, have spin 1/2, while photons and W and Z particles have spin 1.

This seemingly innocuous difference is profound. As the inventors of quantum mechanics discovered, a conglomeration of particles such as electrons or neutrinos, each of which has a half-integral spin, has very different statistical properties than one containing particles such as photons with an integral spin. Since different elementary particles are indistinguishable, any configuration of several particles can undergo a variety of possible unobservable rearrangements obtained by interchanging individual particles. The way the particles behave when we make such rearrangements turns out to depend crucially upon their spin. More important, we find that interchanging two electrons or two neutrinos produces a configuration that, while equally probable, is described mathematically by a function exactly the same as that which describes the initial configuration, except for an overall minus sign. As Wolfgang Pauli was the first to show, this implies that it is impossible to force two electrons to be in exactly the same quantum state—with the same position and momentum—at any one time. This principle has subsequently been called the "Pauli exclusion principle," and it forms one of the fundamental tenets of quantum mechanics.

Unlike many of the postulates of quantum mechanics, the Pauli principle may seem intuitively reasonable. After all, who would expect to be able to put two objects in exactly the same place at one time? Unfortunately, even here our hopes of applying comfortable classical reasoning are dashed. We can show that an arbitrarily large number of particles with integral spins, such as photons—called bosons—can in fact occupy the *same quantum state at a single time*. It is for this reason that bosons can "condense" into the vacuum, or into what is called a "degenerate" ground state configuration, as occurs when *pairs* of electrons (which therefore have an integral spin) condense together in a superconductor.

The statistics that characterizes a distribution of particles with integral spin is called Bose-Einstein statistics; the statistics that characterizes particles with half-integral spin is called Fermi-Dirac statistics. This difference is at the root of most, if not all, the behavior of normal materials; it serves as the basis of all atomic structure and chemistry, as well as nuclear physics. This distinction is therefore *not* a sterile theoretical invention: it has teeth.

Since neutrinos are "fermions"—that is, governed by Fermi-Dirac statistics—one can pack only a limited number of them into any given volume. Beyond that number, quantum mechanical effects exert an incredibly strong pressure that limits any further packing. If neutrinos have a 30 eV mass, then these effects limit the total mass density of neutrinos that can be contained in a given volume. Scott Tremaine and James Gunn, working at the California Institute of Technology in 1979, were the first to apply this limitation to cosmological neutrino clumping.

We observe evidence for dark matter on scales as small as that of dwarf galaxies, containing fewer than 100 million stars. The dynamics of such systems seem to imply that certain dwarf galaxies contain more than fifty to one hundred times the mass that is visible. The arguments of Tremaine and Gunn, refined since their original studies, imply that this cannot be accomplished with neutrinos alone. The most recent estimate I know of maintains that a neutrino would have to have a mass in excess of about 90 eV in order for neutrino clumping to be effective in producing large mass densities on such small scales. This is larger than the maximum limit from cosmology placed on stable light neutrinos if they are not to close the universe. To be fair, I should note that this simply means that the dark matter in dwarf galaxies is not made of light neutrinos; in fact, the large-scale structure arguments I presented earlier suggest that neutrinos would not efficiently clump on such small scales anyway, even without the Pauli exclusion principle to forbid it. Thus, the Pauli principle alone does not suggest that they do not make up dark matter on larger scales.

When the results of the large-scale structure simulations, the CMBR data, and current data on large-scale structure are all taken into account, the situation looks abysmal for neutrinos as dark matter. Yet failure sows the seeds of hope and progress. The very factors that contribute to the apparent demise of neutrino-dominated universe scenarios point toward a solution, described in the next chapter.

CHAPTER EIGHT

Cold Gets Hot

If we examine the problems for the growth of structure in both baryon- and neutrino-dominated universes, a trend becomes clear. Baryons fail in part because the structures they would produce are far too large. Neutrinos, which can begin to collapse at earlier times because they are decoupled from radiation, tend to begin by producing smaller structures, but these still would be much larger than galaxies. What seems to be called for is some material that is weakly coupled, so that radiation pressure does not inhibit the growth of fluctuations early on, and that becomes nonrelativistic, or "cold," at times much earlier than the time that 30 eV neutrinos become nonrelativistic. Particles of such a material would move slowly early enough so that they could not have traversed distances of what would now be galactic dimensions in time to wipe out density fluctuations on such scales before gravity could take over and start the process of collapse. In this case, smaller structures would be able to grow. Moreover, if smaller structures were to grow directly, and not after the fragmentation of larger structures, initial fluctuations on galactic-size scales could begin with smaller amplitude and still have enough time to collapse to form galaxies by the present epoch. This, in turn, suggests that a fainter imprint would be left on the photon background at recombination, avoiding the present stringent limits on the microwave background anisotropy today.

We then ask ourselves the following question: When did the horizon have a size that would have corresponded to the scale we would now associate with galaxies? For if our miracle material was both nonrelativistic *and* decoupled from photons at this time, then initial density fluctuations on these scales would have been neither dissipated due to pressure nor smoothed out due to the outward streaming of the particles of this material. Hence galactic-sized structures could have begun to form directly, early on, as our intuition suggests is necessary. In this way, structure could first begin to form on the scale of galaxies instead of clusters. Such a model, where galaxies form before clusters, is "hierarchical," rather than "from the top down" as in the neutrino scenario.

This simple observation, like our earlier observation that the total mass in the cosmic neutrino background could be cosmologically significant, has spawned many investigations centering on structure formation associated with such hypothetical "cold" (that is, nonrelativistic at early times) dark matter. Even in the absence of candidates for such material, astrophysicists can examine numerically what a cold dark matter–dominated universe might look like, just as they had in neutrino-dominated models. Analytical work on the growth of small fluctuations in a cold dark matter universe can determine good initial conditions for numerical simulations, and these simulations can then follow fluctuations through the period of collapse and galaxy formation.

Davis, Efstathiou, Frenk, and White, among others, carried out simulations in the 1980s, involving a number of "particles" equivalent to the number used in the neutrino simulations, although the scale of the volume used for modeling was half as big, so that smaller structures could be examined. It is harder to simulate the "hierarchical" clustering scheme typical of a cold matter-dominated universe, because clumping takes place almost simultaneously on a wide variety of scales. Resolution on the widest possible dynamical range is necessary. Also, this rapid clumping on different scales makes the results especially sensitive to the decision of what "particles" in the simulation will represent galaxies, and when. Nevertheless, once these details were under control, the results could be compared to the neutrino simulations.

Figure 8.1 displays once again the Center for Astrophysics redshift survey (*upper left*) and the three simulated surveys for neutrino-dominated models. The lower two maps in the figure present similar slices obtained for the cold dark matter–dominated models. In the latter case the amount of dark material is adjusted to account for the amount that we infer from the virial estimates. This estimate is about 20 percent of the closure density (that is, $\Omega = 0.2$). The cosmic fudge factor, h, is set nearly equal to 1. The simulations were evolved until the density and correlations of galaxies generally matched what we observe over a broad range (remember that in the neutrino case this was not possible). It is apparent from these figures that the cold dark matter–dominated universe offers a better approximation of what we actually see. A much wider range of clustering scales is obvious in the hierarchical clustering scheme, and excessive clumps on large scales are not observed.

Over the past decade other independent analyses—analytical as well as numerical—revealed that a cold dark matter–dominated universe consistently reproduces the general features as well as certain fine details of observed structure. Simulations have now progressed forward along with available computer power. The number of particles measured has increased by an order of magnitude, and even physics associated with the hydrodynamics of a baryon gas embedded in a dark matter sea has now been included. In addition, as early as 1986, George Blumenthal and his collaborators Sandra Faber, Joel Primack, and Martin Rees demonstrated that the detailed morphology of galaxies embedded in regions of different densities of cold dark matter is in striking agreement with what is observed. Also, if cold dark matter can successfully cluster on smaller scales than can hot matter, it raises the hope of explaining the observed dark matter abundances in small systems such as dwarf galaxies.

Looks can be deceiving, however, and as good-looking as the numerical simulations displayed in figure 8.1 are, the success of a model—even in cosmology—cannot rest on qualitative agreement alone. Although there is no doubt that cold dark matter simulations, for the aforementioned parameter choice (an open universe with the expansion rate at its maximum allowed value), produced universes that *qualitatively* appear similar to what we see, *quantitative* tests uncovered problems. As usual in such matters, any specific quantitative problem in a model turns up in simulations through a number of different but essentially equivalent statistical measures. Among the various problems first discovered were these five. (1) The relative velocities of nearby galaxy pairs tended to be somewhat too fast, if the velocities were made to fit the observed relative velocities at larger scales. (2) Although the overall level of clustering agreed well with observations on different scales, the clustering as a function of separation—determined quantitatively by calculating the probability of finding a second galaxy within some distance of a given galaxy—did not have the same detailed shape as the observed clustering function. (3) The joint probability of finding three galaxies within some distance for scales less than 5–10 mega-parsecs tended to be much higher than what is observed. (4) The mass to light ratio on the scale of groups of

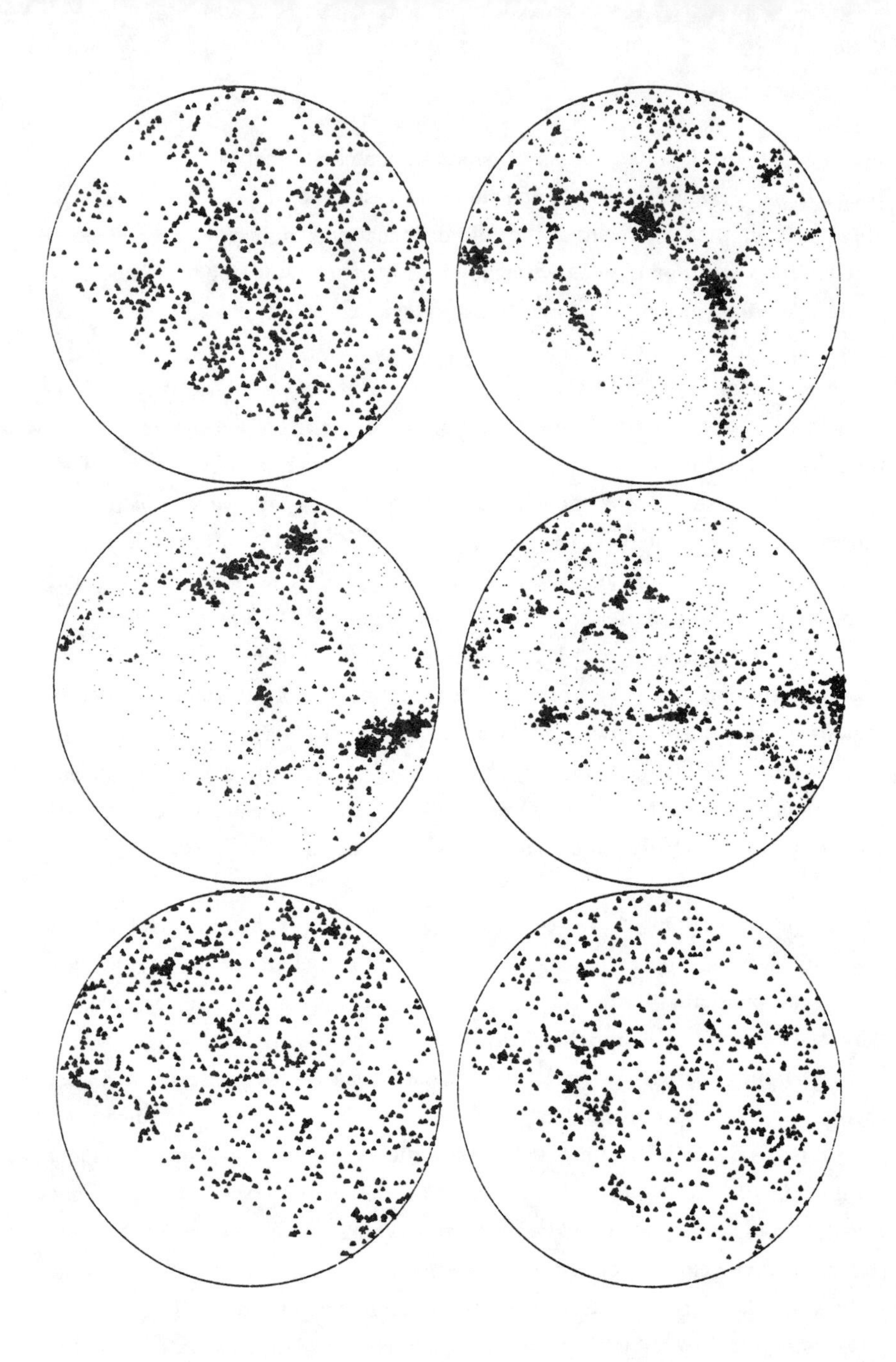

FIGURE 8.1

The bottom two maps shown here are taken from simulations of structure formation in a universe dominated by cold dark matter. The four other maps are identical to those in figure 7.5, with the map in the upper left-hand corner taken from the Center for Astrophysics redshift survey of actual galaxies and the other three showing structure in simulations for a neutrino-dominated universe. (Courtesy of S. White, based on simulations by Davis, Efstathiou, Frenk, and White.)

galaxies was too large. (5) The models predicted remnant fluctuations in the microwave background that were on the edge of, or just marginally ruled out by, our observations. These five "glitches" reflected a single fact: individual clusters of galaxies in these simulations were somewhat more tightly bound than actual clusters seemed to be.

Should these minor quantitative irregularities force us to reevaluate the apparent qualitative success of these cold dark matter models? After all, the simulations did neglect the internal structure of galaxies, and physical effects on these small scales could be important in alleviating at least some of the problems. Alas, the answer is yes. One of the motivations for performing these simulations was to see whether under very simple assumptions a dark matter model might explain observed structure. If we have to rely on unknown dynamics at small scales to solve observed difficulties, this is not immediately satisfactory. More important, however, if we ignored signals revealed by the few quantitative probes we may have had large-scale structure, we risked letting cosmology become devoid of content. A careful examination of the detailed quantitative nature of these potential problems might offer essential clues about how to resolve them.

It is in fact somewhat surprising that a first attempt should have come so close. Moreover, if the simulations showed better agreement with observation, we would have had cause to doubt certain fundamental aspects of our theoretical picture of cosmology. In the first place, the extreme value of h required in the early simulations leads to a universe that today would be quite young, less than about 10 billion years old. As I have mentioned, many independent age estimators, especially analyses based on stellar evolution, suggest that our universe must be *at least* 11 billion years in age. It would be discouraging if all of these estimates had to be tossed out in order to achieve good agreement for dark matter models.

Next, remember that these simulations were for $\Omega = 0.2$, or *open* universes. I alluded to this significant fact earlier without too much hoopla, but you may have heard bells and whistles when you read it. If an $\Omega = 0.2$, *open,* cold dark matter-dominated universe agreed perfectly with our observation, then what about $\Omega = 1$, the flatness problem, inflation, and related ideas so prominent in the preceding chapter?

Any excitement generated by the idea that an open universe may not be in straightforward agreement with the detailed numerical predictions of cold dark matter models is quickly tempered, however. If instead one modeled an $\Omega = 1$ universe by the same kind of treatment, the situation only got worse. Raising the overall density of galaxies only makes clustering more pronounced, and "potential wells" would be deeper on small scales. Since virial estimates based on just such qualities in the observed universe all suggest that Ω is less than 1, it is no surprise that simulations in which these same analyses would offer an estimate in accord with a flat universe produce poor results. In fact, all of the glitches described earlier become exacerbated; in addition, the Hubble fudge factor (h) that made the first simulation agree best with observation was nearly equal to 0.2, and this is much smaller than even the most extreme observational estimates.

What are we to do? Can we reconcile theory and observation in a cold dark matter–dominated universe? Many exotic possibilities come to mind. But a first simple, and many would claim natural, solution lies buried in the simulations themselves. Keep in mind that these simulations only applied to the dominant, dark component of the universe. Identification of this mass distribution with the galaxy distribution is not necessarily the proper thing to do. I earlier alluded to the fact that distinguishing what one means by a "galaxy" amid the clustered particles in such a simulation is quite a delicate task.

In taking statistical "polls," one of the trickiest challenges is making sure that a proper "random" sample has been chosen. As can be seen in many political opinion polls taken during election campaigns, even the most comprehensive statistical analysis of a poor sample will lead to results that probably have nothing to do with reality. Once this Pandora's box is opened, we can think of a hundred reasons why visible galaxies, which represent at most 10 percent of the mass in the universe, might not provide a proper "random" sample of the underlying matter distribution. Indeed, identifying galaxies with the underlying distribution of dark matter, as was done in the simulations described earlier can be justified a priori only on the grounds of simplicity. If we relax this assumption, the results change dramatically.

With this in mind, Davis, Efstathiou, Frenk, and White next numerically examined what would be predicted if galaxies provide "biased" tracers of mass in the universe. Specifically, they investigated the possibility that galaxies are "rare" occurrences, that they would only form where the dominant dark matter background had a significantly greater than average density. One can question what physical mechanisms might cause this to be the case, and I shall return to this later. For the moment, let me merely review what happens if it is the case.

This "biasing" was implemented by taking the same initial conditions as used in the earlier simulations and then smoothing them out to probe the average density on small scales. These averaged regions were examined, and if the density came out to be sufficiently large, say 80 percent greater than the mean density, the nearest "particle" was chosen and labeled as a galaxy. The evolution of this system of "galaxies" was then followed in detail. The number of such galaxies in the simulations turned out to be comparable to the number of bright galaxies that one would expect in an actual volume equivalent to the volume being simulated. A comparison between the underlying mass distribution in such a simulation and the distribution of "galaxies" is shown in figure 8.2.

As the figure shows, if galaxies are "rare" events, they will be more clustered than the overall background. Nick Kaiser, a young British astrophysicist, had already emphasized the importance of this simple statistical fact in explaining the otherwise unexpected observation that very rich, very rare observed clusters of galaxies tend to be more strongly correlated with each other than are average galaxies, most of which are not located in such rich clusters.

This effect can also be used to attempt to reconcile the observed correlations of galaxies with the possibility of a flat universe. Imagine that galaxies are formed from fluctuations in the density of matter on some scale, and that these fluctuations are then embedded inside others that exist on much larger scales, and so on, as is the case in hierarchical models. Then, *if a certain density threshold is necessary before a galaxy is to form,* it will be more probable for the fluctuations on galactic scales to cross this threshold if they themselves lie inside background regions having densities well above the average. This situation is illustrated

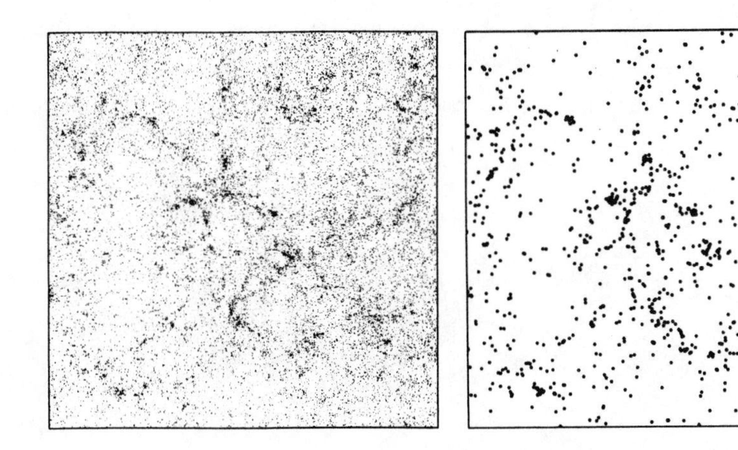

FIGURE 8.2

Shown on the left is the distribution of all particles followed in a cold dark matter simulation performed by Davis, Efstathiou, Frenk, and White. The map on the right displays the distribution of regions with a sufficiently high average density to be identified as "galaxies" in DEFW's biasing algorithm. (From M. Davis et al., *Astrophysical Journal* 292 [May 1985].)

schematically in figure 8.3. Here we can see that the small-scale random fluctuations tend to cross the "threshold" line in regions where the background density is highest. Therefore, galaxies will tend to be clustered together more; we will be more likely to find one galaxy when there are others found nearby. This simple fact has profound implications. It implies that the observed clustering of galaxies in such "biased galaxy formation" models is only indirectly related to gravity. It is also a statistical phenomenon. If gravity is not as intimately tied as we had assumed to galaxy clustering, this means that the gravitational potentials from the large-scale fluctuations in the underlying mass distribution need not be as strong. If these are smaller in value than those which were earlier needed to produce the same clustering level using gravity alone, before we considered the possibility that galaxies were rare compared to the underlying dark matter, then the negative features of the earlier simulations disappear: relative velocities of nearby galaxies would be smaller, joint correlations of galaxy triplets would be reduced, and the overall scale of initial fluctuations that might leave an imprint on the microwave background need not be so large. In other words, agreement with observation would improve. Indeed, a quantitative analysis using the "biased" simulations of Davis's group showed this very behavior.

The fact that a universe dominated by cold dark matter appeared to require statistical biasing can be viewed as a blessing or a curse, depending upon your point of view. One of the most important observational implications of all this is that the level of biasing used in the DEFW simulations suggested that an inferred value of Ω of about 0.2 from virial estimates of galaxy motions in fact would be

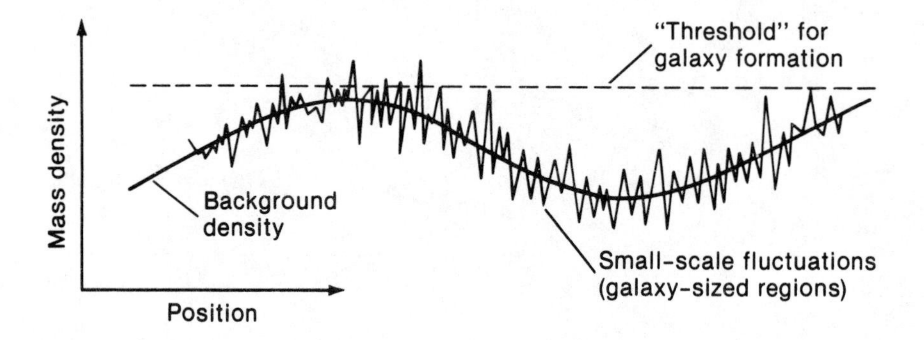

FIGURE 8.3

If local density fluctuations must exceed a certain threshold, as schematically shown here, before galaxies can form, then galaxies will preferentially form only where the background density itself exceeds its average value.

consistent with an actual value of $\Omega = 1$ when the more widely distributed dark matter is included. What a remarkable turn of events: suddenly $\Omega = 1$ would be favored rather than dismissed. On the other hand, for an observer who has spent the better part of his or her career demonstrating that Ω is nearly equal to 0.2 on the largest scales that we can measure, the resort to biasing to explain away one's observations seemed very suspicious.

Moreover the apparent improvements in the biased models should be carefully weighed. These improvements were gained at the expense of introducing more parameters into the models: the biasing scale and the threshold factor. It is always easier to reconcile data with more parameters, so do these improvements justify any optimism? The early answer was yes, based largely on further numerical work by the DEFW group. Once the parameters of the biased models were manipulated to produce agreement on some scale, say between 1 and 10 megaparsecs, there is no more latitude left. But if these models were to explain all structure they must reproduce observed features of galactic "halos" of dark matter on smaller scales, as well as recently observed structures on much larger scales. The next set of simulations carried out in these two extremes, this time with up to 1 order of magnitude more particles in the simulation, yielded very encouraging results. Simulations on lesser scales, with many different particles within an identified "galaxy," yielded predicted "rotation curves" (see chapter 4) of exactly the correct shape and overall magnitude for bright galaxies. The simulated curves for ten bright galaxies are strikingly similar to the curves presented earlier (also in chapter 4) for real galaxies (see figure 8.4).

Alternatively, other simulations for much larger scales showed agreement for such features as the observed abundance of "rich," or populous, clusters, immense empty voids, and filamentary configurations of galaxies. Since it usually happens in physics that detailed studies based on a preliminary hypothesis produce results that disagree with observation more strongly than the initial estimates do, the rare cases in which agreement is *improved* upon detailed study, are all the more compelling.

After all the early approximate agreement between biased, $\Omega = 1$, cold dark matter models and observation really was, in an objective sense, remarkable. It is a simple testable model based on only a few fundamental parameters. That such a simple model that might explain so much remains viable is very encouraging. Indeed, the cold dark matter scenario is now a fundamental part of every current proposed explanation of observed large-scale structure. This does not mean it is correct in detail, or that it faces no serious observational challenges. In fact, one

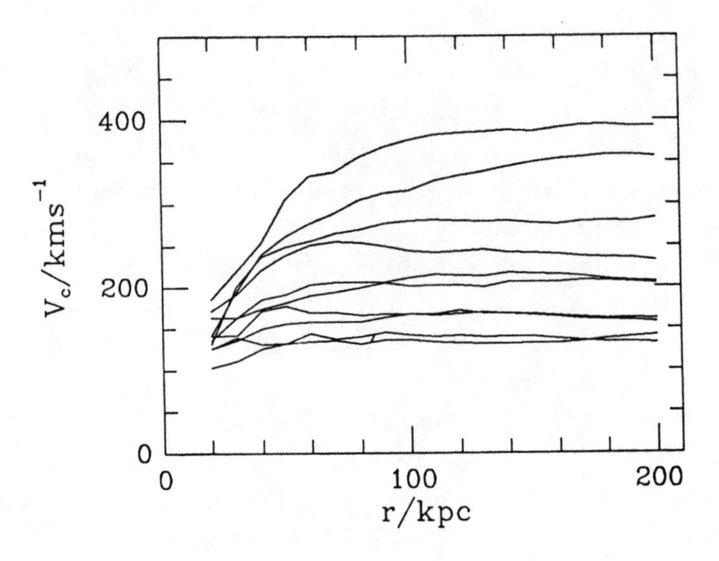

FIGURE 8.4

For the ten most massive clumps present at the end of a numerical simulation for a cold dark matter–dominated flat universe, predicted rotation curves similar to those obtained for real galaxies were derived. Shown here is the predicted circular velocity, V_c, of objects in the system (in kilometers per second) as a function of their distance r from the center (in kiloparsecs [= 1,000 parsecs]), obtained by analyzing the mass contained in spheres of increasing radius r centered on the densest region of each clump. (From *Nature* 317 (1985): 597. Copyright © 1985 Macmillan Magazines Ltd. Reprinted by permission.)

of the chief virtues of this model is that it makes testable predictions—so that we may be able to rule the model out, if necessary. Cosmology tends to be a speculative science. The last thing we need is a proliferation of models with little or no *empirical* implications, or with so many parameters that any observation can be accommodated. If any of the predictions associated with either the notion that the universe is flat or the idea that structure formed due to the gravitational growth of an initial scale-independent spectrum of primordial matter fluctuations turned out to be wrong, we would, in the best tradition of science, be forced to discard or alter the model.

Alas, no sooner said than done! In the decade since the original CDM simulations were done, our empirical and theoretical understanding of structure formation has evolved tremendously. It is now unambiguously clear that *we do not live in a flat, matter dominated universe!* Put another way: *The standard cosmological paradigm of the 1980s is wrong!* All of the evidence from independent cosmological observations now suggests that the total matter density in the universe is less than about 50 percent of the amount needed to result in a flat universe today. Remarkably, this evidence either directly or indirectly suggests that the universe is nevertheless flat, *and that the matter content is indeed dominated by cold dark matter, and that structure formed from the gravitational collapse of an initially roughly scale-invariant spectrum of primordial density perturbations.* I shall address the apparent contradiction of some of these statements in the next chapter. Here I discuss the current status and challenges facing cold dark matter models.

Two sets of observations have completely altered the way we think about large scale clustering of matter since the early heady days of an $\Omega_{\text{matter}} = 1$ CDM–inspired universe. The first involves the remarkable discovery of primordial anisotropies in the Cosmic Microwave Background Radiation. In the first edition of this book, I listed the CMBR as one possible major challenge to CDM models. After all, the nonobservation of anisotropies in the CMBR at the level then currently accessible had already ruled out a baryon dominated universe and had even squeezed the possibility of a neutrino dominated universe. As I then argued, if the resolution could be improved by an order of magnitude, and primordial fluctuations were still not observed, then the entire idea that structure forms from gravitational collapse, not to mention that this collapse was dominated by the gravitational dynamics of CDM, would have to have been discarded.

Within three years of issuing this challenge, the COBE satellite discovered primordial anisotropies in the CMBR. These were within a factor of 2 of *precisely* the level predicted in CDM models of structure formation in which the spectrum of primordial fluctuations was roughly scale-invariant. Besides providing unprecedented information on the nature of the physics of the early universe associated with the generation of fluctuations, this discovery provided clear and direct support for the general CDM picture of structure formation. The general framework I have described thus far, involving structure formation via gravitational collapse of primordial density perturbations in a universe in which the dominant form of matter is cold dark matter, is now on far firmer footing than it was a decade ago.

The COBE result also changed the way we normalize CDM models when comparing with the data. Previously we used the observed magnitude of density fluctuations on galaxy cluster scales, since this was the largest scale structure we knew of, as the basis for evolving the growth of structure on smaller scales. The trouble with this measure is that it still involves visible objects, and it suffers from the problem of the unknown level of biasing of mass versus light, even on these scales. Now, however, the COBE CMBR data yields a direct measure of the overall magnitude of primordial density perturbations in all matter—dark or light. Since the COBE satellite also measured the magnitude of primordial fluctuations in the CMBR on scales that could not have been affected by causal physics after inflation, we are reasonably certain that the fluctuations it measures truly are primordial. For this reason we now speak of COBE-normalized models. This has fixed one fundamental parameter for galaxy formation and clustering which, prior to COBE, had been only loosely constrained.

At the same time as this development, a number of in-depth studies of the real space clustering of huge samples of galaxies have been made, which improves dramatically the statistical understanding of actual galaxy clustering. Surveys, including the Automated Plate Measuring (APM) survey, the Center for Astrophysics (CfA) survey, and the Las Campanas Redshift Survey (LCRS), have probed structures on scales as large as 50 mega-parsec on significant slices of the sky. At the same time, the infrared astronomy satellite (IRAS) has provided a broad sample of infrared galaxies, representing a different class of galaxies from those probed in the other surveys.

These new observational tools have provided a whole new level of quantitative constraints on dark matter models. Cosmology was, as little as thirty years ago, thought to be a field largely devoid of empirical data. Apparently wild speculations were not strongly constrained. At the present time, however, it is difficult to imagine a field of physics that is more data-rich. While this data has provided remarkable support for the general features of the CDM-induced structure formation picture presented thus far, it has also presented several serious quantitative challenges that have forced us to reexamine the details of this picture.

It is worth pointing out that each of the challenges I shall describe below has had the potential of fully invalidating the fundamental assumptions of the CDM picture of galaxy formation. At various times in the past fifteen years, things have looked rocky as preliminary data was announced. However, the fact that CDM is still alive and kicking after almost two decades of intense scrutiny is nevertheless very impressive, even if our prejudice for precisely how much dark matter there is, or should be, in the universe, has radically changed.

THE ERA OF GALAXY FORMATION

One relatively generic prediction of the cold dark matter scenarios described here is that galaxy formation has been fairly recent on a cosmic time-scale. The successes of cold dark matter models stem from the fact that they allow the growth of structure on widely different scales, from galaxies on up. But, in a flat matter dominated CDM model, galaxy formation could not begin too early. If this were the case then we should expect to see more significant clustering on larger scales than we do today. Put another way, if fluctuations on the scale of galaxies had grown sufficiently to collapse at an early time, then fluctuations on larger scales should also have had time to grow to be fairly large by today—assuming that they had the same initial amplitude as galaxy-sized fluctuations when the horizon first encompassed them. The larger these fluctuations are today, the greater will be the clustering on these large scales. Simulations show that structure forms very quickly once fluctuations become big enough to collapse. The earlier such structure forms, the more pronounced it will become today.

The bottom line is that the standard flat universe, cold dark matter scenario, predicts that the bulk of galaxies should have formed more recently than when the universe was about one-fourth its present size, or when its age was greater than one-eighth its present age. The light from stars emitted after this time would have redshifted by a factor of less than about 3 by today. Since we see many galaxies with redshifts on the order of 1, this places the era of galaxy formation between about one-eighth and one-third the present age of the universe.

As our telescopes improve, we are able to see objects at farther and farther distances, or, equivalently, light emitted at earlier and earlier times. Advances in technology now allow us to spot incipient galaxies with a redshift factor of 4 or 5. Already several groups have spotted objects with a redshift factor of between 2 and 5. Determining the redshifts for such objects is sometimes problematic, but, assuming the distance estimates are correct, the fact that such objects have already been seen with limited searches indicates that they are not very rare.

This concern about the era of galaxy formation has recently been dramatically compounded by a series of observations, not of single galaxies, but of clusters of galaxies. As I have said, these are the largest bound objects in the universe, containing up to hundreds of individual galaxies. In a model where galaxies form before clusters, and in which galaxies themselves formed recently, one would not expect to see as many clusters of galaxies early on as now. This is, in principle, a dramatic probe of the CDM scenario. Recently, Neta Bahcall at Princeton University and collaborators have examined the redshift-dependence of the number density of rich galaxy clusters. In a flat CDM matter–dominated model, Bahcall has argued that one would expect to find up to a hundred times fewer rich clusters at a redshift of near unity (about 1/3 of the present age of the universe) than if there were not sufficient matter to result in a flat universe today.

The central point is this: In a flat, matter-dominated universe, density fluctuations continue to grow in magnitude with time. Thus one would expect much less structure at earlier times than one observes today. However, if the density of matter in the universe is not sufficient to make it flat, then either the universe is open, or the remaining energy density required to result in a flat universe is contained in something other than CDM. It turns out that *in either case,* the effect on structure formation is essentially the same. Once matter becomes subdominant, and either the curvature of an open universe begins to become significant, or the energy density in a cosmological term takes over the expansion, the rate of expansion becomes larger than it would be in a flat matter-dominated universe. This increased rate of expansion, tends on large scales, to win out over the local gravitational attraction that causes matter to cluster. Thus, structure formation slows down, or ceases, on large scales.

Bahcall's group claims, for example, that their observations of clustering at high redshift place an upper limit on the density of matter in the universe today of about 50 percent of that required to result in a flat universe. If this matter were predominantly CDM, then observation and predictions agree well. Unfortunately, this observation alone cannot easily distinguish between an open CDM-dominated universe, or a flat CDM/cosmological-constant mixture.

Combining this result with observations of clusters at more modest redshifts, however, can break this degeneracy. The number density of rich clusters in the universe today tells us something about the overall magnitude of present-sized fluctuations on cluster scales. Using the COBE data to normalize the scale of primordial fluctuations one can try to distinguish between open and flat models by the amount of growth of fluctuations on cluster scales between the CMBR era and the present time. Here it appears that flat CDM/cosmological-constant models are preferred over open CDM models. I stress that in both cases the abundance of CDM is between 30 and 50 percent of that required to result in a flat universe today. This abundance is far greater than the abundance of baryonic matter in the universe, so even though the CDM density is a factor of 2–3 less in these models, compared with flat CDM models, CDM is still by far the dominant form of matter.

SCALE-DEPENDENCE OF GALAXY CLUSTERING

The vast amount of new data on galaxy distributions in the universe has provided another powerful new probe of the total amount of matter in the universe. The argument is again simple and is essentially identical to that in favor of CDM over a neutrino-dominated universe. Recall that in the early universe, structure essentially did not begin to grow until slow-moving matter began to dominate. Because in a neutrino-dominated universe this happens at a later time than in a CDM-dominated universe, the first scale to grow is far larger in the former case than the latter. This same distinction can be further refined to constrain the *amount* of matter in a CDM-dominated universe. The lower the overall abundance, the later the time at which it would come to dominate the expansion, and the larger would be the first scale on which fluctuations could grow.

The observable that allows this distinction to be drawn is the "shape" of the observed dependence of galaxy clustering with increasing scale. This function is usually called the power spectrum of galaxy clustering. In 1994, John Peacock and his collaborator, P. Dodds, compiled all the newly available data from the various galaxy surveys discussed earlier, and came to an unambiguous solution. In a standard CDM-dominated universe, *the "shape parameter" of the galaxy power spectrum required the combination* $\Omega_{\text{matter}}\, h$ *to be between approximately 0.25–0.35.* This result has been confirmed by all subsequent refinements in the data, and was perhaps the first definitive evidence that cold dark matter could not result in a flat universe as long as the Hubble expansion parameter *h* were greater than about 0.35, as it certainly seems to be.

One might ask whether this conclusion in fact does force us to either an open universe or a universe dominated by something like a cosmological constant. For example, if one altered sufficiently the scale dependence of the primordial fluctuation spectrum, then one might hope that the resulting inferred scale dependence today, even in a flat CDM model, could agree with observations. However, detailed numerical simulations have been performed in this case, and it has been found that there is no way to simply tilt the primordial spectrum in such a way as to produce the correct overall shape parameter and at the same time produce the observed CMBR anisotropies.

One possibility that was suggested, and which created quite a bit of interest for several years, was to introduce some hot dark matter in the form of light neutrinos into the mix. A neutrino with a mass of a few electron volts would not be sufficient to dominate the universe today, but it would comprise perhaps 10–20 percent of the total mass density of a flat universe. Several groups demonstrated that this possible "mixed dark matter" solution could in principle solve the problems of a flat cold dark matter–dominated universe. The neutrinos would tend to suppress structure formation on smaller scales, but not as dramatically as if they dominated the mass density today. Indeed, the shape function for such a model is reasonably close to the data, as determined by numerical simulations performed by Joel Primack and collaborators, as well as others, for a flat universe with approximately 80 percent cold dark matter and 20 percent hot dark matter.

By adding another parameter one can always get a better fit to the data, and thus this proposal could seem uninteresting, were it not for the fact that in 1995 a group attempting to measure neutrino oscillations at Los Alamos claimed to observe a signal consistent with a 2–6 eV neutrino mass. This caused some excitement for awhile. However, to date this observation has not been confirmed, and moreover, it is extremely difficult to reconcile it with the combination of other neutrino observations, including those of solar neutrinos and neutrinos arising from the collisions of cosmic ray protons in our atmosphere.

Perhaps the factor which is most damning to this scenario, however, is the claimed observation of an accelerating expansion, attributable to a nonzero vacuum energy density in the universe. Since such a vacuum energy would have somewhat of the same effect as the addition of hot dark matter, even the proponents of a mixed dark matter universe have now agreed that the addition of hot matter to the CDM/vacuum-energy mix does not improve the fit sufficiently to consider it interesting.

In any case, when the shape parameter result is combined with the results of the cluster evolution studies, the conclusions appear to be unambiguous. A small admixture of hot dark matter will not resolve the issues associated with cluster evolution. *While a CDM model is the only model consistent with the overall shape parameter, the density of rich clusters, and the CMBR anisotropies, it is only so if the density of matter is less than 50 percent of the closure density. If the universe is indeed flat, as inflation combined with recent CMBR data suggests, something other than matter must make up the difference.*

While these are the main observational factors that have changed our current thinking, a number of other tests of the CDM picture have been carried out, with varying levels of success.

GALAXY FORMATION: ANTIBIASING?

The problems associated with the early CDM simulations involved the inability to discern galaxies amidst the dark matter. The number of particles in the simulations was such that the mass associated with each particle often exceeded a galactic mass. Moreover, visible galaxies are made of baryons, not dark matter. A gas of baryons can emit radiation and cool. Until the processes associated with a realistic gas of particles, called hydrodynamics, were included in the simulations, any detailed quantitative conclusions drawn about actual galaxies embedded in a dark matter sea remained suspect.

In the intervening decade, computer power has improved exponentially, and now simulations are performed with an order of magnitude more particles, and gas cooling and dissipation are routinely included. Here the results are encouraging. Recent simulations by Simon White and colleagues, now performed in the context of a CDM model with a nonzero cosmological constant, suggest that the type of galaxy distribution that evolves in such a universe is in good agreement with the observed luminosity distribution of galaxies.

Moreover, the most recent simulations also address another apparent problem with CDM models, even those with a cosmological constant. In order to reproduce the observed clustering of larger scales on small scales associated with galaxies and small groups, the clustering of the dark matter seemed to be in excess of that observed for galaxies. However, the new simulations suggest that luminous galaxies may be "antibiased" on small scales, compared to the dark matter. The visible matter seems to be less clustered than dark matter on these scales, not more clustered. The authors of these simulations suggest that this antibias is due to the fact that the clustering of the dominant dark matter continues to evolve on these scales until the present time, while the clustering of visible galaxies, located initially in the densest regions, does not. Further work will have to be done to confirm this picture, but the early results are encouraging.

BUBBLES OR NO BUBBLES?

In 1986, Margaret Geller and her colleagues at the Harvard-Smithsonian Center for Astrophysics undertook a systematic redshift survey of galaxies at distances much deeper in space than ever performed. Their results were astonishing. Structures were observable on very large scales, tens of mega-parsecs, which had not been identified previously. Very well defined filaments were seen, and galaxies seem to be located on very sharply delineated, relatively spherical "sheets" surrounding huge voids. Geller and colleagues described it as resembling a "slice through the suds in the kitchen sink." This was not the type of structure that one would imagine could form due to gravity alone. The images suggested that something else had to come into play—perhaps large-scale explosions of undetermined origin. Our picture of the universe on large scales had apparently, in one fell swoop, evolved.

After these results appeared, the DEFW group scurried to their computers to determine how strange this kind of picture really was. They claimed it was premature to argue that something new was necessary to produce the structures we see. They produced simulations from their standard cold dark matter models, but this time on much larger scales. They examined the simulations just as Geller and colleagues examined the real thing: qualitatively, things looked pretty similar. Judge for yourself (see figures 8.5 and 8.6).

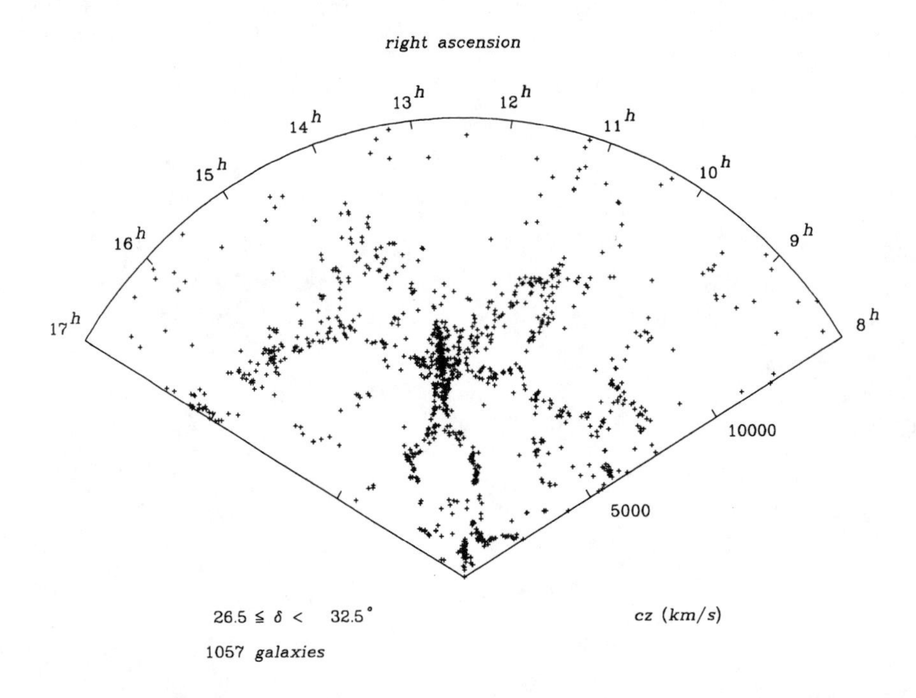

FIGURE 8.5

A "wedge" diagram from the Center for Astrophysics redshift survey showing the distribution of galaxies along a slice of the sky. The radial direction represents the redshift, and hence roughly the distance to the galaxies shown. The velocities of galaxies at the edge of the map relative to an observer on earth (located at the lower triangular vertex) are over 10,000 kilometers per second, indicating a distance of over 100 mega-parsecs. Huge circular voids with no galaxies can be seen. In addition, most galaxies seem to lie along thin shells at the edges of the voids. If slices of the sky slightly above and slightly below the wedge are examined, one can see that the circular pattern seen here in two dimensions extends to a "foam-like" pattern of galaxies distributed in three dimensions. (From V. de Lapparent, M. J. Geller, and J. P. Huchra, *Astrophysical Journal* 302.)

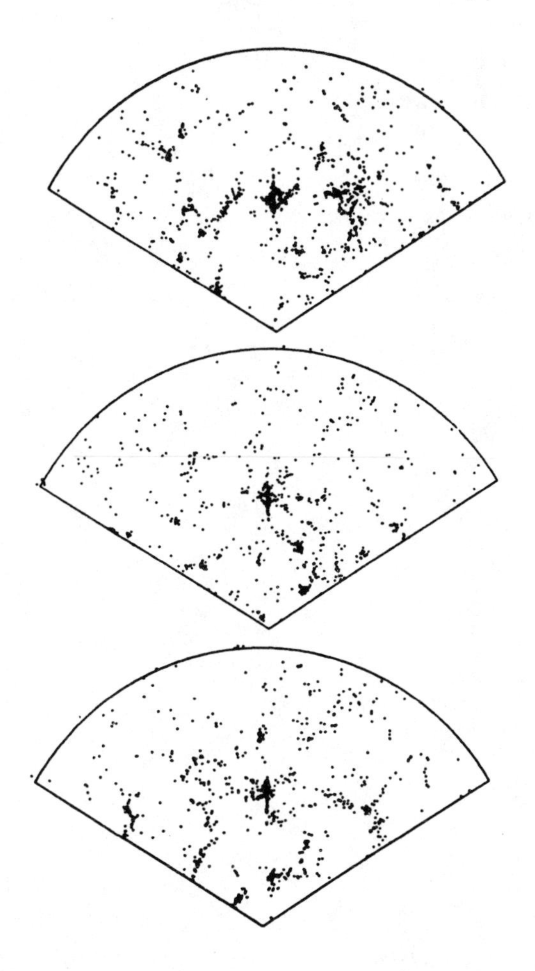

FIGURE 8.6

The "wedge" diagram distribution of "galaxies" for three hypothetical redshift surveys designed to mimic the conditions of the actual Center for Astrophysics survey, but in this case obtained from simulations of a cold dark matter–dominated flat universe with biasing. One can see filamentary structures and voids of comparable size to those observed in the actual data. Adjacent slices seem to show that the structures seen here continue across several slices. (From M. Davis et al., *Astrophysical Journal* 313 [1987].)

The DEFW group also pointed out a key attribute that may allow us to distinguish between different models. The large voids seen in the simulations do not seem to be continuously surrounded by sheets of galaxies, as in the case of soap bubbles, but seem rather to percolate throughout the volume, as in the case of a sponge. The use of these crude analogies is telling. Early on we had little more than a qualitative handle on our data. Until quantitative analyses of the data and numerical simulations were performed, it was too soon to choose between gravitational collapse in a cold dark matter–dominated universe or explosions as the seeds of large-scale structure.

Since these initial results, a host of observations have now come down firmly in favor of the gravitational collapse picture. The COBE data demonstrates that primordial fluctuations existed at precisely the level required for gravity to do the job. Numerical simulations have also indicated that gravitational collapse produces the correct general statistical features of galaxy clustering. Moreover, the

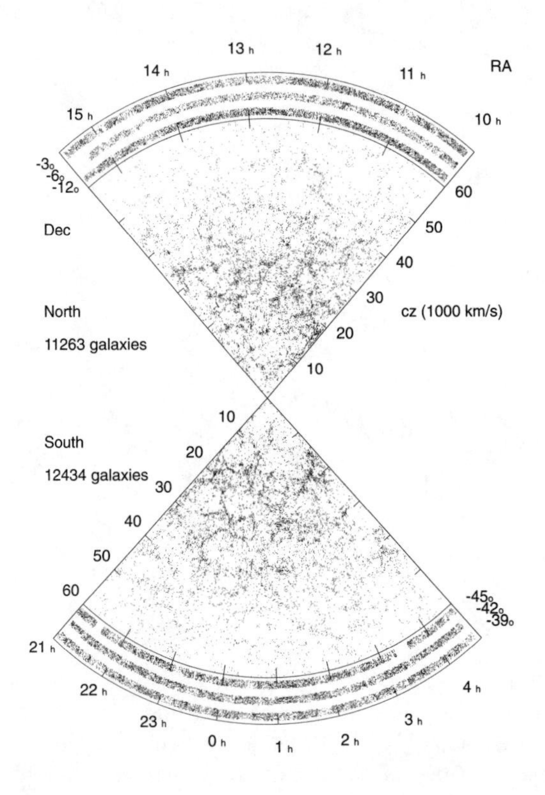

FIGURE 8.7 Las Campanos Redshift Survey

The Las Campanas Redshift Survey (LCRS), carried out by an international collaboration at the Carnegie Institution's Las Campanas Observatory in Chile, consists of 26,418 redshifts of distant galaxies. The survey covers more than 700 square degrees in six strips, each 1.5 degrees x 80 degrees, three each in the North and South galactic caps. The typical redshift in the survey is 30,000 km/s. Two-dimensional representations of the redshift distribution (as above) reveal many repetitions of voids, on the scale of about 5000 km/s, sharply bounded by large walls of galaxies as seen in smaller, more nearby samples. Most important is the fact that while the voids and walls observed in earlier surveys are repeated throughout the sample, larger walls and voids are not seen. The earlier CFA redshift survey (figure 8.5), for example, would be contained in the innermost corner of this figure. (Figure courtesy of Paul Schechter, MIT) early CfA redshift studies have now been supplanted by studies going out to redshift velocities in excess of 60,000 kilometers/sec, over 6 times deeper than the CfA study. Shown in figure 8.7 is a slice of the universe from the Las Campanas Redshift Survey taken of over 7,000 square degrees in the sky in both the northern and southern hemispheres. In this figure, the entire CfA "stick-man" survey would be contained within the first tick mark.

As can be seen, the behavior observed by the CfA survey of voids surrounded by walls persists. But more important, the scale of such clustering remains the same. No larger coherent structures are seen in the data. Again, this is precisely what one would expect in CDM clustering scenarios.

LARGE-SCALE DRIFT IN THE UNIVERSE

The observations of "large-scale drift" in the universe are still somewhat controversial, as is the issue of an "edge" to galactic halos discussed in the next section. Both pose very significant questions in assessing the evidence for dark matter, and in fact I have mentioned each of them earlier. Both involve apparent direct observational challenges to the cold dark matter structure-formation scenario I outlined; both stir intense interest. What differentiates the two issues is the fact that the large-scale drift observation was totally unexpected, and like the Geller group's observation of "soap suds" in the distribution of galaxies, it forced us to reevaluate our picture of what our region of the universe looks like on large scales.

I noted earlier the fact that the microwave background provides a frame by which we can measure our local motion relative to the general expansion of the universe. If we are moving in some direction relative to the frame of the microwave background, the photons coming from the direction into which we are moving will be blueshifted, and those coming from behind will be redshifted. This creates a dipole anisotropy in the background which is measurable. Measurements suggest that we are moving with respect to the microwave background with a speed of about 600 kilometers per second in a direction about 30 degrees away from the Virgo supercluster of galaxies.

We can perform another measurement in addition to determining our own motion with respect to the microwave background. We can examine the local motions of as many nearby galaxies as we can see, so we can learn what their net motion is relative to ours. I have said that on a scale of up to about 10 mega-parsecs in distance we and nearby galaxies seem to be falling toward the Virgo cluster. However, if we examine a sufficiently large volume, we would expect that any local motions would cancel out and that this volume would on average be essentially at rest with respect to the microwave background frame, that is, this large region would not be coherently drifting with respect to the microwave background. Thus, we would expect that our motion relative to the large conglomeration of galaxies should be of the same order of magnitude and direction as our motion relative to the microwave background.

In this context, the significance of the results of the so-called Seven Samurai group of astrophysicists, mentioned earlier, becomes clearer. This group undertook a very ambitious and careful survey designed to settle this issue of large-scale drift. The data were quite perplexing. Our motion relative to the local group of galaxies was comparable in magnitude to our motion relative to the microwave background, as expected, but in a *different direction*.

Taken at face value, this finding would mean that our whole local group is not at rest with respect to the microwave background, but moving at a very large coherent velocity with respect to it, at about 500–600 kilometers per second. This came as a shock to the astrophysical community. No one had ever anticipated such large-scale coherent motions in the universe. Certainly such motions were not predicted to result from cold dark matter simulations.

This result was so unexpected that it came under a great deal of scrutiny, by other observers and even within the Seven Samurai. The immediate impact was to change the interpretation of the results, but not the results themselves. The new interpretation, which I mentioned previously, suggested that the data agreed better with a large-scale infall of the local group of galaxies into an apparent huge mass overdensity region, called the "Great Attractor," located near the Perseus-Pisces cluster perhaps some 200 mega-parsecs distant. When viewed on the plane of the sky, no huge mass agglomeration stands out to the eye, but analysis of the data is not inconsistent with such a possibility.

I should note that the measurements done by the Seven Samurai were very difficult to perform consistently, and even harder to interpret directly. If some galaxies were missed in the sample, and if these systematically had velocities in other directions, it could reduce the effect. Also, since they were measuring velocities and not mass densities directly, they require some theoretical input in order to translate the measurements into assumptions about the mass distribution in the universe.

In any case, these findings about large-scale drift, even with their new interpretation in terms of an infall into some large mass overdensity, are astonishing. It is easier, although by no means a snap, to accommodate infall than it is to accommodate large-scale coherent random drift in existing cosmological models. But the magnitude of the infall is surprising. The probability of finding a mass overdensity on such a large scale in the cold dark matter structure-formation simulations is small. Of course it is very dangerous to do statistics on a sample of one. This has been pointedly demonstrated in the last few years in the field of particle physics, where exotic and baffling events have been reported with alarming regularity, only to end up stemming from rare and improbable background sources upon reexamination.

HALOS: TO BE OR NOT TO BE?

This whole business began with measurements of galactic dark matter halos, and perhaps fittingly it could end with them.

I spoke in chapter 2 of two independent results that challenge the idea that galactic dark matter halos extend out with smoothly diminishing densities so that mass increases with distance well out into the intergalactic medium. These results suggest instead that halos might fall off to zero outside some distance, say five to ten times the visible size of galaxies.

But if this is the case, some extra physical process other than gravity should be responsible for such a cutoff. Visible matter accumulates quickly in the cores of galaxies rather than in the outer regions because baryonic matter can dissipate or lose energy by emitting light. In this process, stars and gas can fall deeper inside the potential well of galaxies. But any material, such as cold dark matter, made up of weakly interacting stuff should not be able to dissipate its energy nongravitationally. Accordingly, it should form halos that extend out smoothly without a sharp cutoff. And here is the apparent paradox. If halos have a fixed size, then something is amiss with the standard cold dark matter hypothesis.

The consensus of observers is that the early measurements done by Tyson on the lensing phenomenon of distant galaxies by foreground galaxies which suggests finite halos, while very ambitious and creative, were very difficult to interpret. More recent results find extensive lensing by foreground galaxies, but the fact that galaxies lie in extend clusters makes it difficult to untangle the effects of single galaxies. The other analysis, by Tremaine and collaborators, of the motion of satellite galaxies around our own galaxy, is also perplexing. Moreover, it involved just one system, our own Milky Way. Perhaps some violent encounter in the past with another system caused our galaxy to shed its outer halo. Until firmer data demonstrate this cutoff of halos to be generic—and the data on rotation curves suggest that it is probably not—we should not jump to any conclusions that suggest halos do not have the form predicted in cold dark matter models.

DETAILS, DETAILS, DETAILS

When all is said and done, the new "standard" CDM model—involving 30–50 percent of the closure density of the universe in cold dark matter—does a surprisingly good job of explaining the general features of the growing body of data on the nature of large scale structure.

Still, it is not perfect. Even if supplemented by vacuum energy or by a small tilt in the spectrum of primordial fluctuations, perfect agreement with observations has not been achieved. Perhaps this is telling us that there is some new physics we are missing in our simulations. Or perhaps it is a reminder of a maxim in cosmology: "No theory should agree with all of the data on cosmology, because some of the data is *wrong!*"

The exciting aspect of these potential small quantitative discrepancies is that the observational situation is improving with time. The launch of new X-ray satellites, the Next Generation Space Telescope, and the next generation of CMBR satellites, along with new ground-based computerized redshift surveys of millions of galaxies, will provide a hitherto unimagined wealth of data on the nature of large scale structure.

This chapter, perhaps more than any other, has underscored the volatility of our picture of the large scale structure of the universe. Nevertheless, cosmology is now a fully empirical science. And cold dark matter has survived a host of challenges over the past two decades. While a flat matter-dominated universe is now clearly ruled out, a universe whose matter content is dominated by cold dark matter, at between 30–50 percent of the closure density—a density perhaps ten times that of baryonic matter–is remarkably consistent with the general features of data ranging from observations of galaxy motions to observations of the entire visible universe. Moreover, it is a model that can be falsified, and one that will be subject to the tests of increasingly precise and comprehensive observations.

Finally, the initial conditions that are assumed in the model are well motivated by our ideas of the physics of the early universe. Even more remarkable is the fact that the central ingredients of what is now the standard cold dark matter picture—the inferred existence, abundance, and distribution of dark matter; the general notion that the universe is flat and perhaps dominated by a cosmological constant; and the calculations of primordial fluctuations during a possible period of inflation in the early universe—these are all relatively recent developments. Until the early 1990s, what is now the standard cosmological model had no firm theoretical or empirical basis; even the most ambitious cosmologists could not have guessed the progress that has been made. We now have a model, based on the fusion of ideas from elementary particle physics and astrophysics, that *might* be correct. More important, we have a model that we can *test*.

It is too early to judge the outcome of the confrontation between the model and our observations. The points of contention involve areas at the forefront of research, where data are not yet firmly established. Nevertheless, work is proceeding at a rapid pace. At this point, however, many astrophysicists and particle physicists find it extraordinarily encouraging that a single simple model with definite predictions—a universe with cold dark matter more widely distributed than visible matter, and structure resulting from a scale-free spectrum of primordial fluctuations predicted by the theory of inflation—fits so closely the universe we actually see about us. The aroma of a grand synthesis floats temptingly in the air.

Future astrophysical measurements can test and even ultimately firmly establish this scenario, yet only one direct observation would be needed firmly to establish it immediately: the discovery of the nature and identity of the dark matter itself. With such a direct discovery, the whole picture would jell immediately. On the other hand, if all candidates are disqualified by experiment, we may be forced to review our present cosmological prejudices. Of course, it is easy to make a grand synthesis when we are allowed to invoke anything we want to explain our observations. While I have made every attempt here to judge in some depth the viability of the astrophysical aspects of the cold dark matter scenario, I have not yet touched on perhaps the most fundamental question. Just how realistic, from a microscopic viewpoint, is the very idea of cold dark matter, and how can we test this idea?

The answer to this question is stimulating in itself—that is why I have chosen to write about it. While all of these astrophysical findings about dark matter were being amassed, the developments in particle physics in the 1960s, 1970s, and 1980s have been independently leading to several bold proposals about the fundamental structure of matter on very small scales. In the process, a number of natural candidates for cold dark matter have arisen, along with mechanisms by which they could become abundant in the universe today. It turns out that each possibility leads to a potentially observable direct signature. The final sections of this book are devoted to a discussion of the candidates for dark matter and the experimental search that is now under way to discover the nature of the dominant matter in the universe today.

CHAPTER NINE

COLD, COLDER, COLDEST: THE NEW IMPROVED STANDARD MODEL

Before proceeding to discuss the candidates for dark matter, dark energy, and the exciting search going on today in terrestrial laboratories, it is only appropriate to end this penultimate section of the book—one that has provided the observational motivation for our current picture of the universe on large scales—with a short summary of the observational and theoretical issues associated with perhaps the biggest surprise of all: that even dark matter itself may still be a sideshow as far as the ultimate fate of the universe is concerned.

In one sense, it is very surprising that the possible discovery of an unknown energy in empty space, which may in turn dominate the current expansion of the universe, has not dramatically altered the story of dark matter. Luckily for me as I was rewriting this book, the possible existence of a such a uniform background energy only marginally affects the many cosmological and astrophysical arguments associated with the existence, distribution, abundance, and detection of dark matter. It is true that the abundance of cold dark matter in the universe seems to be perhaps 1/3–1/5 of what we had earlier expected in order for us to live in a flat universe today, but aside from that reduction, most of the other arguments, including the claim that dark matter dominates the gravitational dynamics on the scale from galaxies to clusters, remain unchanged.

Nevertheless, perhaps nothing in the history of physics resembles more the quintessence of Aristotle than the possibility that empty space may contain the seeds of our own ultimate destiny. Early on in this book, I argued that the suggestion that most of the mass in the universe might be dark could be termed as the *ultimate* Copernican revolution. This was an understatement. We have now come, at the end of this millennium, to a new and even more surprising paradigm. The dominant "stuff" of the universe may not even be matter at all, no matter *how* exotic! The new "standard model" of cosmology, against which we test our observations and experiments, now includes three components: normal matter, comprising perhaps 5–10 percent of the energy density of a flat universe; dark matter, made from something other than baryons and comprising 30–50 percent of the energy density of a flat universe; and vacuum energy, comprising 50–70 percent of the energy density of a flat universe.

On the other hand, it is worth stepping back and reviewing how truly preposterous is the notion that empty space provides the dominant energy driving the overall dynamics of the universe. Remember first that if empty space has energy, fundamental arguments from particle physics suggest that this energy should be immense, more than 120 orders of magnitude larger than that associated with the energy of normal matter today. The question then becomes, how can one manage to find a mechanism to reduce the energy of empty space by 120 orders of magnitude and not make it precisely zero? To date, no good theoretical idea exists that gives even the slightest inkling of how this might be possible.

There is a second bit of craziness, however. If the energy density of truly empty space has remained constant, as it must in standard General Relativity, then the fact that this energy is today comparable with (if slightly larger than) the energy density associated with matter in the universe is also completely inexplicable! After all, the energy density of matter and radiation gets diluted as the universe expands. If one starts out with a certain total abundance of matter, for example, then as the universe expands, the number of particles per unit of volume will decrease. In this way, the density of matter on average decreases with the expansion of the universe, and thus the energy per unit of volume associated with matter decreases. A similar argument applies to radiation. However, if the energy density in empty space has remained constant over time (after all, in empty space there is nothing to dilute as the universe expands), then *this is the first time in the history of the universe that the energy density in empty space is comparable to that in matter and radiation!*

From every possible theoretical vantage point, then, a cosmic vacuum energy at the level required to result in a flat universe today seems completely ludicrous! If this is in fact true, then it is extraordinary! And extraordinary claims require extraordinary evidence, so it is worth reviewing precisely why we believe this absurd possibility. I will list the issues that have led to this belief in historical order, first focusing on the factors that led Michael Turner and me, as well as several other groups, to make this claim, and then focusing on the discoveries since that have reinforced our view.

(1) The Age Problem. I alluded to this puzzle very early in chapter 3. Over the course of the past hundred years in cosmology, the nagging problem that the universe appears to be younger than the age of objects within it has continued to crop up. In the last century, for example, the British scientist Lord Kelvin and the German scientist Hermann Helmholtz made a famous estimate of the time it would take to cool the sun (without knowing that it was powered by nuclear reactions) as less than 100 million years. This suggested that the sun was less than 100 million years old, but even at that time it was clear from geological records that the earth was far older than this.

Of course, this discrepancy was resolved once it was recognized that the sun has an energy source in addition to the gravitational energy released as it slowly collapses inward, namely nuclear energy. Following on the recognition by Hans Bethe in 1939 that nuclear fusion of hydrogen into helium could power the sun for billions of years, detailed modeling now suggests that the sun is about 4.5 billion years old. In approximately 5 billion years, it will exhaust its hydrogen fuel. As it does so, it will slowly grow to become a red giant, engulfing the earth. We can observe many stars that have already undergone this fate. The rate at which a star consumes its hydrogen fuel depends upon its mass. More massive stars burn their fuel far more quickly. If one can observe a group of stars of differing masses that formed at the same time, and determine which have already exhausted their hydrogen, one can estimate the system's age.

Such systems are called globular clusters—compact groups containing thousands of stars. Some, located in the outskirts of our galaxy, are believed to have formed before the bulk of the galaxy collapsed into a disk and are thought to be among the galaxy's oldest objects. As I earlier mentioned, estimates of their age ranged between 15–20 billion years. This can be compared with an independent age estimate of the universe, based on the measured Hubble expansion. In the absence of any gravitational attraction, the universal expansion would continue with a constant speed. Then by measuring galaxies' recession velocities, one could determine how long it took them to arrive at their present positions, assuming they all started out at the same place: Simply divide their distance traveled by the speed at which they have been traveling. This ratio is nothing other than the inverse of the Hubble constant—the rate of expansion of the universe. The faster the expansion rate, the younger the universe. A Hubble constant of 100 kilometers/second/mega-parsec would imply a present age of about 9.8 billion years.

Of course the gravitational attraction of matter slows the expansion. Objects recede at a faster rate early on, taking less time to get to their present distances than they would if their speed had been constant. A flat matter-dominated universe with H = 100 kilometers/sec/mega-parsec today would be only about 6.5 billion years old.

Indeed, Hubble's first 1929 estimate of the expansion rate (about 500 kilometers/sec/mega-parsec) suggested the universe was less than 2 billion years old, again making it less than the known age of the earth. While the estimates of the expansion rate have changed by almost an order of magnitude since Hubble, this tension between the estimated age of the universe and the age of objects within it has persisted.

In the past decade, with the launch of the Hubble Space Telescope and other refined observational techniques, measurements of the Hubble constant are finally beginning to converge. Recently, Wendy Freedman and colleagues at the Carnegie Observatory compiled an estimate of approximately 70 kilometers/sec/mega-parsec. This is based on determining the distance to distant galaxies by using the luminosity of a certain type of variable star in these galaxies. The uncertainties in the estimate could, with reasonable confidence, make *h* as small as 63 or as large as 80. What is particularly exciting is that this is now in reasonable overlap with a competing effort to determine the distance to distant galaxies by using the brightness profile of exploding stars—supernovae—in these galaxies. Assuming *h* as small as 63, one finds an upper limit on the age of a flat universe of about 10.5 billion years. Is this *old* enough? Not if the globular-cluster age estimates of the 1980s, of about 15–20 billion years, held up.

To determine if this age problem was real or an artifact of stellar modeling, in 1995 I assembled a group of colleagues, Brian Chaboyer, then at the Canadian Institute of Theoretical Astrophysics, Pierre Demarque at Yale University, and Peter Kernan at Case Western Reserve University to assess the uncertainties in stellar evolution estimates of globular cluster ages. We modeled the evolution of over 3 million different stars, in each of which important parameters including nuclear reaction rates, diffusion, etc., were chosen from a range that we felt spanned the existing uncertainties. We then compared our model stars to the distributions in observed globular clusters to estimate the allowed range of ages.

The major uncertainty turned out to arise from comparing models to observation. In order to model a star, you must have a good estimate of its brightness. However, unless you have a good estimate of its distance, you don't know its brightness. The estimated uncertainty in the distance to globular clusters implied that the oldest could be as young as 12.5 billion years old—still too old to exist in a flat matter-dominated universe with a Hubble constant greater than 60.

However, results in 1997 from the Hipparcos Satellite launched by the European Space Agency to measure the distance to over 100,000 nearby stars using parallax (how much nearby objects appear to shift relative to far away objects when one shifts the line of sight) suggested our estimated uncertainty in the distance to globular clusters was too small. We recently concluded that globular clusters could be as young as 10 billion years old. Allowing time for our galaxy to begin to form, one derives a *lower* limit of about 10.5–11 billion years on the age of the universe, with a best estimate of about 12.5 billion years.

Thus, while our results reduced the apparent discrepancy, they did not completely eliminate it. Both the Hubble constant measurements and the globular cluster estimates would have to be pushed to their extreme limits even to get marginal agreement. However, everything changes if we consider the possibility that the universe is dominated by a vacuum energy. Recall that in this case, the expansion rate of the universe can *increase* with time, due to the universal repulsive force induced by the vacuum energy. In this case, if the universe were earlier expanding at a *slower* rate than it is today, it would have taken longer for galaxies to reach their present distances than one would imagine otherwise. When the estimates are performed, there is no problem reconciling a flat, vacuum energy–dominated universe with an age of 12–13 billion years.

(2) The Baryon Problem. I have described how recent observations of the absorption of light from distant quasars have allowed us to determine, with good accuracy, the primordial abundance of deuterium. From that, we now know that the actual abundance of baryons in the universe today is limited to be less than about 8 percent of the density of matter needed to result in a closure density.

Elsewhere, I have also pointed out that measuring X-ray emission from large clusters of galaxies gives us a handle on the total mass of these systems, as well as the total mass of gas in these clusters. Since most baryons exist in the form of hot gas between galaxies rather than in the visible stars in galaxies, if we measure this total gas mass and the total mass of the cluster, and take the ratio of these two quantities, we get an estimate of the fraction of the mass in baryonic matter.

Now here is the problem. Clusters are the largest bound objects in the universe. They should therefore provide a good sample of all of the gravitating mass in the universe. In 1993, Simon White and colleagues demonstrated that the fraction of baryonic matter to the total mass in rich clusters was large, perhaps 20 percent. If clusters sample all the matter in the universe, and if the universe is flat, then this suggests that the fraction of the closure density is 20 percent. But this disagrees with the Big Bang Nucleosynthesis estimate above!

Since this early result, Gus Evrard at the University of Michigan and colleagues have refined the estimates based on X-ray observations and numerical simulations of galaxy clusters, and the problem has, if anything, gotten worse.

There are two solutions. Perhaps the universe is open, and the total matter in clusters is far less than the closure density, and thus the ratio of baryons to total matter in clusters is larger than the ratio of the baryon density to the closure density. However, an open universe is ugly! Even before there were any suggestions from the CMBR that the universe might actually be flat, it was pointed out by several groups, including our own, that a uniform energy density, like a vacuum energy, could resolve this problem. Such an energy density would be spread out between clusters, would not affect the internal dynamics of the gas in clusters, and thus would not be probed via X-ray measurements. Thus the total mass in galaxy clusters could fall far short of the closure density just as in an open universe, but in this case, the remaining energy density needed to make the universe flat would exist in empty space, hidden from such observations.

(3) The third problem is one that I have already alluded to in the last chapter. The clustering of matter as a function of scale provides a powerful probe of the precise time that the universe became matter dominated. This time, in turn, depends on how much matter there is today. There is definitive evidence now, based on galaxy clustering data from a variety of sources, that the density of matter lies in the range 0.3–0.5. Once again, in order to rescue a flat universe, one can add vacuum energy to the mix. Since the vacuum-energy density remains constant in the early universe, the density of matter would have been far larger than it is today while the vacuum-energy density would have remained the same. Thus, the time that the universe became matter dominated would not have been affected by what was then a miniscule addition of vacuum energy. One need not change the inference that matter contributes at best 50 percent of the closure density today in order to have a flat universe today.

This was the situation in 1995, when we, as theorists, suggested that the best-fit flat universe involved a nonzero vacuum energy. However, as I have described in the past chapters, there is now much stronger empirical evidence that this is the case, particularly points 4 and 5, below.

(4) COBE and the CMBR: As I have discussed, the currently observed peak in the "power spectrum" of CMBR anisotropies, while still tentative, agrees precisely with the angular scale predicted in a flat universe. If the universe is indeed flat, then the X-ray data above, combined with the data on the evolution of galaxy clusters also described in the last chapter, makes some form of energy beyond that stored in matter, including dark matter, essential.

(5) Finally, the recent claimed observation, based on type 1a supernovae, that the Hubble expansion has been *accelerating* gives direct confirmation of the picture espoused here. In fact, it is almost eerie that the value for the vacuum energy inferred from this measurement is precisely that predicted based on resolving the problems 1 to 3, above, with the addition of a vacuum energy (see figure 9.1). Of course, while on one hand this is heartening, one should recall the caveat I mentioned earlier: If a cosmological model agrees with *all* the data, one should begin to worry, because some of the data *might* be wrong!

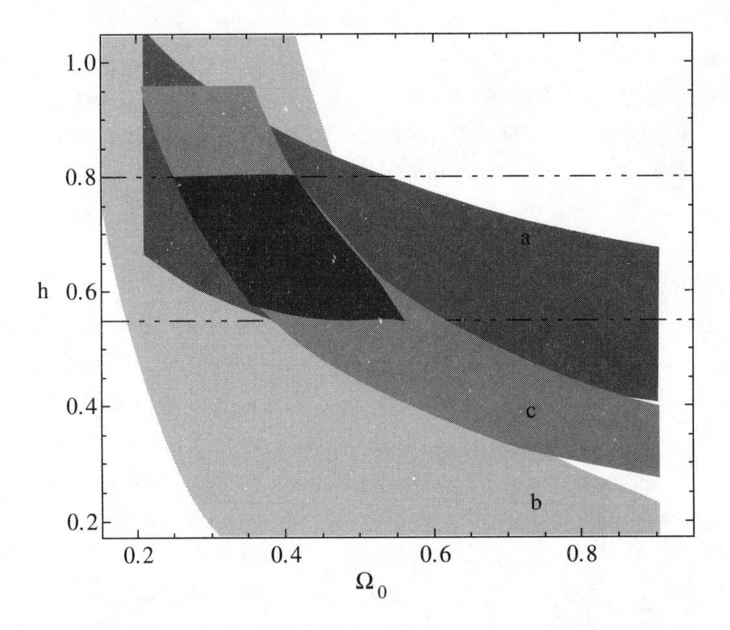

FIGURE 9.1

Shown here are constraints, assuming a flat universe, on the Hubble parameter, h, and the fraction of the density of the universe in matter, Ω_0, including dark matter. Shaded areas show allowed regions; (a) based on the age ofthe universe, (b) based on the baryon density of the universe, and (c)based on large scale structure observations. The dark region is that which is allowed by all constraints. Note that the allowed region of Ω_0 coincides almost exactly with that suggested by recent supernova measurements of the acceleration of the universe.

Indeed, I believe it is presently premature to conclude, as *Science* magazine did in 1998 when it called the discovery of a cosmological constant the "breakthrough of the year," that in fact the supernova data is definitive or unimpeachable. Attempts to measure acceleration and deceleration of the expansion rate have been repeatedly plagued by the fact that if systems evolve with time, one cannot assume that the parameters derived from measuring these systems today can be directly applied to these systems at early times.

However, the other arguments given above, each pointing to the possible existence of a nonzero vacuum energy, are based on the three independent fundamental observables in cosmology: age, matter content, and large scale structure. Taken alone, each might not be compelling, but taken together, I find them highly suggestive at the very least. If we are to believe that we live in a flat universe today, there is no other option than to endow empty space with energy. While from a fundamental theoretical perspective there is nothing crazier one can imagine, the beauty of cosmology at the turn of the twentieth century is that it has become an empirical science. Whether we like it or not, nature is as nature is. The challenge to us is to figure out why.

We are thus apparently left with two great, probably intertwined, puzzles in cosmology to be solved in the next millennium: What is the nature of dark matter, and what is the possible source of vacuum energy? These are the issues I shall turn to in the remainder of this book. It will probably be easier to solve the first than the second, and I shall concentrate largely on the search for dark matter. But as a theorist, I fully expect that the resolution of the mystery of dark matter will provide vital clues for solving the mystery of the darker energy of empty space.

PART FIVE

The Candidates

CHAPTER TEN

All Roads Lead to Dark Matter

. . . and this time it vanished quite slowly, beginning with the end of the tail, and ending with the grin, which remained some time after the rest of it had gone.

—Lewis Carroll, *Alice in Wonderland*

The 1970s witnessed an extraordinary transformation in our picture of the fundamental structure of matter. At the beginning of the decade, particle physics as a field was "data rich" and "theory poor." In spite of mountains of data made possible through the development of mammoth particle accelerators and huge particle detectors that have become the hallmark of the field today, of the three nongravitational forces in nature, only one—electromagnetism—had been satisfactorily described in terms of a fundamental theory. The weak interaction and the strong interaction were both treated using purely phenomenological models. Some physicists were even beginning to doubt that the whole formalism of quantum field theory would allow a successful description of the microphysical world.

By the end of the 1970s, the situation had completely reversed. All three nongravitational forces had been explained in terms of fundamental "gauge" theories. The many theoretical ideas born of despair in the 1960s had borne fruit beyond the dreams of many of those involved in their development. Rarely in the history of physics has so much fallen into place so quickly. Since about 1978, only one reproducible experiment has produced a result that signaled the need for any new physics beyond what has since become known as the "standard model," and this result involved solely the weakly coupled neutrino sector.

Particle physics is not free of theoretical enigmas, however. The standard model is by itself logically incomplete. Within it a number of crucial questions about why the world is the way it is cannot be answered. For example, while spontaneous symmetry breaking as an idea had been phenomenally successful in leading to the unification of the weak interaction and the electromagnetic interaction, the precise mechanism of symmetry breaking has not yet been confirmed. The simplest model predicts the existence of one or several new fundamental particles with spin value equal to 0, called "scalar" particles, although none have yet been observed. Moreover, any theory of a fundamental scalar particle has a serious shortcoming that is associated with a severe "naturalness" problem. The scale of weak interaction symmetry breaking is 17 orders of magnitude below the scale where quantum gravity becomes important. Furthermore, if Grand Unified Theories are correct, new physics concepts are likely to be necessary at a scale some 13 orders of magnitude larger than the weak symmetry-breaking scale. The existence of such vastly differing scales in physics is innocuous unless fundamental scalar particles do exist. In this case, the effects of physics at large energy scales can be "communicated" to the interactions governing such scalar particles at lower energy scales. The result is that the energy scale of weak symmetry breaking should then naturally be near the GUT scale or the Planck scale. The only way to stop this from happening in the standard model is to "fine-tune" parameters to many decimal places, so the observed hierarchy of scales that vary by many orders of magnitude can be maintained. Without some further explanation of why this should occur in nature, the situation is clearly unsatisfactory.

Other problems with the standard model beg for new physics. One glaring flaw is that the model does not possess a fundamental symmetry (discussed shortly), which is apparently strongly manifest in the real world. There have been some proposals for resolving this problem, but we still have no idea what the right answer will be.

Finally, one of the gravest omissions in the standard model relates to Rabi's original question about the muon, "Who ordered that?" We have discovered three different copies of the electron, electron neutrino, and the light quarks. We have no idea why there are three such "families" of elementary particles; we are not sure whether there are more, and why they vary in mass. Until we understand why the spectrum of elementary particles is as it is—why the proton is lighter than the neutron by just enough so that the neutron can decay, so that the sun can operate, so that our planet could evolve—we are a long way from unraveling all the secrets of nature on its smallest scales.

Unfortunately, little experimental direction has guided theorists in responding to these inadequacies of the standard model. It may be that the resolution of these problems will be forever beyond the range accessible in terrestrial experiments. Perhaps the solution merely requires pure thought—Theories of Everything, or the like—although I bet not. Indeed, at least one of the problems discussed here demands that some new discovery occur just around the corner. Whatever breaks the weak gauge symmetry—be it a new scalar particle or something more fancy—it must produce observable effects at a scale that the next large collider in Geneva, to be operating in the first decade of the next century, should be able to probe.

The absence of direct experimental constraints, however, has not discouraged theorists from proposing a variety of solutions to the fundamental logical problems I have described. Indeed, they have been free to roam largely unfettered over vast plains of ideas, retaining those that seem especially elegant or powerful. If the history of physics is any guide, some subset of the most elegant of these proposals might actually be true. From the perspective of this book, however, what makes these proposals especially exciting is that almost all of these extensions of the standard model result in candidates for cold dark matter.

Thus, just at a time when cosmology has been crying out for a new kind of matter in the universe, particle theory has begun supplying new candidates. It would be naive to suggest that this is pure coincidence. But it does illustrate the noteworthy parallel developments in the fields of particle physics and astrophysics since the 1960s. In any case, the prime particle physics candidates for dark matter are "honest" ones—at least as honest as Anaximander's "indefinite" or Aristotle's "quintessence." Each candidate has been proposed to solve an outstanding problem in particle theory, independent of any possible relevance to cosmology. And like the Cheshire cat, each may leave behind some subtle but detectable signature of its existence in the universe today.

I want to spend this chapter introducing three such "honest" candidates for cold dark matter. Each is of intrinsic interest primarily because of its implications for our understanding of the fundamental structure of matter, independent of whether it also forms the dark matter. The properties of the candidates can vary considerably, however. They may have silly names, such as axions and WIMPs, but they are serious: each potentially could solve a pressing problem in particle physics. Of course, by the time this book has become dog-eared, one of three things may have happened: any one of these candidates may have been seen, some of them may be ruled out on other grounds, or we may still not know whether any of them exists. In view of these possibilities, I will next discuss the general mechanisms by which it has been proposed that elementary particles might become dark matter. Conveniently, I will claim that these can be categorized in three classes, one for each of the candidates I describe here. In this way, I am safe: even though the specific candidates I list might not survive the test of experiment, one of the mechanisms might. Alternatively, independent of their possible relevance to the dark matter question, any one or all of the candidates might actually exist, and thereby resolve questions essential to our understanding of the elementary structure of matter.

My choice of candidates to describe here is somewhat subjective. I am guided by four factors: (1) I believe that these three contenders are on the soundest theoretical footing; (2) they span the three generic production mechanisms I shall discuss later; (3) the candidates differ by some 30 orders of magnitude in mass and interaction strength, and thus give a fairly representative idea of how diverse are the possibilities; and (4) experiments are under way to search for each of them.

Table 10.1 outlines these various elementary particle candidates for dark matter. I include the old-fashioned light neutrino for comparison purposes. Along with the name of the particle, I indicate what mass range would be expected to result in an approximately flat universe ($\Omega \approx 1$) today. I use eV units throughout for mass. For orientation purposes, recall that the mass of the electron in these units is about 500,000 eV, while the mass of the proton is about 1 billion or 10^9 eV. The table also shows the mass range expected for these particles based on existing theoretical models, and where appropriate I show the mass range allowed by existing experimental constraints. Finally, I note the motivation, from particle physics, for postulating the existence of such particles. *The specifics are not important;* they will become clearer as I discuss each candidate in detail.

It is worth pointing out several general features first. Note that the candidates differ widely in possible mass, from *axions*—some billion times less massive than electrons—to *magnetic monopoles* possibly billions of times more heavy than a proton. Of course, this does not exhaust the range of yet more exotic alternatives that have been proposed, but it gives some idea of the variety. Next, note that with the exception of supersymmetric (SUSY) WIMPs (weakly interacting massive particles), the mass range predicted independently by theory does not necessarily single out that region relevant for dark matter or a flat universe. This is not a damning blow, however. Since we have no idea why the particles we see have the masses they do, we know even less about the masses of particles we have not seen. It would not surprise us and, more important, it would not be unnatural should any of these particles turn out to have masses in the requisite range of dark matter. All this means is that at this point in time, if you locked a particle

Particle	Mass Range for $\Omega \approx 1$	Predicted Mass Range (allowed masses)	Motivation
Light neutrino	≈ 30 eV	1,000,000 eV–10^{-8} eV (same)	they exist
Axion	$\approx 10^{-5}$ eV	100,000 eV–10^{-8} eV (10^{-3} eV–10^{-6} eV)	strong CP problem‡
WIMP*	$\approx 10^{10}$–10^{12} eV	10^{10}–10^{12} eV†	mass scales–hierarchy problem
Magnetic monopoles	$\approx$ (>10^{28} eV)	$\approx 10^{24}$–10^{31} eV (> about 10^{12} eV)	(a) must exist if GUTs are true; (b) charge quantization

*WIMPs: weakly interacting massive particles.
†Existing direct (that is, nondark matter–related) experimental constraints are model dependent.
‡CP: charge conjugation and parity symmetries, discussed later in this chapter.

TABLE 10.1 "STANDARD" EXOTIC DARK MATTER CANDIDATES

theorist in a room and asked her to invent any of the theoretical models that predict these particles, she would not, with the possible exception of certain WIMPs, necessarily choose values in the ranges that would make them dark matter contenders. These points stated, let me get on with the story. . . .

AXIONS

I must admit my partiality to axions. They are a very pretty theoretical construct, they have some remarkable properties, and finally, I like their name. Axions originated through a consideration of a familiar notion: symmetry. As I have described, we have come to recognize that the symmetries of nature determine everything from the form of the equations governing fundamental interactions to the shapes of such crystalline materials as diamonds or snowflakes. Understanding the fundamental laws of nature has become synonymous with understanding the symmetries of nature.

The breakthroughs in particle physics since the 1960s have been associated with the symmetry called gauge symmetry. As I described briefly in chapter 6, in its simplest form this symmetry is merely a reflection of the conservation of electric charge in nature—the so-called gauge symmetry of electromagnetism. In its more complicated forms, culminating in the Yang-Mills type of gauge symmetry, this single principle has led, by a roundabout series of steps, to the development of theories describing all of the nongravitational forces we know of: the strong, weak, and electromagnetic interactions. The theory of the strong interaction, called *quantum chromodynamics* (QCD), describes the interactions of fundamental quarks that combine to make up protons and neutrons. The term "chromodynamics" refers to the fact that the gauge symmetry in this case is associated with a new quantity that quarks have in addition to electric charge. In the absence of a better name, physicists called this new property "color." It has absolutely nothing to do with real color; it is just a property of these particles which characterizes their strong interactions with each other and is conserved in these interactions, just as electric charge is conserved in electromagnetic interactions. It could have just as easily been called "quarkiness" or any other term. The quantum theory of interactions related to "color," then, in analogy to quantum electrodynamics, is called quantum chromodynamics.

Comprehending the interactions of quarks via QCD was a great triumph for particle physics. Quark interactions at ordinary energies had seemed so powerful and so complicated that they defied explanation. In particular, the fact that the fundamental particles of the theory—quarks and the QCD analogues of photons, called *gluons*—are apparently never observed directly, but instead are strongly bound together inside observable particles such as protons, made it seem virtually impossible to probe their identity through direct experiments. However, a remarkable discovery was made in 1973 by David Gross, Frank Wilczek, and David Politzer. They found that QCD was "asymptotically free"—that is, it gets weaker at short distances—allowing for the possibility of performing direct tests of the theory. If probes such as electrons or neutrinos, which are not subject to the strong interaction, can scatter at very high energy and very close range when they collide with individual quarks inside protons, the interactions of the quarks with each other should be weak enough so that simple estimates could be used to make reliable predictions. To date, the results of all such high-energy scattering experiments continue to be in agreement with the predictions of QCD. The experimental discovery of this fact was awarded a Nobel Prize several years ago. I am personally hoping that the theoretical breakthroughs will also be similarly recognized sometime soon.

Now, if QCD is the proper theory of the strong interactions of quarks, then the symmetries of the theory should be reflected in the symmetries of observed particles and their interactions—unless some of these symmetries are "spontaneously broken." Here again QCD is triumphant. The entire spectrum of all particles that are built of quarks, called *hadrons,* including protons, neutrons, and their more exotic cousins, can be explained in terms of the symmetries of QCD.

Or so it seems. Although the particles we have observed fit very well with the symmetries of QCD, it transpires that there are other particle states that should occur based on these symmetries but that apparently are not observed. This caused serious concern until the very deep and thoughtful work of the Dutch physicist Gerard 't Hooft resolved the difficulty. Like his predecessor Huygens, 't Hooft's name is impossible for an English-speaking person to pronounce properly without hearing it first. (I remember the spelling by recalling a remark made by Harvard physicist Sidney Colemen. He stated that if 't Hooft had monogrammed cufflinks, they need only have apostrophies on them!) Gerard 't Hooft has had his hand in almost all of the developments leading to the use of gauge theories to explain the strong and weak interactions. He was the first to show that Yang-Mills theories, when spontaneously broken, make sense as quantum theories. It was this result that first caused the models of Glashow, Weinberg, and Salam to be taken seriously. Gerard 't Hooft also came very close to discovering "asymptotic freedom" in QCD, and indeed he may have independently discovered the mathematical relations leading to this result, although he did not publicize or publish them. He is a brilliant, somewhat shy individual who plays piano for relaxation and derives his deep insights in a more individual way than almost any other physicist I have known. And I am pleased to say that he was awarded the Nobel Prize in 1999 in recognition of his work on guage theories.

In any case, 't Hooft made an important discovery in about 1975 concerning QCD. He was able to show that the quantum theory did not possess all of the symmetries of the classical theory due to the presence, in the "vacuum" state, of complicated configurations of gluons, the particles that hold quarks together. This complex result was one of the most significant applications of mathematical techniques beyond perturbation theory to our understanding of particle theory. Because of the intricate nature of the vacuum state, an extra term must be introduced in the equations governing the interactions of quark and gluons in the theory to achieve consistency; it reduces the symmetry expected of the physical ground state in the theory so that the observed particle spectrum then agrees in form exactly with the predicted spectrum.

So far so good. Unfortunately, this wonderful, and correct, solution of a long-standing problem in particle theory introduced yet another, perhaps more troubling, problem that has yet to be resolved. The extra term present in the equations of QCD may have removed unwanted symmetries, but it unfortunately also removed a symmetry that many had believed was a necessary part of the theory.

Since the 1950s we have become accustomed to rude surprises regarding apparent "discrete" symmetries of the world. By "discrete" I mean that they involve transformations that are not continuous. For example, reflection in a mirror involves a discrete symmetry that interchanges right and left (the left hand of your mirror image is really your right hand). It is an all-or-nothing affair. But the symmetry of physical theory with respect to translations in space—namely, that experiments give the same results no matter where they are performed—is continuous. We can make an arbitrarily small translation to test this symmetry.

One traditional discrete symmetry is an invariance of electromagnetism under interchange of "positive" and "negative" charges. In any physical situation when we reverse the "sign" of all charges, that is, change positives to negatives and vice versa, exactly the same behavior will result. A negative test charge will be repelled from a negatively charged object, and similarly, a positive test charge of the same magnitude will be repelled by exactly the same force from an object with positive charge equal to the negative charge in the previous case, for example.

This is simply a reflection of an obvious fact. The labels "positive" and "negative" are arbitrary conventions that have no objective physical significance. The person who first labeled an electron charge "negative" could have just as easily called it a "positive" charge, and physics would have been the same. It would simply mean that the charges we now call positive, such as the proton, would be called negative. It was Benjamin Franklin who first labeled the flow of current in a certain direction in a wire "positive." This turned out afterward to be opposite to the flow of actual particles—electrons—in the wire. The only way to resolve this was to call the charge on the electron "negative." The flow of negative particles in one direction is then mathematically equivalent to the flow of positive charge in the opposite direction. In any case, stating that "positive" and "negative" have no *objective* significance is equivalent to stating that no experiment we can do in electromagnetism would behave differently if we reversed the signs of all charges within it.

A similar dualism in nature comes to mind. Think of the quantities "left" and right." If you interchanged "left" and "right," how could you tell the original universe and its "mirror image" apart?* Try to imagine how you would explain, by telephone, to some alien being on a different planet how to perform an exper-

---

*Strictly speaking, I am actually imagining here a "three-dimensional" reflection, obtained by reflecting all objects about a single point. Namely, if an object has a position defined by the coordinates x, y, z in three dimensions, imagine reflecting this object to now have coordinates $-x, -y, -z$. Of course, for any single point object, one can always imagine placing a mirror at a specific location so as to effect this reflection. However, a single mirror will not necessarily reflect two different points properly in this way. This more general kind of reflection is called a "parity" reflection, and it is really parity of which I speak here. Left-right symmetry is just a special case of this more general symmetry.

iment that would distinguish between left and right in an *absolute* sense, that is, if you had no reference points. These conventions too appear to be purely *subjective;* in other words, the macroscopic laws of physics seem to be symmetric under left-right interchange.

As sensible as this sounds, however, it is wrong. It was one of the most startling discoveries in physics in the 1950s that the weak interactions *can* distinguish between "right" and "left." This bold possibility was proposed in 1956 by T. D. Lee, and C. N. Yang (of Yang-Mills fame) who pointed out that there really was no direct empirical evidence that left-right symmetry or "parity" is a manifest property of the weak interactions, and if one gave up this idea, it would be possible to explain otherwise inexplicable data. Within a year their hypothesis had been experimentally verified, by experiments on the prototypical weak interaction process—beta decay of a neutron.

The neutron as an elementary particle has a nonzero intrinsic angular momentum, or spin. I remind you that it is a fundamental property of quantum mechanics that pointlike particles, which classically cannot have any spin since there is no physical way to picture an object with no spatial extent as "spinning," can and do have such angular momentum. The intrinsic spin of elementary particles is a purely quantum mechanical notion. The situation is similar to describing the motion of electrons around atoms as "orbits," even though quantum mechanics tells us that the electrons are not really whizzing around the atom the same way the planets orbit around the sun. Again, in this case much of the electron's behavior can be explained as if the electron were indeed physically orbiting around the atom.

A spinning top, like a neutron with spin, distinguishes a certain axis in space, namely, the axis about which it is revolving. Now, along this axis, there are two different directions in which one could choose to label the orientation of the spinning top: that is, whether it is spinning clockwise or counterclockwise. Say it is spinning counterclockwise, as shown in figure 10.1. Then, if you wind your right hand around the direction of rotation, your thumb will point upward. If you wind your left hand around the direction of rotation, your thumb will point downward.

Now, it should be a matter of mere convention which direction I choose to associate with the spinning top. I can call the first convention the "right-handed rule" and the second the "left-handed rule." If the laws of physics are invariant under left-right interchange, then no physical process should distinguish one choice from the other. There should be no way I can unambiguously associate the direction "up" with the spinning neutron by doing any experiment *which does not also* depend on the same choice of conventions.

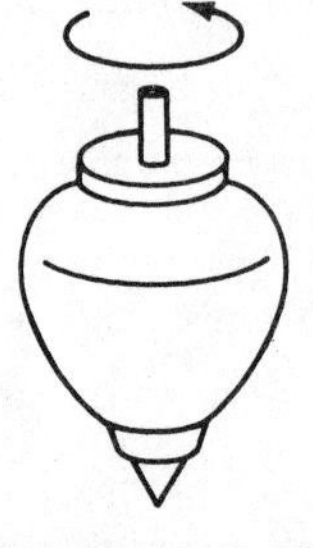

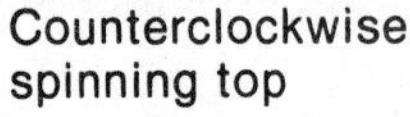

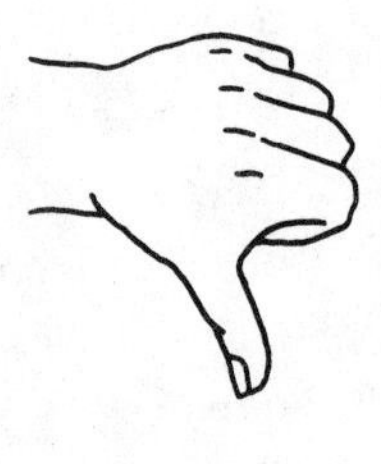

FIGURE 10.1

If a top is spinning in a counterclockwise direction around some axis, one may associate a direction along that axis with the spinning motion of the top. Winding your right hand around in the direction of the spinning motion makes your thumb point upward along the axis about which the top is spinning, while winding your left hand around in the same direction makes your thumb point downward along this axis. Thus, the direction along the spin axis that is associated with the spinning motion depends upon which hand you choose.

Within a year after the proposal by Lee and Yang, an experiment was performed that demonstrated that an unambiguous choice was indeed possible in the case of beta decay. In this case, a nucleus, composed of protons and neutrons, was "polarized" in a magnetic field. This means that the net spin axis of the nucleus was aligned along the magnetic field axis. The nucleus used, cobalt 60, is radioactive: this means that one of the neutrons inside the nucleus is unstable and undergoes beta decay. This neutron decays into a proton that remains in the new nucleus that is formed, and an electron is observed to be emitted (a neutrino is also emitted in the decay, but is not directly observed). When many decay electrons were observed in the polarized sample of cobalt 60, a curious result was obtained. Electrons were not emitted uniformly in all directions. Instead, the electrons were preferentially emitted in a direction *opposite* the spin axis as defined by the "right-handed rule" (see figure 10.2).

Since the direction of motion of the electron is a direct observation that does not depend on our conventions of right- or left-handedness, it should not be correlated physically to something that does. The weak decay of the neutron therefore picks out a certain "handedness" as physically measurable. *It violates left-right symmetry, as predicted by Lee and Yang.*

The discovery that left-right symmetry, or "parity," was not a universal property of the laws of nature, was astounding (so astounding that Lee and Yang won

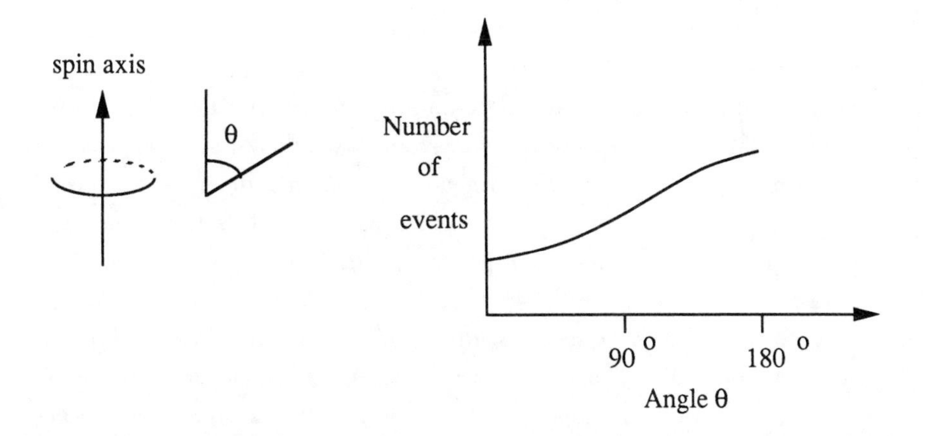

FIGURE 10.2

If the nuclei in cobalt 60 have spins directed primarily upward (defined by the "right-handed rule") by the application of a strong magnetic field, then electrons observed in the beta decays of these nuclei are seen to be emitted primarily in the opposite direction (that is, near an angle of 180° to the direction of the nuclear spins).

the Nobel Prize within a year of their work). Because electromagnetism and gravity are left-right symmetric, all the phenomena that govern our daily lives in an observable way do not distinguish between left and right. To suddenly find that this cherished symmetry was not a fundamental property of nature, when the weak interactions—responsible for the radioactive processes powering the sun—were included, turned all of our conventional physical notions upside down.

After losing this sacred discrete symmetry, physicists were soon to find that others came tumbling down as well. There are two other closely related discrete symmetries in nature besides left-right parity. The first, *charge conjugation symmetry,* involves interchanging particles and antiparticles. One of the precepts of quantum field theory is that every elementary particle has an antiparticle with the same mass and opposite value of all other quantum numbers such as charge. Thus, this symmetry is similar to the charge reversal symmetry described earlier but somewhat more general since it applies even to particles that are not charged (the neutron has a distinct antiparticle called the antineutron with which it would be interchanged under this symmetry, even though both are neutral objects). This is again an apparent symmetry of the world, since if all particles were replaced by their antiparticles, they could form "antiatoms," and "antimolecules," out of which presumably viable living things could be made. Since electromagnetism and gravity apparently do not distinguish between particles and antiparticles, the macroscopic world would appear virtually identical to the way we see it now.

The last of the three discrete "space-time" symmetries is called *time-reversal invariance.* Nowhere do the classical macroscopic laws of physics, namely electromagnetism and gravity, appear to distinguish a direction of time. For example, if we run a film of billiard balls moving on a table backward, the action of each ball appears equally possible in terms of physics. Similarly, if time were reversed, so that all the planets moved in the opposite direction around the sun, this would entail an equally acceptable solution of Newton's law of gravity. Time appears to have a directional "arrow" only when one considers the motions of many objects at the same time, when statistical laws become important. On a fundamental level governing the motion of each individual object, however, reversing the direction of time causes no problems whatsoever.

Now, there is an essential mathematical property that must be true if the world is to be described at a microscopic level by quantum mechanics, and by relativity. One can show rigorously that if one makes a simultaneous exchange of left and right (parity), exchanges particles with antiparticles (charge conjugation), and changes the direction of time (time reversal), physical laws must remain invariant, independent of the detailed form of the interactions between particles. Because such a transformation comes from the successive application of each of these symmetries, parity, charge conjugation, and time reversal, we call it a PCT transformation, after the first letters of the separate transformations. If PCT is not really a valid symmetry of the world, then one could show that probability would not be conserved—specifically, the total probability that any one of all possible mutually exclusive outcomes of an experiment actually occurs would not be equal to unity. It is clearly inconsistent with reality to imagine a world in which, among all the allowed possibilities in nature, there is a finite probability that none of them actually occurs. In other words, "something" must happen, even if that "something" means that things stay the same. Obviously, we regard PCT as a very sacred symmetry whose violation would indicate something very wrong with all our notions of physics.

Keeping in mind this theoretical baggage, then, we can see that the violation of parity (P) by the weak interaction suggests that this interaction must also violate one of the other symmetries; if both transformations occur together, the violations of each symmetry cancel out, so that PCT itself remains a viable symmetry. In fact, this turns out to be the case. The weak interaction also violates charge conjugation (C) invariance. We can see this in the following example. One manifestation of parity breakdown in the weak interaction is that the neutrino angular momentum "spin" axis always turns out to point along the direction of its motion. This picks out one specific "handedness," so we call the neutrino a "left-handed" particle. If we looked at a neutrino in a mirror, it would appear right-handed, since its direction of motion would reverse but the *direction of its spin angular momentum would not*. Since no such particle is observed in the real world, we can distinguish the real world from its mirror image, and this signals the breakdown of parity as a symmetry. Now, the antiparticle of the neutrino, the antineutrino, turns out to be right-handed. This is a signal of C violation, for if we were to switch the identity of neutrinos and antineutrinos we would end up with left-handed antineutrinos, which are not seen in nature. Finally, if we looked at a neutrino in a mirror and at the same time switched the identity of particle and antiparticle, we would see a right-handed antineutrino. Since this is a perfectly acceptable particle seen in the real world, we could not distinguish between the original world and a world in which parity had been reversed and particles had been interchanged with antiparticles. Another way of saying this is that while parity (P) alone is not a symmetry, and charge conjugation (C) alone is not a symmetry, the combination CP appears to remain a symmetry of the weak interactions. This of course implies that time reversal (T) should also be a valid symmetry, since we believe PCT together must always be a symmetry of nature.

So far so good. The world of the weak interaction is kind of screwy, but not too screwy. At least CP and T are apparently valid symmetries. Well, as you may have guessed, even this turns out not to be true. Within a few years of the discovery of parity *noninvariance* of the weak interaction in 1956, a startling experiment in 1964 showed that even the combination CP could be violated. This experiment involved one of the oddest systems in nature, involving a particle called the *kaon* and its antiparticle. These particles are made up of quarks, as are protons and neutrons. But kaons contain a new quark, called the "strange" quark. The name is apt, because the kaon system is unique in nature, so far. It turns out that one can form kaons in two different types: a long-lived type and a short-lived type. The long-lived kaon is observed to decay into three particles called *pions,* while the short-lived kaon decays into two pions. It is harder to decay into three particles than two, since it is more difficult to apportion the energy and momentum of the original particle among three objects than it is among two. This is why the long-lived kaon lives longer. Moreover, based on symmetry arguments it can be understood why the long-lived version is forced to decay into three objects and not two. A configuration of three pions behaves differently under simultaneous charge conjugation and parity transformations than does a configuration of two pions. Thus, if CP were a universal symmetry in the world, then one could understand why the long-lived particle would undergo only the three-particle decay mode by assuming that its internal CP properties allow it to decay only into this configuration. The short-lived kaon, on the other hand, would have different CP properties that allow it to decay into two pions.

Theory and experiment coincided until an experiment led by Val L. Fitch and James Cronin in 1964 produced a preposterous result. Their careful experiments on the long-lived kaon showed that about one in a thousand times it would decay into not three pions but two. Since the two different decay modes have different CP invariance properties, there was no way that the weak interaction that mediated the decays could preserve CP. But this was tantamount to saying that in the kaon system, time-reversal invariance had to be *violated* (in order for PCT to remain a valid symmetry). Understandably, there was a strong reaction to this result; many were tempted to dismiss it as incorrect. It has since been verified in great detail, however; in 1980 Cronin and Fitch were awarded the Nobel Prize for showing that once again, nature still held surprises. To date, the kaon system is the only one in which CP violation is clearly manifest. New experiments on heavier particles, involving heavier, more esoteric quarks, should soon be able to distinguish CP violation in these systems, as theory now predicts. Until then, the kaon system remains the only "laboratory" in which to study this obscure but fundamental property of nature.

We now realize that the weak interaction violates P and CP. I have emphasized that both electromagnetism and gravity, which govern the macroscopic behavior of the world, preserve these symmetries, however. Indeed, this is one reason why their violation by the weak interaction seems so nonintuitive to us. What about the only other force in nature we know of—the strong interaction described by quantum chromodynamics? Since QCD is really a direct generalization of electromagnetism, on a classical level it also retains all the symmetries of its simpler counterpart. Thus, it should preserve P and CP, and the equations describing the theory appear to do just that. But when 't Hooft made his breakthrough explaining why certain particles that were apparently predicted to exist in QCD were *not* observed, he discovered that a new term had to be added to the equations governing the theory. This term turns out to violate CP symmetry, and thus it predicts that CP is not a valid symmetry in the strong interaction.

There is nothing in principle wrong with this. After all, CP is not a valid symmetry in the weak interaction, so why should it be respected by the strong interaction? Unfortunately, this logical statement is not supported by experiment. If CP is not a symmetry of the strong interaction, it should show up somewhere in the properties of particles whose interactions are governed by QCD. Among these particles are the familiar proton and neutron.

Using the PCT invariance of nature—which I remind you is so sacred that its violation would require us to devise a whole new physics where probability was not conserved—we can equate violation of CP invariance to a violation of T invariance. How can we search for such a violation in the properties of objects such as protons and neutrons? Well, in addition to having spin angular momentum, these particles can possess several other properties. In particular, they possess, associated with their spin, a *magnetic orientation*. They act like little magnets with two opposite poles, north and south. Because of this they are said to behave like magnetic "dipoles." It is also possible that they might act like electric "dipoles"; they might behave as if they had an asymmetric charge distribution inside so that one could pick out a direction representing a positively charged "pole" and another with a negatively charged "pole." Now, because the proton and neutron are elementary particles, their structure is only described by "quantum" observables such as mass, charge, and spin. The only such observable that is associated with a direction is the spin angular momentum of these objects. Therefore, if they are to have an electric "dipole" direction, one can show that it too must point along the spin axis, either up or down.

But what happens under a time-reversal transformation? The spin of these particles "flips" in direction. Classically, a counterclockwise rotation would be changed into a clockwise rotation if the direction of time changed. This reasoning holds up for elementary particle spins as well. However, the static charge distribution would not change, so the direction of any electric dipole would not flip. By this argument, we can see that in order to have a permanent electric dipole in these particles, time-reversal invariance *must be violated,* since if the dipole is aligned with the spin initially, it will point in an opposite direction after a time-reversal transformation.

Very sensitive experiments searching for an electric dipole contribution in both neutrons and protons, and even in electrons for that matter, have been performed throughout the 1970s and 1980s. To date no positive signal has yet been observed. From the upper limits on the magnitude of such a dipole contribution, one can designate upper limits on the magnitude of any CP- and hence T-violating effects associated with the strong interaction. One finds that the T-violating part of the strong interaction, theoretically predicted by 't Hooft, must be much smaller than the T-conserving part of the strong interaction in order for its effects to be consistent with these upper limits. In fact, the ratio between the respective terms in the equations governing the theory must be smaller than about 10^{-8} (that is, one part in 10^{8}).

Why should such a small ratio appear in the theory? Like the flatness problem in cosmology, this is another example of a "naturalness" problem. One might hope to set the coefficient of 't Hooft's T-violating term to be zero, by fiat. But that is really an illusory solution. If we do this, we can show that T violation is likely to crop up somewhere else in the theory. Indeed, since CP symmetry is violated by the weak interaction, the weak interaction of the quarks themselves would be expected dynamically to produce, à la 't Hooft, such a T-violating effect in QCD.

This dilemma, called the "strong CP problem," stands as one of the more glaring inconsistencies in our theory of fundamental interactions. In 1978, a solution was proposed by physicists Roberto Peccei and Helen Quinn at Stanford. Shortly thereafter, Frank Wilczek and Steven Weinberg, working independently, pointed out that an inevitable consequence of the Peccei-Quinn proposal was the existence of a new light particle, which Wilczek dubbed the "axion." For some time, Wilczek had been amused by the name of a popular laundry detergent and had hoped that he would one day have the chance to name a new particle after it. The name was particularly appropriate, since the Peccei-Quinn proposal involved a symmetry called axial symmetry, and the interactions of the new particle predicted by Wilczek and Weinberg therefore would be of a special "axial" type.

The Peccei-Quinn proposal was to suggest a dynamical mechanism by which the same effects that would normally ensure that T violation occurs in QCD would instead conspire to minimize such a violation. Since the weak interaction of quarks posed a potential problem for maintaining CP symmetry in the strong interaction, they suggested tying the symmetries of these two interactions together somehow so that a single dynamical mechanism could redress this problem in principle. To do this, they proposed to extend the symmetry present in the equations governing the weak interaction. At the same time as the weak symmetry was broken, this additional symmetry would also be broken. Most important, this "axial" symmetry (different for left- and right-handed particles) was also violated explicitly by 't Hooft's T-violating term in QCD. One could then show that the dynamics that determined how the weak interaction symmetry breaking occurred would also make the effective coefficient determining the strength of the CP-violating term in QCD acceptably small.

This idea was very elegant; many physicists would say that it is "too pretty" not to be adopted somewhere by nature. Of course, in physics whether pretty ideas are actually adopted by nature is something only experiment can tell. Wilczek's and Weinberg's demonstrations showing that associated with this new symmetry breaking there must be a novel light particle opened the avenue for experimentalists to probe the Peccei-Quinn mechanism. Because the axion is the product of symmetry breaking, and of the existence of 't Hooft's new term in

QCD, its properties are almost entirely fixed once the symmetry-breaking scale of the new axial symmetry is given. If this scale turns out to be the same as the weak symmetry-breaking scale, the axion interactions, while almost as weak as neutrinos, would still be in principle directly measurable.

Soon after Wilczek and Weinberg announced their theoretical predictions, experimentalists began searching for this new particle. They looked at accelerator data and at the decay debris of various heavy particles; they looked outside nuclear reactors, under rocks, everywhere. While the theory led to definite predictions in most cases for what they should see, all results were negative. By the early 1980s it looked as if the axion issue was dead.

The advent of Grand Unified Theories resuscitated axions. By suggesting a possible new scale in physical theory, GUTs offered a new scale for the Peccei-Quinn mechanism. Several different groups recognized that the Peccei-Quinn symmetry-breaking scale could be raised to the GUT scale and the important Peccei-Quinn mechanism for removing strong CP violation could still be operative. The new, "improved" axion that resulted would still be viable. Because it transpires that the axion mass and couplings to ordinary matter are inversely related to the Peccei-Quinn symmetry-breaking scale, both its mass and couplings would be drastically reduced by raising this scale by some 10–15 orders of magnitude. The new ultralight axion was coined the "invisible" axion, because it would not be observable in the experiments that had previously searched for the weak interaction scale object, and also because its couplings were so small that it seemed it would remain forever unobservable. The new, improved, invisible axion was an experimentalist's nightmare!

Freeing the axion from the shackles of the weak interaction scale, and the experimentalist's detectors, moved it into the realms of physics accessible only in the very early universe. As such, it became fair game for the cosmologists. The GUT scales with which the axion was now associated had revolutionized our models of the early universe. Physical processes occurring at the GUT scale could potentially explain the baryon number of the universe and provide a basis for the inflation theory and all its wonders. Within a year or so after the "invisible" axion was proposed, it was shown that axions might not be so innocuous after all. Indeed, these little beasties, which had faded from notice and were now reborn, might form almost all matter in the universe, except for the piddling little bit of it that we see. Rather generically, by a new type of production mechanism, which seemed inevitable and which I shall discuss later, a very small energy density in axions at early times could grow until a uniform background axion "field" would eventually dominate the energy density of the universe.

Suddenly the "invisible" axion had potentially observable consequences. It could form dark matter that dominated the expansion of the universe. One could argue that if the Peccei-Quinn scale were too high, then axions would close the universe today.

Soon other astrophysical analyses placed lower bounds on the Peccei-Quinn symmetry-breaking scale. As that scale is lowered, the interactions of axions become stronger. Just as neutrinos are created in stellar processes and in supernovae, one could show that axions, which in these models are very light, could also be produced at the temperatures available in stellar cores. So that axions would not take away excessive amounts of energy from stars such as the sun, one could derive upper bounds on the axion's coupling strength to ordinary matter, and thereby derive lower bounds on the scale of Peccei-Quinn symmetry breaking.

Supernova 1987A also had an impact on these developments. The observation of neutrinos from the supernova told astrophysicists that their models for the energetics and timing of the supernova explosion were largely correct. This hinted that axions probably would not have an unexpectedly strong influence. Turning this into a bound on the Peccei-Quinn scale is not easy, but the most recent comprehensive calculations suggest that if the energy scale at which the Peccei-Quinn symmetry is broken is less than about 10^{10} times the mass of the proton, then axions would severely alter the explosion dynamics in such a way as to make theory inconsistent with observation. The simplest cosmological arguments maintain that the Peccei-Quinn symmetry-breaking scale should be about 2 orders of magnitude larger, if axions are to compose the dark matter in the universe today. Astrophysical probes are getting closer and closer to directly constraining such an axion.

The completion of the transformation of invisible axions from theoretical toys to potentially observable particles was accomplished by a valuable insight of a young physicist named Pierre Sikivie. He pointed out that if axions formed the dark matter in the universe, one could envisage experiments that might directly detect them. The work of Sikivie has since been refined, and new ideas have been proposed—by my colleagues and me, among others. I shall discuss the stimulating prospects for detecting these objects, if they exist, in the final chapter of this book.

Before leaving axions, however, I want to add a historical footnote that warns us not to be too cavalier about our ability to rule out new physics, even at scales well below the scales I have talked about here. In about 1985 an unexpected observation once again revived the possibility of weak interaction scale axions. Experimenters at a nuclear physics facility in Germany, under the direction of Jack Greenberg of Yale, had been analyzing the collisions of beams of heavy atoms—uranium and thorium, for example—as they impinged upon thin target foils made of similar materials. They were searching for a signature of a rather esoteric phenomenon predicted in quantum electrodynamics which might be observable in the strong fields near very heavy atoms. What they saw was apparently something very different.

They had analyzed the outgoing positrons, and then later the outgoing electrons in coincidence with these, emitted from the collisions of these heavy atoms. They first observed a peak in the positron spectrum roughly where theory had predicted it. But when they examined the spectra resulting from the collisions of six different combinations of pairs of heavy atoms, to their great surprise they discovered that the same peak was observed, in the same position, and with roughly the same magnitude. This remained true when they looked at the coincident electrons as well. Neither signature was characteristic of the phenomenon they were probing, and both signatures were notably consistent with the existence of a new light particle produced in the collisions which could decay into an electron-positron pair.

Now nothing could smell more like a weak interaction scale axion! Animated conversations were heard in the halls of physics departments as soon as these results were announced:

> Simplicio: Why, a new light neutral particle . . . it must be an axion!
> Sagredo: But such a particle is ruled out!
> Salviati: Yes, but a new light neutral particle . . . it must be an axion!

The results achieved by Greenberg and his colleagues caused people to rethink the existing bounds on axions from previous terrestrial searches. A potential loophole was soon spotted that caused experimentalists quickly to reanalyze their five-to-ten-year-old data—with no luck.

After the standard axion's revival seemed doomed again, however, it was proposed by Wilczek and me, and independently by Peccei and collaborators, that some of the assumptions in the simplest axion models could easily be relaxed to produce more general models. In this case, no data seemed to rule out such a variant axion. This axion was predicted to be extremely short-lived, and to decay predominantly into electrons and positrons. Moreover, we even mumbled some words about why one should expect to see such an object in the observed collisions.

Our axion theory was itself short-lived. My own work, and that of others, over the course of the next six months demonstrated convincingly that our hypothetical axion should have been observable in various experiments performed concurrently. Moreover, I believe that we demonstrated convincingly that any elementary particle interpretation of the Greenberg group's data seems inconsistent. The Greenberg experiment has been repeated at another accelerator facility, and the earlier result has not held up. Indeed, I think it is fair to say that once the axion explanation was ruled out, stock in this experimental result decreased dramatically.

This story has a moral. The fact that a claimed observation in Germany by Greenberg and his associates could spark theorists to invent models of new physics—on scales accessible at laboratory energies—which survived for some time before being ruled out, suggests that there still is room for extraordinary discoveries at ordinary scales of energy. Surprises may still lurk just around the corner. . . .

SUPERSYMMETRY

If you are a pessimist, you will insist that to date there is no experimental evidence whatsoever for the notion of supersymmetry. If you are an optimist, you will claim that exactly half of the particles predicted by supersymmetric theories have already been observed—namely, all the particles we have discovered thus far. In any case, supersymmetry as a theoretical construct has flourished in the last two decades among particle theorists. Here is the reason why.

In an age when symmetry serves as the guiding principle for much of theoretical physics, supersymmetry (SUSY) is one of the most elegant of all symmetries, standing next to gauge invariance in power and mathematical beauty, and perhaps next to scale invariance (or "conformal symmetry," associated with "superstring" theory) in promise. As a symmetry principle, supersymmetry relates two otherwise disconnected but fundamental manifestations of matter: bosons and fermions. I have already noted that matter comes in two different types, classified according to the magnitude of the spin angular momentum of particles. Objects with half-integral angular momentum quantum numbers (for example, 1/2, 3/2) are called fermions, and objects with integral values (0, 1, 2, and so on) are called bosons. This categorization determines many of the important properties of conglomerations of these particles. As I have described, using neutrinos in dwarf galaxies as an example, fermions obey one kind of statistics, so that no two identical particles can occupy the same quantum state. Bosons, on the other hand, have no such restriction, and in fact they "like" to occupy the same state. A good example of this preference is the fact that in a beam of photons (spin 1 particles and therefore bosons) if all of the photons are in the same quantum state, the probability of producing yet more photons in the same state becomes strongly enhanced when the light beam traverses matter. This is one of the guiding principles of lasers.

These two categories of particles—fermions and bosons—apparently could not be more different; certainly it is a tribute to the ingenuity and sophistication of theorists that they have uncovered a mathematical symmetry capable of relating them. Indeed, supersymmetry as developed by physicists has opened up whole new areas of mathematics and has led to a renaissance of collaboration between mathematicians and physicists.

One of the properties that make SUSY so unique is the way it incorporates both symmetries associated with the structure of space and time, such as rotational symmetry and hence spin angular momentum, with symmetries associated with the internal structure of particles, such as electric charge and gauge invariance. Before the advent of supersymmetry, such harmony was generally regarded as impossible. Indeed, in 1967 Sidney Coleman and Jeffrey Mandula established a theorem stating that any internal symmetry that is mathematically like gauge symmetry—or all other internal symmetries we know of—can only connect particles of the same spin. The beauty of supersymmetry is that it circumvents this theorem by generalizing the mathematics associated with symmetry transformations in such a way that particles of different spin, integral or half-integral, can be related together at the same time as particles of different charge or "color."

The end result is that supersymmetry makes a one-to-one correspondence between every "fermionic" particle in nature and its supersymmetric partner, which must be "bosonic." Put another, simpler way, SUSY predicts that for every particle we observe today, there exists a particle with identical quantum numbers, but different spin. For every boson we observe, there should exist a corresponding fermionic partner:

$$\text{bosons} \leftrightarrow \text{fermions}$$

Because supersymmetry emerged from the hazy boundary between mathematics and mathematical physics, its history is not easy to trace. Unlike axions, SUSY was not developed initially to solve some specific problem in particle theory, nor was there a well-defined logical progression of results that led to its development. The mathematical formulation of a supersymmetry transformation entered the physics literature in the early 1970s, when physicists were still attempting to explain the then poorly understood strong interaction in terms of "stringlike" excitations rather than particle-like excitations.* These "string" models possessed a sort of supersymmetry invariance. Later it turned out that these theories would be replaced by quantum chromodynamics as the correct theory of the strong interaction. This early, inchoate effort to introduce both supersymmetry and strings into physics eventually led to the development of the "superstring" theories of the 1980s and the "M" theories of the 1990s—which some physicist-mathematicians view as candidates for a unified theory of all elementary particle interactions, including gravity.

Once introduced, the mathematics and physics of supersymmetry were refined. By about 1975 it was recognized that because supersymmetry relates to space-time symmetries, it might be relevant to a quantum theory of gravity. Thus, "supergravity" was born. Because supersymmetry enlarges the symmetries of pure gravity, it held out hope for canceling out or otherwise removing the apparent disastrous phenomena that occur at the Planck scale in classical gravity. Indeed, preliminary results in this regard were encouraging, and a small industry of physicists set about investigating supergravity theories.

Supersymmetry as a theoretical idea did not catch on widely until the advent of Grand Unified Theories. It was only after the development of GUTs that a well-defined question arose to which supersymmetry appeared to provide a generic and plausible answer. Once physicists began thinking about possible new physics at very high energy scales, a disturbing incompleteness in the standard model began to rear its ugly head. This is the so-called hierarchy problem—the presence of vastly differing scales in nature—to which I referred at the beginning of this chapter.

The successful unification of electromagnetism and the weak interaction depends on a viable mechanism for spontaneous symmetry breaking. One of the remarkable features of the model, however, is its lack of dependence on many of the details of the mechanism by which symmetry is broken, as long as it is broken. In the original formulation of Salam and Weinberg, spontaneous symmetry breaking is, as I have described, accomplished by the introduction of a new set of elementary particles in nature. Some of these particles "condense" into the vacuum, thereby changing its properties so that the weak interaction symmetry

*Supersymmetry may have actually crept into physics earlier. However, I am tracing the version of SUSY that made its impact in particle theory in its present form. I have not made efforts to trace methodically the history of this idea, and I apologize to any seminal figures whose work I do not mention here.

appears broken at low energies. In order to avoid destroying other symmetries in nature, most notably rotational and translational invariance, these elementary particles must have no spin angular momentum. They are called scalar particles, and since this mechanism for gauge symmetry breaking was first described by the Scottish physicist Peter Higgs, they are called Higgs scalars.

Sheldon Glashow has referred to the Higgs particles as the "toilets" of modern particle theory, because while they are very useful, they are also the guardian of dilemmas that physicists would rather not bring out in the open. Introducing new particles in nature is always a little worrisome, especially when they have not yet been seen. Unfortunately, however, in the case of the Higgs scalar particles the situation is even more disturbing. The standard model as it stands does not pin down very closely any of the properties of the Higgs particle. The mere existence of such a particle comes close to being all that is needed. Experiments over the last decade have begun to significantly constrain the parameter space for this particle, but no positive results have yet emerged. Moreover, on a deeper level, the theory describing such a scalar particle, or particles, suffers jarring formal problems. The worst of these is the problem I alluded to earlier: scalar particles are notoriously sensitive to new physics that may occur at very high, presently inaccessible energy scales. If somehow the strong, weak, and electromagnetic forces are unified at some energy scale, and the Higgs particle becomes part of a larger structure, it is virtually impossible to keep the scalar particle light while its relatives become heavy. We cannot preserve the hierarchy of scales in any natural way within the standard model. This is where supersymmetry comes in.

How does supersymmetry solve this problem? Recall that scalar particles are sensitive to physics at high energy scales because of the effect of virtual particles. Virtual particles in the vacuum, especially those of very high mass, interact with scalar particles, providing an effective mass for these particles which can be very large, regardless of the "bare" mass they would have in the absence of interactions. Now, I have already said that the important thing about SUSY is that it implies that for every boson there exists a fermionic partner in nature. One of the many significant differences between bosons and fermions is how they interact as virtual particles. It turns out that each boson and fermion in a supersymmetric pair gives the same contribution to the effective mass of the Higgs particle, but their *contributions come in with opposite sign* (that is, if one gives a positive contribution, the other gives a negative contribution). Thus, as long as supersymmetry is an exact symmetry of nature, the effects of all virtual particles—fermions and their bosonic partners—cancel out, and the mass of the Higgs scalar is *not affected by physics* at higher energy scales.

This miraculous cancellation, which occurs in supersymmetric theories in calculating the Higgs particle mass, is generic, and it affects many other physical processes as well. It is one of the reasons that these theories modify much of the "bad behavior" of gravity at high energies, and why "superstring" theories may be entirely free of any such unmanageable effects above the Planck scale.

There is one rather obvious fly in the ointment, unfortunately. Supersymmetry *cannot* be an exact symmetry of nature. If it were, every particle we see should be accompanied by a partner of opposite spin type, with the same mass and same interaction strength. The predicted particle content of physics should include the particles and spins listed in table 10.2. Observed particles and their spins are shown in ordinary typeface, and the predicted partners are in italics.

Fermions	Spin	Bosons	Spin
electron	½	*selectron*	0
photino	½	photon	1
neutrino	½	*sneutrino*	0
quarks	½	*squarks*	0
wino	½	W	1
zino	½	Z	1
Higgsino	½	Higgs	0
gravitino	3⁄2	graviton	2

Note: Italics indicate unobserved hypothetical SUSY partners. The names used here for predicted hypothetical objects are those most often used in physics literature.

TABLE 10.2 PREDICTED PARTICLES AND THEIR SPINS IN A SUPERSYMMETRIC THEORY

The vast number of particles not yet seen (exactly half of the total) indicates how poor this prediction is. The solution? Simply allow supersymmetry to be spontaneously broken! In this way, the mass of all the supersymmetric partners of ordinary matter might be raised until they are outside the range that our present accelerators can probe. Although this option keeps the idea phenomenologically viable, if supersymmetry is spontaneously broken we risk losing the key factor that gives SUSY its major appeal—the possible resolution of the hierarchy problem. If ordinary particles are split in mass from their supersymmetric partners, then the mechanism by which both cancel out in calculating the effects of virtual particles on the Higgs mass fails. We can show that the cancellation is now no longer exact but instead leaves a residue on the order of the supersymmetry breaking mass scale—that is, the order of the mass difference between particles and their SUSY partners.

Hence, if supersymmetry as an idea is to be motivated by a desire to solve the hierarchy problem in particle physics, the scale at which the SUSY partners of ordinary matter must exist cannot be much higher than the scale of weak interaction symmetry breaking. This scale is, in turn, just slightly above the scale accessible in today's accelerators, and the next generation of machines should surely see something. This impending development stirs great excitement among theorists. Supersymmetry makes many predictions about observable processes at accelerators. Confirmation (or invalidation) may be just within reach.

Over the past decade two other pieces of indirect evidence have strongly increased the theoretical motivation for considering supersymmetric models in particle physics. In the late 1980s, the large electron-positron collider at CERN in Geneva began to make precision measurements of the physics associated with the W and Z particles. In the process, the strength of both the weak and strong forces at these energy scales was probed with high precision. These new measurements were then used as inputs in order to determine the possible features of GUT that might unite the strong, weak, and electromagnetic forces.

Remember the remarkable discovery of Gross, Wilczek, and Politzer that the strong force gets weaker at shorter distances (or equivalently at higher energies). The weak and electromagnetic forces also vary in strength with distance, and it was this result that motivated the original work of Georgi, Quinn, and Weinberg, showing that the three forces could merge together with the same strength at some high energy scale associated with a GUT. At the time, the values of the three interaction strengths were not accurately known. However, in the intervening years, the quantitative improvement in our knowledge has been dramatic. So dramatic in fact that it was possible to extrapolate the predicted strengths of the forces to very high energy scales using just the known particles in nature (including the Higgs particle) and discover that the three interaction strengths would *not* come together at a single point, as one would expect if the forces became unified. However, if one adds into the mix the expected plethora of supersymmetric particles above the weak scale, it turns out that the same calculation remarkably reveals that the three forces do merge at a single energy scale, as first pointed out by Wilczek and Dimopoulos! This not only increases the stock in GUT theories, but it is strongly suggestive that SUSY might actually exist.

Another development that supports the theoretical belief in SUSY models is the surprising discovery in 1995 of the long-awaited "top" quark, the sixth and final quark predicted to complete the structure of the standard model of particle physics. The top quark has a mass of 175 times the mass of the proton. This is a huge and rather inexplicable mass. However, it turns out that in SUSY models, if the top quark mass is sufficiently large, one has a natural mechanism for the

Higgs particles to condense in the vacuum. Thus, in SUSY models, one might say that one *predicted* a heavy top quark in order to explain spontaneous symmetry breaking associated with the weak interactions.

In any case, doubling the number of particles that exist in nature is not without other implications for particle physics, and for cosmology. One of the most severe constraints on supersymmetric models of elementary particles is the requirement that they do not cause the proton to decay. Because of the plethora of new particles and processes available once supersymmetry is introduced, we find that processes involving intermediate SUSY particles can enhance the rate of proton decay well beyond that which is predicted in standard GUT models, or which is now empirically acceptable, based on data from proton decay detectors. However, many realistic models contain a symmetry (called R parity) that might be termed *supersymmetric partner conservation* (I define a supersymmetric partner as any one of the particles in italics in table 10.2: those particles that are not part of the standard model *sans* supersymmetry). This symmetry ensures that the only interactions that are allowed are those for which the same number of supersymmetric partners of ordinary matter leave any interaction as enter it. Hence, in a given situation, if one begins with only ordinary matter, one must end up with only ordinary matter. This implies something very important. The lightest supersymmetric partner (LSP) of ordinary matter must be absolutely stable. There is nothing for the LSP to decay into if R parity is to be conserved in the decay of the LSP.

The existence of such an absolutely stable LSP has pointed implications for detection in accelerators, and also for cosmology. This very object is a promising contender for dark matter. Indeed, of the exotic candidates that have been proposed for cold dark matter, the LSP has a distinct advantage. The simplest supersymmetric models predict the SUSY-breaking scale to be approximately equal to the weak symmetry-breaking scale. These models in turn predict an LSP with a mass that will then *automatically* result in close to a closure density today of such particles as leftover remnants from the Big Bang. Of course, as I stated earlier, this is not a required feature for any model. We know very little about what sets the scale of any observed particle masses, so it is no surprise that most models do not immediately predict in advance the mass scale of as yet unobserved phenomena. It is, however, nice when a model, such as the low-energy supersymmetry models mentioned here, make predictions on the basis of one set of arguments, which are then favored on the basis of their impact on other independent phenomena.

Before you rush out to vote for the LSP as your candidate for dark matter, I should note that these same supersymmetric models are not without other cosmological snags. For example, the supersymmetric partner of the graviton (the particle that presumably mediates the force of gravity), called the *gravitino,* tends not to be the lightest particle in most models. As such, it can decay into lighter LSPs plus ordinary matter. Yet because it is the partner of the graviton, its couplings are fixed to be of gravitational strength. Since gravity is so weak, this implies that the gravitino tends to live a long time before it decays. If even a tiny fraction of the mass density of the universe was in gravitinos after the universe cooled down to the temperature at which nucleosynthesis began, then the results of gravitino decay afterward would have been generally disastrous, in that they would have tended either to alter the abundance of primordial elements in a manner inconsistent with observation, or to distort by an unacceptable amount the spectrum of the microwave background. Physicists can always fiddle with their models to make sure these problems do not happen, but such manipulation is difficult at best. Of course, if the LSP is *discovered* in an accelerator or in a dark matter detector, then theorists will happily find a mechanism to explain why the gravitino drawback is no problem!

Before leaving supersymmetry, I want to discuss what kind of object the LSP would probably be. In the simplest SUSY models, the lightest particle in the spectrum is perhaps about ten to one hundred times lighter than the scale of SUSY breaking. If this latter scale is say 10 times as large as the weak symmetry breaking scale, then the LSP would be expected to be about fifty to five hundred times the mass of the proton. Thus it would be a respectably but not extraordinarily heavy object—it would be comparable in mass to the nuclei of atoms. How powerfully would it interact with matter? If R parity is a valid symmetry, then any interaction with normal stuff—which takes place by the exchange of some virtual particle—must be mediated by the exchange of a supersymmetric partner of ordinary matter, so that the same number of SUSY particles enter and leave each interaction. For example, the scattering of a *photino* by an electron can be sketched in a Feynman diagram (see figure 10.3). The intermediate particle is the *selectron,* the SUSY partner of the electron.

The photino is constrained to have the same coupling strength to the electron as the photon does, so one might expect its interactions with matter to be electromagnetic in strength. However, because the particle exchanged is a SUSY particle that is very heavy, the interaction is instead not only short range but also very weak. How weak? Well, the selectron and most SUSY particles are expected to have masses on the order of the SUSY-breaking scale, which we assume is perhaps slightly higher than the scale of weak symmetry breaking. Thus the mass of these particles is expected to be slightly larger but of roughly the same magnitude as the W and Z bosons that mediate the weak interaction. If this is the case, then the interaction strength of the LSP with ordinary matter will slightly weaker than the strength of the weak interaction, which also involves the exchange of heavy intermediate virtual particles. Compare figure 10.3 to figure 10.4, which diagrams the scattering of a neutrino by an electron.

What can we derive from this comparison? If the LSP forms the dark matter due to SUSY breaking around the weak symmetry-breaking scale, then the LSP should interact almost as strongly as would a very heavy neutrino, with a mass fiftyfold to five hundredfold that of the proton. This will be important in the discussions that follow. Also, since it is fairly "massive" and "weakly interacting," the LSP is a prime example of so-called WIMP candidates for dark matter (weakly interacting massive particles). (I regret the nomenclature, but it has become ingrained.) If the LSP is actually observed, particle physicists will be assured of exciting times to come as the other supersymmetric partners of ordinary matter come rolling along. Until then, however, supersymmetry must remain an elegant mathematical notion with an intriguing physical motivation.

MAGNETIC MONOPOLES

Since the time when Maxwell first developed his famous four equations governing the behavior of electric charges, currents, magnets, and electromagnetic waves, physicists have been aware of a noticeable asymmetry in electromagnet-

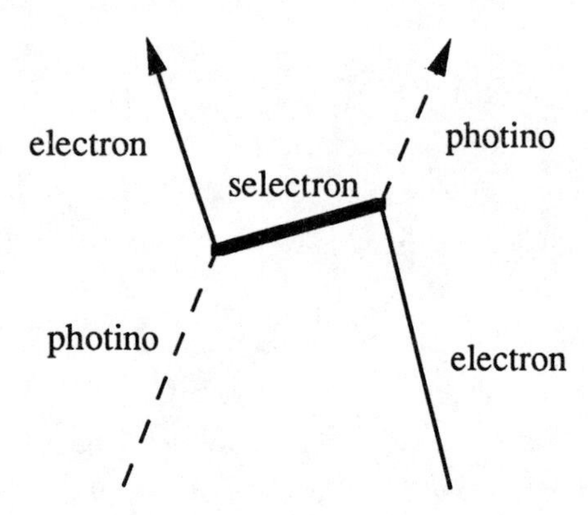

FIGURE 10.3

A Feynman diagram illustrating the scattering of a photino off an electron. The interaction between the photino and the electron is mediated by a heavy supersymmetric partner of the electron, called a "selectron."

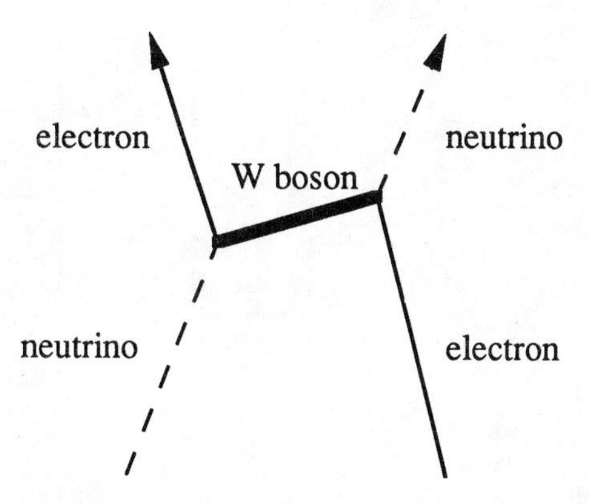

FIGURE 10.4

A Feynman diagram illustrating the scattering of a neutrino off an electron. The interaction between the photino and the electron is mediated by a heavy W boson, whose mass is comparable to that which might be expected for SUSY partners of ordinary particles.

ism. Electric charges act as the source of electric fields. What acts as the source of magnetic fields? The source is not magnetic charges, but rather currents. Indeed, it is a consequence of Maxwell's theory as it stands that no single elementary magnetic charge exists in nature. If we take a magnet with a north pole and a south pole and cut it in half, we do not find two isolated magnetic poles. Instead we produce two smaller magnets, each with a north pole and a south pole (see figure 10.5).

This process could continue indefinitely until we slice the magnet up into all of its elementary "spins" comprising the protons, neutrons, and electrons that make up the material. Each spin acts like a little magnetic dipole, and one can go no further.

If instead we take an electric dipole, with a charge separated at the ends of a barbell, say, and cut it in half, we will produce two separate charged objects, one positively charged and one negatively charged (see figure 10.6).

One could remedy this asymmetry between electricity and magnetism if somehow there could exist an isolated magnetic charge, a "north pole," if you wish. Now this alone is not sufficient reason to expect such an object to exist. However, in 1931, one of the originators of quantum mechanics, the laconic but brilliant British physicist Paul Adrien Maurice Dirac, made a startling discovery. He pointed out that when electromagnetism was coupled to quantum mechanics, the existence of a single "magnetic monopole," as it became called, could explain

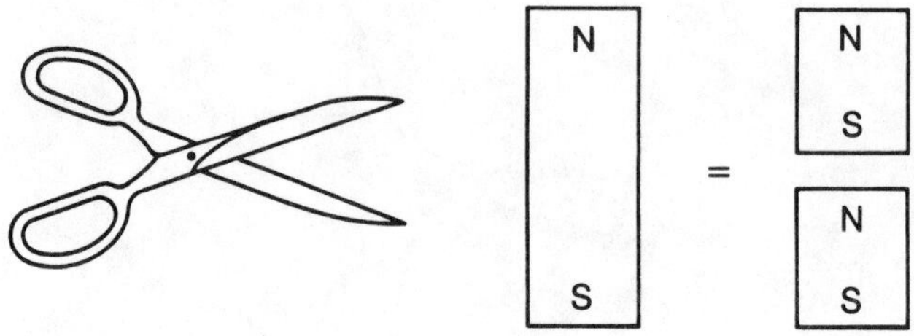

FIGURE 10.5

Cutting a dipole bar magnet in half does not produce an isolated north or south pole, but rather produces two smaller replicas of the original magnet, each containing both a north and a south pole.

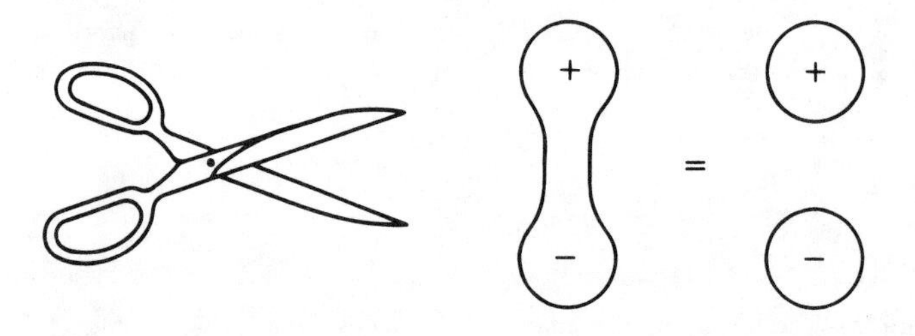

FIGURE 10.6

Cutting apart an electric "dipole," that is, a system in which the positive and negative charges have been separated, will produce an isolated negatively charged object and another positively charged object.

something that bewildered physicists at the time, and to some extent still does. All electric charge that exists in nature seems to come in integer multiples of the charge on the electron. The proton, for example, is 2,000 times more massive than the electron, and resembles it in no particular way, except charge. Its electric charge is exactly equal and opposite to that on the electron. Nowhere do we observe charges that are, say, $5/3$ or 1.335678 times the charge on the electron. Even quarks, which are fractionally charged, cannot exist as isolated objects, but only in combinations that result in an integral charge. What Dirac showed in his succinct and elegant piece of work was that the existence of a single magnetic monopole would not only restore complete symmetry between electric and magnetic charge in Maxwell's equations, it would force all isolated electric charges to come in integer multiples of a fundamental charge.

This observation was duly noted by the physics community and resulted in periodic efforts to search for magnetic monopoles in cosmic rays, for example. But if their mass was comparable to that of, say, the electron, they should have been relatively easy to see, and their absence was notable. Dirac's observation remained a provocative theoretical curiosity.

This situation changed dramatically with the advent of gauge theories, and in particular after the development of Grand Unified Theories. In another of his classic papers, 't Hooft (and independently the Russian physicist A. M. Polyakov) demonstrated not only that a monopole solution of the classical equations governing these theories existed, but that any unified theory that breaks down to the standard model at low energies must result in magnetic monopoles. These objects would be quite different from the electron-mass monopoles that were envisaged earlier. Their mass would be characteristic of the Grand Unified scale, or higher, so that they might be expected to weigh at least 10^{16} times as much as a proton. A single monopole alone would weigh about one-hundred millionth of a gram. In normal matter, such a mass of material contains about 10^{16} particles.

Such a heavy monopole would interact quite differently with normal matter than would a light monopole. Because of its large mass, and the fact that a monopole at rest would undergo no attraction or repulsion from matter, a slowly moving monopole falling to earth, for example, would fall right through the surface, since its interactions with light atoms would not be sufficiently strong to impede its motion. It would be like trying to stop a Mack truck by throwing popcorn at it. Our searches for monopoles in normal matter, such as ocean water, would be expected to come up with zilch.

Monopoles went from curiosities, however, to real concerns when it was first shown by a graduate student at Harvard, John Preskill (now a professor at Caltech), and independently by the renowned Soviet astrophysicist Y. B. Zel'dovich, that monopoles could cause difficulties for standard cosmology. I have mentioned the problem already. At the time in the early universe when a GUT symmetry breaks down into the separate strong and "electro-weak" theories, many monopoles should be produced. So many, in fact, that they would easily close the universe today. The problem becomes not one of how to produce a single monopole, but rather how to get rid of so many.

In 1982, a young experimentalist named Blas Cabrera, motivated to perform an experiment by the suggestion that monopoles might be cosmologically significant, announced what appeared to be an unambiguous discovery of a magnetic monopole in his basement lab at Stanford. Suddenly cosmic monopoles were the stuff of headlines. Cabrera, a careful experimenter, comes from a long line of physicists. His grandfather, of the same name, was one of the most famous Spanish physicists of his time. Cabrera's father is also a physicist, who has returned to Spain to work. I have become both a friend and collaborator of Blas and can attest to what his experimental colleagues knew at the time of his monopole announcement: when he made the claim, it was taken very seriously by the scientific community. Cabrera has since become the chair of Physics at Stanford, and is occupied in the modern search not for monopoles, but WIMPs.

The circumstances of the discovery were a newspaper reporter's dream. Sometime around 2:00 P.M. on Valentine's Day in 1982, in a deserted laboratory, a detector that had been set up to search for monopoles in the cosmic rays bombarding the earth recorded a signal that could not have resembled more closely the trace that would have been left by a monopole traversing the ringlike detection region. It had been anticipated that such a traversal would cause the magnetic field inside the ring to jump instantaneously by a discrete amount proportional to the magnetic charge on the monopole. This is exactly what the chart recorder showed. Moreover, in controlled experiments during the previous several months, no "false alarms" had been detected. On top of that, the density of cosmic monopoles inferred from the observation was exactly the density that would give 10^{25} eV monopoles the density of dark matter in galactic halos. After the fact, to demonstrate the insensitivity of the detector to jarring blows, or other possible student-induced pranks, Cabrera demonstrated that even hitting the device directly would not make the chart recorder jump in a similar manner.

Unfortunately, since that time, Cabrera and others have built much larger and more complicated devices but have not detected a single comparable signal. If that original occurrence was just a random event, the probability that another such event would not have been seen since in any of these detectors is about 1 in 1,000, that is, perhaps possible, but certainly not likely. To date, no good explanation of the Valentine's Day event has been given, but perhaps it was bad luck, one of those cases where a variety of extremely rare background noise sources conspire to reproduce a signal. Nevertheless, experimentalists have been undaunted in their continued search for these objects. Even theorists provided moral support. Exactly a year after the original event, Sheldon Glashow, a closet poet, sent Cabrera a Valentine's Day telegram:

> Roses are red, violets are blue,
> The time has come for Monopole II.

I will return later to the experimental search for monopoles. Since Cabrera's announcement, however, the field of cosmology has been revolutionized in response to the problem first posed by Preskill and Zel'dovich. It was exactly the monopole problem that first caused Alan Guth to start thinking about Grand Unified Theories and cosmology. The result was his inflationary universe cosmology. Besides all of its other advantages, one of the things that inflation theory does well is get rid of monopoles. Indeed, if the GUT transition in which monopoles are produced is the same one that results in inflation, then Guth's theory assures, by inflating all the rest of the monopoles in the universe outside our present horizon, that there would be at most one monopole left in the visible universe today. Perhaps that was the one seen by Cabrera.

Yet the inflation theory does not always preclude remaining monopoles today. It is quite possible that inflation took place *before* the monopole-producing transition took place. Various models of transitions at lower energies result in much lower, and perhaps even acceptable, levels of primordial monopole production—maybe even enough to just close the universe today. While these scenarios are by no means universal, magnetic monopoles, at least in theory, are. They are on such solid theoretical footing, in fact, that experiments are still under way in the hope of detecting a closure density of monopoles. If their mass is greater than about 10^{29} eV, their predicted event rate falls below the sensitivity of today's experiments, as I shall discuss; thus magnetic monopoles are still at least plausible contenders for dark matter.

In my review of the three "best bet" candidates for cold dark matter arising from the field of particle theory, I hope I have convinced you that each has a good reason to exist. It would not be too surprising if at least one, if not all, were discovered in the years to come. What still needs to be done, however, is to demonstrate convincingly how any of these objects could have been produced in the early universe with sufficient abundance to become the dark matter that we infer on large scales today. All of the candidates I have described are very weakly interacting. With the exception of the axion, which I will discuss again later, they are all very heavy and therefore would be expected to be nonrelativistic at late times, when fluctuations in their number density could easily grow to form the seeds of galaxies and clusters. While they thus fit the bill for cold dark matter, why should we believe that any of them actually populates the heavens?

CHAPTER ELEVEN

Three Modest Proposals

Given the obvious differences among the various dark matter candidates I have described, it may seem unlikely that there should exist plausible mechanisms by which all of them could be produced in the early universe. Nevertheless, I can describe three possibilities that are sufficient to allow any of these candidates to become the dark matter we are searching for. Specifically, elementary particle candidates are either *born to be dark matter,* or they *achieve "dark matterhood,"* or they *have "dark matterdom" thrust upon them*. It is the purpose of this chapter to convince you of both the generality of these options and their plausibility. More important, the mechanisms of each option lead to signatures that guide the experimental search for these elusive objects.

BORN TO BE DARK

The most well known prototype of a natural dark matter candidate is the good old-fashioned light neutrino. Recall that these particles need not do anything active to dominate the universe today. Simply by virtue of the existence of thermal equilibrium conditions early on, light neutrinos would have been as abundant as photons in the early universe. After they decoupled from the Big Bang expansion, all they had to do was wait patiently and eventually the energy density in radiation would fall below that stored in the rest energy of the neutrinos, and they would come to dominate the expansion. As I earlier showed, if their mass happens to be in the range of 30 eV, then the neutrinos would have just the right density to result in a flat universe today.

It turns out that axions, too, are born to be dark, but by a different gestation process that results in a very different distribution of dark matter particles today. The characteristics of this production mechanism play a crucial role in providing the key properties of the axion background which could allow it to be detected.

Cosmic axions are born out of the darkest of all energy sources, the vacuum. We have seen earlier that the vacuum in particle theory can be far from innocuous. Indeed, in the inflationary universe scenario the vacuum itself can store enough energy to alter and eventually dominate the whole expansion of the universe. Things start out much more innocently in the case of axions. Axions, as I have related, are theoretical offspring of our considerations about the symmetries associated with the vacuum state of the world. Their whole existence is attached to the notion of the spontaneous breaking of symmetry—the Peccei-Quinn symmetry to be exact. To understand axion dynamics, therefore, we shall have to spend a little more time describing the dynamics of the vacuum itself.

How can the vacuum—that sophisticated version of empty space—have dynamics? The answer is pretty straightforward. The vacuum is supposed to be the ground state of matter. The ground state is supposed to be the lowest energy state of matter. Hence it is dynamics that determines the conditions of the vacuum state. We modify those degrees of freedom which are variable until we find the combination that results in the lowest energy density; that combination labels what we call the ground state.

In field theory, the fundamental degrees of freedom are the elementary particle excitations themselves. We have seen that one way to characterize the vacuum state is by the number density of real particles that exist, or that have "condensed," into this state. This number density is most often zero, in which case only "virtual" particles populate the vacuum state. However, we have also seen that a nonzero density of particles can "condense" into the vacuum to initiate spontaneous symmetry breaking.

In the language of dynamics, we may now try to understand, at least pictorially, why this kind of phenomenon happens. Recall that it usually takes *energy* to make real particles out of the vacuum. If the particles do not interact with each other, then the more particles we create the more energy it takes. Figure 11.1 presents this situation diagrammatically. On the horizontal axis I have let vary the number density of real particles existing in the ground state. Particle physicists call this the "expectation value" of the "quantum field" describing the particles in question. As shown in the figure, it is obvious what will be the energetically preferred ground state of matter. The vacuum will be devoid of real particles.

Imagine that, instead of the situation depicted in figure 11.1, I allow the particles to attract one another such that as I increase the number density of real particles I actually *lower* the total energy until I saturate the system with particles and the energy once again begins to increase, as depicted in figure 11.2. This diagram repro-

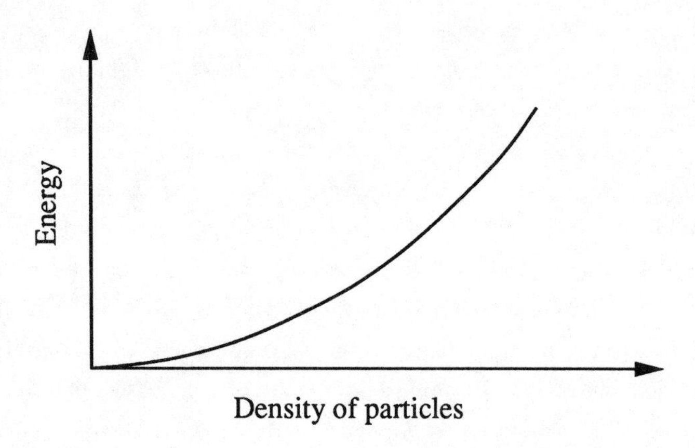

FIGURE 11.1
Normally the "vacuum state," which is the state of minimum energy of a quantum system, is that state in which one expects to find no real particles, because the energy of the system increases as you add real massive particles.

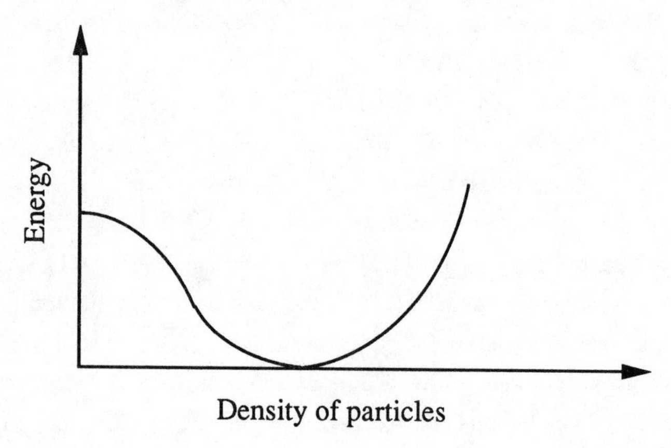

FIGURE 11.2
If the interactions between particles are attractive, one might imagine that the minimum energy configuration is not one with an absence of real particles, but rather one in which a nonzero density of real particles "condenses" into this vacuum state.

duces exactly what I earlier described in words (in chapter 2) as leading to spontaneous symmetry breaking. In this case, it is clear that the energetically favored ground-state configuration of matter, because of particle interactions, contains a nonzero density of real particles. Particles can "condense" into the vacuum.

How does such a particle condensate actually behave? Assume, for simplicity, that such particles have a nonzero mass. Clearly, it will cost energy for the particles in the condensate to move around, so the lowest energy configuration of such particles will be one in which they are at rest. Remember that all of these activities are really happening in the context of "quantum field theory," whose basis is quantum mechanics. And this has even stronger implications: all the quantum mechanical *wavefunctions,* which describe the probability of finding each given particle at some point and time, of the different particles in the vacuum will be correlated in a precise way. Such a correlated combination of particles is called a *coherent state.*

Coherence means that the quantum mechanical wavefunctions of all the particles are coupled in such a way that the net combination acts uniformly together, or "in phase." Coherence is not a purely theoretical notion, but a significant and familiar phenomenon. Compare laser light—which involves a coherent amalgamation of photons—with light from a normal light bulb. Holograms, three-dimensional pictures made with laser technology, are possible only because the "phase" relationships among the individual photons in a coherent beam are stored to provide much more "visual information" than is possible in a normal two-dimensional picture.

Now, back to axions. Axions are closely related to the scalar particles introduced into the Peccei-Quinn theory which actually condense into the vacuum and break the Peccei-Quinn symmetry. It is a general property, first discussed by British physicist Jeffrey Goldstone, that every time scalar particles condense in the vacuum and break a continuous symmetry, there should be associated with these particles another degree of freedom manifested as a massless particle. In the Peccei-Quinn theory, this particle is the axion.

If cosmic axions begin life as massless particles, it does not cost any energy also to populate the vacuum with a nonzero number density of these particles. Any massless particle acts like light insofar as its kinetic behavior is concerned. In particular, its energy is not determined by its mass, but by its frequency (or wavelength) as it travels along at the speed of light. Massless particles with an arbitrarily large wavelength have arbitrarily small frequencies associated with them, and hence they possess arbitrarily small amounts of energy. If they do not interact with each other, or do so very weakly, I can place as many of them as I want in a very large box. It will not cost any energy. Pictorially, I may represent this situation as shown in figure 11.3.

If figure 11.3 describes the situation for axions, then we can see that dynamics allows a large number of real axions to populate the vacuum. Actually, for technical reasons it turns out that the number density of axions in the vacuum cannot be arbitrarily large but instead must lie in some fixed range around zero. In appropriate units, the expectation value of axions in the vacuum can be scaled to be between zero and 1, say. This value is indicated in figure 11.4. Physically, what do we expect to happen in this case? Since it costs no energy for the axion "expectation value" to take any value in this range, the actual density of axions in space which result after Peccei-Quinn symmetry breaking can also take any value in this range. Now, in the early universe, different regions of space are not initially in causal contact (that is, they are outside of the horizons of the other nearby regions). So one would anticipate that in different regions the axion expectation value would take different random values in the allowed range. If we had to guess what that value would be in some region, we would say that it would be about 0.3, or 0.5, rather than, say, 0.0000001. Nothing can guarantee this, however, except for the laws of probability. With equal probability of picking any value between zero and 1, there is a probability of one in a million or so that we should find ourselves in a region of the universe where the axion expectation value was initially between zero and 10^{-6}. The important point here is that nothing really favors the value zero above anything else.

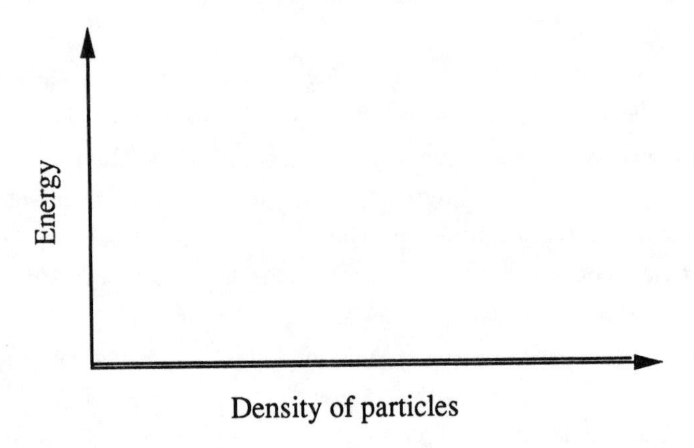

FIGURE 11.3

Massless particles in a sufficiently large volume can have an arbitrarily small amount of energy associated with them. Thus, if one ignores any possible interactions between such particles, one can pack as many of these particles as one wishes into such a volume and not increase the energy density. In this case, therefore, the preferred ground state of the system cannot be determined by energy arguments alone.

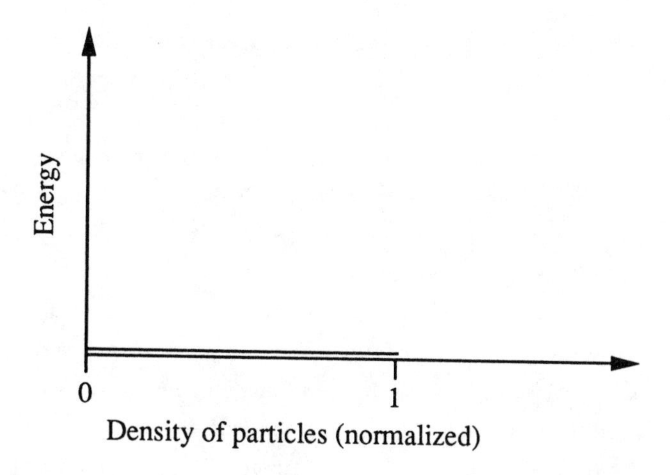

FIGURE 11.4
Axions, as long as they remain massless, can have a number density in the vacuum state which is not determined dynamically. Because of the particular nature of axion physics, it turns out that the actual number density of axions that are present in the vacuum state is limited to lie in some fixed range, from zero up to some maximum quantity. By appropriately normalizing the relevant quantities, one can then imagine the number density of axions present in the vacuum state to take any value between zero and 1.

Now, say that the axion background has a nonzero expectation value in some small region of space, which then inflates so that this region encompasses completely what will evolve to be our observable universe.* Then, after inflation, we should find ourselves living in a universe with a constant background density of axions. These would be in a coherent configuration of minimal energy. Of course, since the axions are massless, this configuration would store no energy, and it would not affect, or be affected by, the subsequent expansion of the universe.

But once the universe cools sufficiently so that the strong interaction begins to become important, things change. The term in the quantum chromodynamic equations which otherwise would lead to charge conjugation-parity (CP) violation produces interactions that explicitly break the Peccei-Quinn symmetry (see chapter 10). The effect is that interactions are induced among the axions which make it cost energy to produce a real axion. In other words, these interactions effectively give axions a mass, where before they had none. Now, the dynamical picture that would determine the number density of axions in the vacuum resembles the good old standard picture, which would be minimized by having no axions in the vacuum (see figure 11.5).

*The argument does not change significantly if this is not the case . . . trust me.

But wait a minute—I have just indicated that we may find ourselves in a universe with a nonzero density of axions condensed in the vacuum. What gives? Well, the state in which we found ourselves earlier, with, say, an axion density of 0.5 in the units that I introduced earlier, is no longer the minimal energy configuration. It is no longer the favored ground state. This state must then dynamically "relax" and reduce the magnitude of the axion background until it approaches the new minimum energy configuration.

On what time-scale will the axion background relax? Processes that can alter the axion background's coherent wavefunction happen on a time-scale *inversely proportional* to the axion mass. The more steeply the energy changes as we increase the number of axions in the vacuum (the greater the axion mass), the more quickly the axion background will respond. The coherent configuration of many background axions acts for this purpose like any other classical object, say a ball, when it finds itself in a "potential well." The time it would take to slide down a valley floor, if you found yourself suddenly about to start down on skis from the top, is inversely related to how steeply the valley floor is curved. In the case shown in figure 11.5, the curvature is proportional to the energy that it takes to produce axions, that is, their mass.

If axions are very light, then it means that the characteristic time it takes for the axion coherent wavefunction to change can be very long indeed. In the early universe, it would have been quite possible for this time-scale to exceed the age of the universe at that time. What happens in this case? The axion coherent background remains constant and stores a constant energy density until the back-

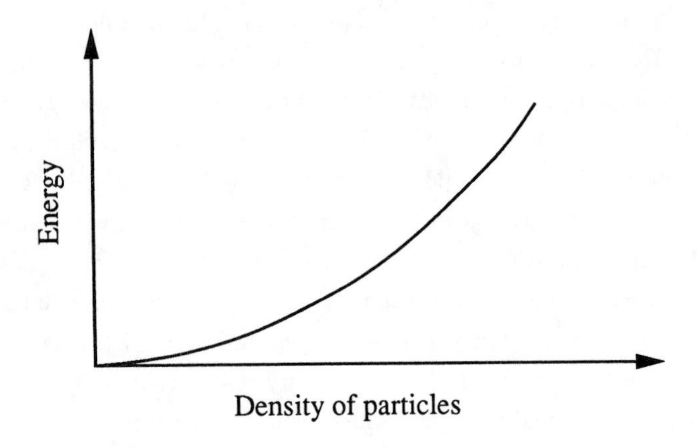

FIGURE 11.5
Once the axion mass becomes dynamically significant, the favored lowest energy configuration becomes one in which no axions are present.

ground has time to relax dynamically. In this sense, the axion background behaves, for some time, precisely as we would expect a vacuum energy associated with a fundamental cosmological constant to behave. The difference, however, and one we will return to later, is that the axion background can relax to zero dynamically. Once it does begin to change, its magnitude, and hence the energy density of the axion background, will diminish as it dynamically dilutes while the universe expands.

During the time it takes before this axion background reacts to the new state of affairs, the rest of matter and radiation in the universe *will* be responding to the background expansion of the universe by becoming more dilute. As a result, the energy of other matter and radiation decreases. One can easily imagine that if the axion mass is small enough, the energy that is stored in the initial constant axion background will remain constant long enough so that even if this energy density were miniscule compared to the energy density in ordinary matter initially, the axion background could eventually catch up. By today, the amplitude of the axion background that still remains, even after dilution, could easily contain enough energy to dominate the universe. If we assume that the initial magnitude of the axion background was on the order of 0.5 (as compared to, for example, 10^{-6}), in units where it was constrained to lie between zero and 1, then a simple calculation reveals that if the axion mass is about 10^{-5} eV, or about 1 ten-billionth of the mass of the electron, the axion background in the early universe would have remained constant long enough so that its remnants could result in a sizable fraction of a critical density, or a flat universe, today.

I am sure that this discussion has struck many readers as a bit otherworldly, and to a certain extent it is. It is worthwhile to realize that very similar processes occur in much more familiar settings: the formation of ice crystals on a window pane in the winter, the magnetization of a nail when you hit it with a hammer. We have no reason to believe that such occurrences should not also happen on cosmic scales. It just takes some faith in the basic laws of physics.

If you buy this mechanism, several implications are salient. First, as promised, no specific axion dynamics were involved. The axion background simply must initially have a nonzero expectation value at early times, and everything else happens automatically. Consider what a miraculous production mechanism this really is. After it is finished, we are left with a coherent configuration of axions—something like being bathed in laser light—containing a very large density of otherwise very light particles. This will be very important when we consider ways to detect this stuff. Furthermore, because these particles are initially created quantum mechanically in a minimal energy configuration, and not through a thermal process, this means that they would behave nonrelativistically at early times, even when the temperature of the radiation bath (with which they are not thermally coupled) far exceeds their mass. This is the chief requirement to be met for dark matter to be cold, so that an initial density fluctuation will have no problem collapsing due to gravity at the right time. This is of great consequence. It means that we do not require cold dark matter to be made of heavy particles, even though a more naive analysis might have led one to expect this to be the case. As long as nonthermal production mechanisms, such as this one, are allowed, the range of possibilities is much richer. Axions are a prime example of this new freedom.

DARK MATTER THE HARD WAY: EARNING IT

The second mechanism by which a large remnant density of exotic weakly interacting particles could have been produced in the early universe is identical to the way a remnant abundance of light elements is created in Big Bang nucleosynthesis. One starts with a thermal equilibrium configuration of matter and radiation and lets it evolve, following the course of specific reactions as they go out of thermal equilibrium as the universe cools. These calculations show that particles with the interaction strength of neutrinos would, if they have a mass of about one to one thousand times the mass of the proton, be produced in sufficient abundance in the early universe so that they may dominate it today. This argument was first applied to the possibility of very massive neutrinos, but its range of applicability is much broader. It goes as follows.

At a very high temperature, the density of matter and radiation in the early universe was so great that even neutrinos, which are weakly interacting, were able to interact quickly enough so that they remained in thermal equilibrium. In other words, the rate of their interactions with other stuff was faster than the rate at which the universe was expanding. The fact that they were in thermal equilibrium implied that they were about as abundant as photons at that time. Now, if neutrinos are very light, we already know what would happen: they would decouple, and redshift, and remain today with roughly the same number density as photons. However, if neutrinos are very massive, then events would be slightly different. If neutrinos remain in thermal equilibrium with matter when the temperature drops below their mass, then their abundance begins to drop. This is a reflection of the simple fact that in thermal equilibrium the probability of finding an object with mass/energy much larger than that which is thermally available at the time is small. Accordingly, the probability is reduced of finding massive neutrinos still around when the available thermal energy drops below their mass. Physically this occurs for an equally simple reason. Particles can be destroyed by mutual annihilation with their antiparticles. Alternatively, particle-antiparticle pairs can be created out of collisions occurring in the background radiation bath. Now, if the thermal energy in the radiation bath falls below the mass of the particles that one wishes to create, then on average there will not be enough energy available in the bath to create these particles. The opposite process, however, particle-antiparticle annihilation, is not so restricted, since the particles are already around and they merely have to collide with each other to annihilate. Thus, more particles will annihilate at low temperature than will be created, driving the number density down to its thermal equilibrium value.

This process could continue ad infinitum, leaving in the end no massive particles, just radiation. Yet as particles annihilate upon collision with their antiparticles, and their net density is reduced, it becomes harder for a particle to find an antiparticle with which to annihilate. Thus, the rate of annihilations will itself also fall. At some point, the annihilation rate will be less than the expansion rate of the universe, and particles will be moving apart faster than they can find partners with which to annihilate. And at this point, essentially no more annihilations will take place and the number density of these massive particles will be "frozen out," that is, it will not change further due to annihilation.

If we use the known interaction strengths of neutrinos, we find that if their mass is approximately the mass of the proton, or slightly greater, then the remnant number density of particles that will be frozen out would be sufficient to result in a closure density of neutrinos today. Since their interaction strength increases with their mass, if they were lighter they would freeze out earlier, and too many would be left over. If they were heavier, more would annihilate, and their remnant number density today would be smaller.

This suggested to some cosmologists that a new neutrino, if it were very heavy (remember, the three types of neutrinos that have been observed so far all have masses constrained to be much less than that of the mass of the proton), would naturally be a candidate to be cold dark matter. That neutrino became the prototypical WIMP. I say this with something less than wild enthusiasm, because no one necessarily expected to discover a new neutrino with the mass required by this argument, and such particles are now ruled out by a combination of constraints from accelerators, and direct searches of the type I shall shortly discuss.

What really brought WIMPs to the top of the heap, however, was the discovery that the lightest supersymmetric partner (LSP) of ordinary matter not only would have an interaction strength comparable to neutrinos, but naturally would be expected to have a mass in this range. Of course, since the parameters of supersymmetric models are not as strongly fixed as the parameters of the standard electro-weak theory, there is still room for play. Things work out naturally in the right ballpark, and that alone is reason to be encouraged.

Weakly interacting massive particles, if they exist, would be produced by the same type of thermal mechanism that has produced all the light elements we observe in the universe today. If we believe the results of the incredibly successful Big Bang nucleosynthesis calculations, we should have no doubt at all that if a WIMP candidate such as a light supersymmetric particle is observed in accelerators with a mass of tenfold to five hundred fold the mass of the proton, then this particle can also exist in the universe with an abundance sufficient to explain all the dark matter we infer to exist today. Of course, WIMPs may be detected first as dark matter.

DARK MATTER BY FORCE

We have already discussed the prototypical example of a particle that can be forced to be dark matter: the magnetic monopole. By "forced" I mean that normal dynamical arguments suggest that these objects would not be produced in the requisite numbers to be significant today. However, just like "defects" found in crystals of normal material, it is sometimes impossible to avoid the creation of objects that are energetically not favored. This is often associated with the occurrence of a phase transition. Earlier I discussed the can of ice crystals on a window. Here let me describe another, somewhat hypothetical, possibility. Consider the case of a very large magnetic system in two dimensions made up of lots of individual magnets. We have seen that at high temperature the favored configuration is for all the magnets to point in random directions. As the temperature is lowered, it becomes energetically favorable for them to align in one direction. If the system is very large, small regions may form with magnets lined up in one direction, and these regions begin to grow as further magnets align along with the prevailing direction in that region.* At some point the different regions nearly merge (see figure 11.6).

---

*In an actual magnetic system in two dimensions, it turns out that near the temperature at which the phase transition occurs, the microscopic magnets are correlated over an entire sample volume, so the kind of defect I describe here does not form. However, for the purpose of this argument, imagine that the temperature is changed very fast or that the system is very large, so that distant regions do not have time to align.

What happens next? At the interface of the regions, the magnets begin to adjust so that the direction makes a smooth transition from one region to another. Eventually, this boundary grows outward so that the two regions end up with their magnetic fields aligned (see figure 11.7).

But what about the region about the point marked with an X in figure 11.6? As one moves in a circle about that point one finds that the magnetic fields in the different regions also change direction by one complete rotation.

Now imagine what happens when the microscopic magnets try to line up. They can alter their directions smoothly at all points except near the point X. Here there is no way for one magnet to align closer to a nearby one without at the same time misaligning more with the magnet on its opposite side. The only way to proceed continuously from a magnetic field pointing down below point X to one pointing up above it is for a small region around point X to form where there is no preferred magnetic field direction, that is, where the magnets are not aligned. Then, as we travel through point X from below, we begin in a magnetic field that is pointing downward. As point X is approached, the magnetic field gets weaker, and then it has zero value at point X. Finally, as we continue moving upward, the field begins once again to intensify, but this time it points upward. This pattern is the same from whichever direction we approach X (see figure 11.8).

What results from this smoothing process? As can be seen in figure 11.8, a small region of space, where no net magnetic alignment has taken place, has been "trapped" amid a background where the magnets are aligned over large

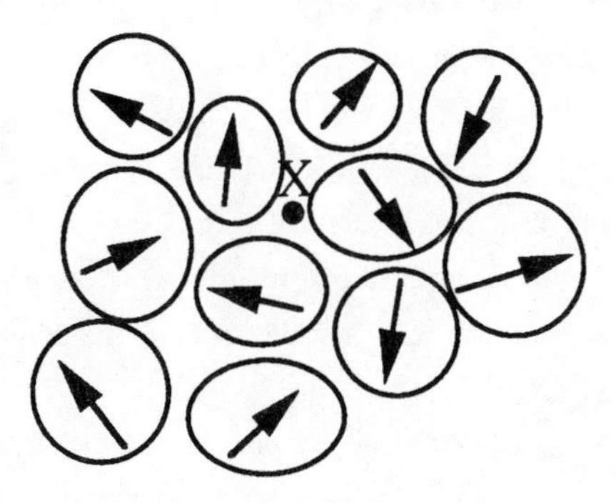

FIGURE 11.6

Imagine some extremely large hypothetical system of magnets in two dimensions, so large that separate regions act independently (that is, they are not in causal contact) as the temperature is quickly cooled. In this case one expects that separate magnetic domains will form, inside of which the magnetic field is uniformly aligned along some random direction. These regions will grow in size until they encounter neighboring regions.

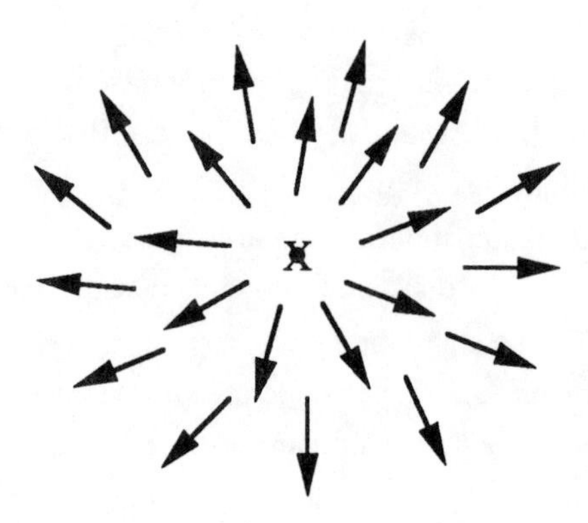

FIGURE 11.7

At the boundaries of adjacent domains, the magnetic fields will tend to align to minimize their energy. In general this will result in magnetic fields that are uniformly aligned over large volumes. However, in regions such as those around point X, it is impossible for the magnetic fields in each region to align with one adjacent region without becoming more misaligned with the adjacent region on the opposite side. The region around point X is called a "defect."

regions—the energetically favored configuration. Why does this configuration develop, and not one where the magnetic field configuration discontinuously changes direction at the point X? Because locally the smoothed configuration is energetically favored. The magnetic field configuration has relaxed to make the best out of a bad situation. Clearly, the final configuration is not as energetically preferred as in the case where all magnets are aligned, but since changes can only be made locally, by one magnet moving in response to its nearest neighbors, a global realignment is not likely. In the process of relaxing to the final configuration, a small amount of energy is stored in the "defect" that has resulted. In figure 11.9, I show the direction and magnitude of the magnetic field at each point in space around the defect, and, in the vertical direction, I display the energy stored in the configuration.

It transpires that this kind of behavior is universal. These kinds of defects are not necessarily restricted to ice crystal lattices or to hypothetical configurations of microscopic magnets. A very similar phenomenon could have taken place during a phase transition in the early universe. Associated with some scalar particles that can condense into the vacuum and break some symmetry, there can be a property that behaves mathematically exactly as if it describes some direction in a hypothetical internal space. As with the little magnets in the foregoing example, during a phase transition, defects associated with this internal degree of freedom

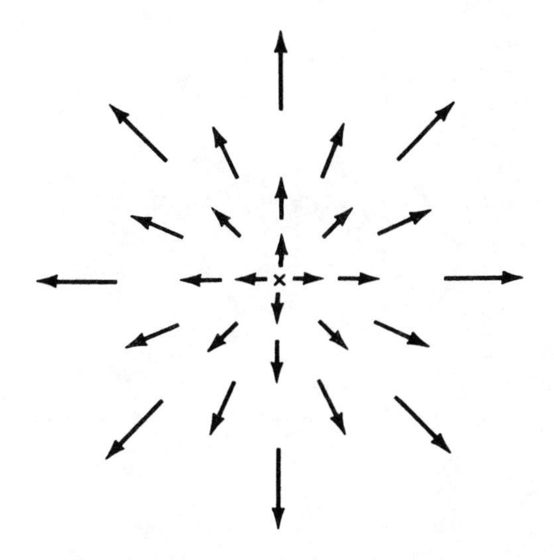

FIGURE 11.8

The only way for the magnetic field to interpolate smoothly above and below point X is for the average magnetization to vanish in a small region around X. Locally this is the energetically favored configuration, even though a globally preferred configuration would be for the field to be aligned uniformly throughout the volume. Thus a small region containing a little bit of extra energy is trapped around point X.

can and will be formed. One such defect, which is universal in any phase transition associated with Grand Unified Theories, is a configuration that relaxes to become a magnetic monopole. One could expect that monopoles would form fairly often, perhaps one defect at the intersection of one to ten regions in which the phase transition has occurred independently (that is, regions that were not in causal contact at the time of the transition). It was just this type of argument, first framed by the British physicist T. W. Kibble, that led Preskill and Zel'dovich to postulate the population explosion of monopoles produced during a GUT phase transition in the early universe. If there is some Grand Unified Theory that governs the interactions of matter, then monopoles were probably abundant at one time in the universe. Whether or not they are abundant today remains to be seen.

The arguments leading to the monopole production scenario outlined here are "topological." That is, they have to do with the impact of the global properties of symmetry breaking upon local configurations of matter. Another, much more straightforward, example shows how particles can be forced "against their will" to remain around long enough to dominate the universe today. In fact, it is an argument we all depend upon for the very existence of all that we can see!

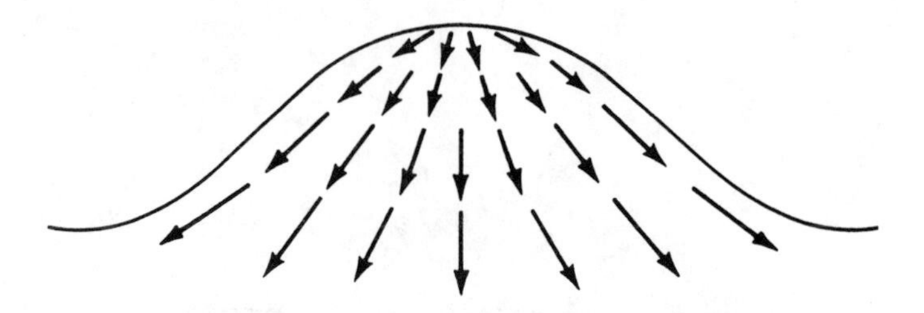

FIGURE 11.9
The energy density in space at each point is displayed in the vertical direction, while the alignment of the local magnetic field in space at each point is also shown; here the trapping of energy in the defect is clearly seen.

If, after the initial Big Bang explosion had cooled to temperatures at or near the proton mass, there were an equal number of protons and antiprotons in the universe, then I would not be around today to write this and you would not be around to read it. Protons have strong interactions with each other, and by the same arguments used in the previous discussion of WIMPs, one could easily show that essentially all protons in the universe would have annihilated with their antiparticles by the present time. Why has this not happened? Why are we here? At some time (perhaps during the GUT era, around 10^{-35} seconds after the Big Bang), physical processes must have produced an asymmetry between protons and their antiparticles. If this occurred, leaving perhaps 10,000,000,001 protons for every 10,000,000,000 antiprotons in the universe, then at late times all the antiprotons could have annihilated with an equivalent number of protons. There would have been one extra proton left over for every initial 10,000,000,000 protons in the universe. Because there were initially about as many protons in the universe as photons, when both were in thermal equilibrium, this would explain today why we find about 10,000,000,000 microwave photons left over in the cosmic microwave background for every proton we see in the universe. It also would explain why there are enough protons still left over to form all the elements in all the stars in all the galaxies that we can see.

Since we know that there must have been an asymmetry in protons and their antiparticles in the early universe, or else we would not be here, it is reasonable to speculate that a similar asymmetry might exist between other particles and their antiparticles. If this were the case for WIMPs, for example, one would be able to avoid the detailed relationship between their mass and abundance today derived on dynamical grounds. If there had been an asymmetry then, try as they might, they would not have been able to annihilate down below a certain level.

One might question whether it is just a numerical coincidence that the density of dark matter is only about ten to one hundred times that of protons and neutrons left over from the Big Bang today. After all, if the production mechanisms of the two kinds of matter were different, why should they end up with such similar energy densities now? If the dark matter is in some form of WIMPs, and if there was exactly the same cosmic asymmetry in the WIMP population as we know there must have been between protons and antiprotons, then this might explain today's ratio of energy densities. It would in fact just be equal to the ratio between the WIMP mass and the proton mass. The abundance of each type of particle that remains would be determined by the initial asymmetries, if these were equal, so that the ratio in their mass densities in the universe would be exactly the ratio of their masses. Since WIMPs might be expected to be ten to one hundred times the mass of the proton, the numbers work out.

Of course, this argument has primarily numerical coincidence going for it. While equal asymmetries in different populations of particles are possible in some models, they are not required. Not only that, but WIMPs are interesting in the mass range that was discussed earlier precisely because dynamical arguments imply that they would have cosmologically significant abundances today. Hence, a comparable ratio in energy densities between protons and WIMPs today would result regardless of whether the latter particles also had a primordial asymmetry. An asymmetry is not necessary to explain the numerical coincidence, although it does offer another way to force a remnant abundance of such particles.

If the dark matter that surrounds us is in fact made of something new, it is likely that one of the mechanisms (or something very like them) described in this chapter will explain why. Nevertheless, one should be very wary of cosmological reasoning. As the preceding numerical coincidence suggests, one can fall into traps when one gives the imagination free rein to reinvent creation. Although axions, WIMPs, and magnetic monopoles are all valid theoretical candidates for dark matter, with real reasons for existing and natural cosmological production mechanisms, the proof of the pudding will be in the tasting.

PART SIX

Desperately Seeking Dark Matter

CHAPTER TWELVE

THE MUSIC OF THE SPHERES?

The sound given forth by the circular movement of the stars is a harmony.

—PYTHAGORAS, quoted by ARISTOTLE

The Pythagoreans wondered why the sound that they were convinced must be produced by the movements of the heavenly bodies was not easily heard. Their explanation was remarkably simple: this sound is in our ears from the moment of birth, and hence we cannot distinguish it from silence. In the words of Aristotle, "What happens to men, then, is just what happens to coppersmiths, who are so accustomed to the noise of the smithy that it makes no difference to them."[1]

Alas, life has gotten more difficult. What distinguishes modern science from that of the ancients is that we are no longer content merely to postulate why we cannot hear sounds that should be there. Instead we search for other ways to detect them and thus prove their existence. When everything is said and done, physics remains an empirical science. All the fuss about exotic dark matter candidates, inflation, primordial fluctuations aside, until we actually experimentally determine what makes up the dark matter in our galactic halo, everything else—however well founded—is not good enough. If we are to be convinced of the existence of elementary particle candidates for dark matter, nothing short of directly detecting them in the laboratory will do.

The poetic notion of an audible "music of the spheres" is long gone. We now know that no sound emanates from the planets and stars as they move through the heavens. Sound waves, resulting as they do from vibrations in some ambient medium, are not transmitted through "empty" space. Yet this Pythagorean legacy lingers on in a much more subtle form today. If we believe that the dark matter in the galactic halo is distributed in the form of diffuse elementary particles, then interstellar space is not entirely empty after all. Dark matter particles move throughout the galaxy, in the space between the stars, even at the surface of the earth. As I suggested in the preface, the dark matter is not only "out there," it is "in here" as well. The interactions of this material with normal matter are so feeble that the motion of the earth through this aetherlike background does not produce anything like sound waves per se, but perhaps we can disturb this medium just enough so that, if we are very clever, we can detect its presence by other means.

It is this notion that makes the search for dark matter so enticing to theorists and experimentalists alike. Only if the dark matter in the galactic halo is made of elementary particles can we hope to detect it directly in terrestrial laboratories, without resorting to indirect telescopic data. The particles that compose the halo of our galaxy should be whizzing around us all the time. Most of these particles manage to penetrate the entire earth unimpeded, so they should have no problem traveling through laboratory walls and through waiting detectors. Just as neutrinos emitted by the sun are streaming through your body by the billions as you read this, so might cosmic WIMPs or axions. It is no longer acceptable science simply to marshal arguments to explain why we do not notice them. The challenge is to find some way by which we can apprehend them, if they do exist.

This mission is no easy task. Dark matter is invisible precisely because it interacts so weakly with normal matter. This interaction translates into at best a miniscule signal in even the most refined detector. Consider the following. One can calculate, for a given particle candidate, and a given halo density of particles, the theoretical rate at which such particles should interact with material in a detector. What happens when such an interaction takes place? Either the dark matter particle "bounces" off something in a detector, preserving its identity in the process and proceeding merrily along its way, or else it scatters and in the process loses its identity, converting into something else that may or may not be easily observed. In either case, the net effect is to deposit a tiny bit of energy into the detection apparatus. We can calculate how much energy will be deposited in an ideal detector, and at what rate. By any standards, for all particles under consideration, this number is exceptionally small. Table 12.1 presents the rough values for expected energy deposit per event, and total average power predicted to be produced by halo axions or WIMPs interacting in an ideal detector on earth. Contrast this with the average power emitted by one's big toe due to the sparse traces of radioactive materials inside it.

Axions or WIMPs would produce a signal, even in a large detector, whose power is between one-trillionth and one ten-millionth the power emitted by the radioactivity inside your big toe. Put another way, in order to build a detector that can register a signal equal in magnitude to that produced by a 100-watt light bulb, we would need, in the case of cosmic axions, for example, a detector volume equivalent to the size of the sun!

This staggering problem spotlights the central difficulty we face in trying to detect dark matter in the laboratory. *How can we detect such minute energy deposits that themselves happen very rarely, in large volumes of material?* Because this dilemma is universal, it is no accident that the detection techniques proposed to discover particles as different as axions, WIMPs, and monopoles share certain features. To detect such faint signals amid a background of other "noise" makes two demands on detectors: there must be ingenious mechanisms for amplifying the signal, and extraordinary measures must be designed to reduce any extraneous signals. Surprisingly, both of these requirements point clearly in a single direction: *cryogenics*—the technology of the supercold. Most experimentalists looking directly for monopoles, axions, WIMPs have all been forced to work at temperatures within 5 degrees of absolute zero—more than 450 degrees below zero on the Fahrenheit scale.

When we consider what is required to attempt to detect dark matter, we may wonder whether the effort is worth the rewards. Let me remind you of some. Dark matter would provide a totally new window on creation. At stake are not only our notions of how the universe evolved, and the origin of everything we can see

Candidate	Energy per Event	Average Power in Detector	Radioactive Power Produced in Big Toe
axion	0.00001 eV	10^{-24} watts/cubic meter	10^{-12} watts
WIMP	10–10,000 eV	10^{-19} watts/kilogram	10^{-12} watts

TABLE 12.1 PREDICTED ENERGY DEPOSITED BY DARK MATTER PARTICLES IN DETECTORS COMPARED TO TRACE AMOUNTS OF RADIOACTIVITY

about us today, but also what its future will be. We stand to make important discoveries about the most fundamental forces and particles that govern the way the world works. The discovery of dark matter would be one of the most stunning experimental feats of modern times. In any case, as long as the identity of the most dominant form of matter in the universe today remains unknown to us, we cannot claim that our scientific picture of the world even approaches completion.

If there is one lesson I have learned from my own work on dark matter detection it is this: never underestimate the abilities of modern experimentalists. It is fascinating to discover not only what is possible in a modern experimental research laboratory, but what is routine. Technology really does go hand in hand with pure research. New applications of technology continually challenge what is possible. And what is possible is often determined in research laboratories where practical applications may seem worlds away. As long as we continue to probe physics, at even the most esoteric scales, this interplay will persist. Dark matter detectors can have an impact on areas as diverse as enhanced nuclear reactor safety, the manufacture of computer chips, or measuring the magnetic field produced by an individual human brain as it thinks.

The validity of the estimates presented in table 12.1 for the power generated by dark matter interacting in a detector will determine whether we shall ever see a pulse registering the discovery of this elusive stuff. These estimates depend solely upon the "flux" of dark matter particles impinging upon the earth's surface. This in turn depends on the value of a single astrophysical observable: the density of dark material in the galactic neighborhood in which our solar system is located.

Fortunately, our observations of galactic dynamics fix the value of this all important quantity with relatively good accuracy, *independent of specific particle models.* First, in our galactic neighborhood, detailed studies of the dynamics of stellar systems suggests that the dark matter to light matter ratio is on average about unity. This is in accord with what one might expect from larger-scale models of the halo based on the observed galactic rotation curve. This translates into a specific inferred mass density of the halo in our region of the galaxy. In units useful for future discussions, the specific value of this mass density is about 3×10^8eV per cubic centimeter. A single proton weighs about 10^9eV (1 GeV); this suggests that there is nearly one proton's worth of dark stuff per cubic centimeter in the region of space in which our solar system is located. To put it another way, there is on average about one pound of dark matter at any one time in a region of space comparable to the volume of the earth, anywhere in our solar system. This estimate of the mean dark matter mass density is expected to be accurate to within a factor of about 1.5–2.

How much of this material crosses any given surface at any time depends on its average velocity. Fortunately, this too we can simply and reliably estimate. Since the solar system is moving around the galaxy at a speed of about 200 kilometers per second, and all other luminous objects in our region of the galaxy have comparable circular velocities, it is reasonable to assume that any dark matter particles will also move with a velocity of this magnitude. It certainly will if the motion of dark matter is in fact determined by the gravitational dynamics of the galaxy, as one expects. Now, if this material moves under gravity in a three-dimensional, roughly spherical halo, then one can show in a straightforward manner that the mean velocity of dark matter particles relative to the earth as their orbits intersect that of the earth's around the galaxy would be about 1.5 times as large as the sun's orbital velocity around the galaxy, or about 300 kilometers per second. This number is somewhat better estimated than the mass density itself and is probably accurate to within about 50 percent.*

The "flux" of dark matter impinging on earth's surface can now be determined as follows. An average mass density of 0.3 GeV per cubic centimeter composed of particles moving with an average velocity of 300 kilometers per second suggests that about 10 million GeV (10^7 GeV) worth of dark material (equivalent to about 100 million protons) will cross a surface of 1 square centimeter each second, on average. *This is the magic number.* Plug in the mass of your favorite dark matter candidate, and you can turn this mass flux into a particle flux. For example, if the dark matter is in WIMPs, each with a mass of about 100 GeV, then about 100,000 of them from the galactic halo population would be passing through each square centimeter of your body each second. If dark matter consists of magnetic monopoles, on the other hand, each of which may weigh 10^{20} GeV, then this would translate into a flux some 10^{20} times smaller. These numbers set the goals that any experimentalist must try to meet if he or she hopes to become the first person to discover what makes up most of the universe.

Obstacles remain, yet it is awesome to ponder how close we may be to finally answering the questions first posed by the philosophers of Miletus more than

*Pierre Sikivie has argued that dark matter which is only just now falling into the galaxy could have much larger velocities, and thus produce a somewhat different signal than other estimates would suggest. If we detect dark matter directly we can then test this hypothesis.

twenty-four centuries ago. What makes up the universe? How did it originate? What might its future be? It is likely that within our own lifetimes many of these questions will be resolved. This section is devoted to describing how modern experiments designed to meet the challenging requirements discussed above are closing in on the answers.

HUNTING THE MAGNETIC MONOPOLE

Ever since Dirac first demonstrated the profound theoretical significance of the existence of even a single magnetic monopole in the universe, experimentalists have continued to search for some evidence that magnetic monopoles might actually exist in nature. As described in chapter 10, the nature of these searches was radically altered when a fundamental particle physics theory finally was developed that actually predicted the existence of monopoles. After Blas Cabrera, using the first apparatus uniquely sensitive to magnetic monopoles, detected his startling Valentine's Day event, the worldwide search for magnetic monopoles began in earnest.

Monopoles were probably the first exotic cold dark matter candidates to be sought directly in the laboratory. We can learn a great deal about the general issue of dark matter detection by reviewing this search, both because the effort is mature and because it is ongoing. To be honest, however, I must note that theoretical interest in the possibility that monopoles actually make up the dark matter of the universe has subsided since the development of inflationary cosmologies. There are no very compelling arguments why the cosmic abundance of monopoles today should lie between a level so large as to be ruled out on cosmological grounds and a level infinitesimally small, which is as might be predicted by inflation. Nevertheless, nothing rules out this possibility. Arguments about the physics of the very early universe must always be taken with a grain of salt. It is certainly true that if someone sees a gold-plated monopole event tomorrow, at a rate that is cosmologically influential, all these statements about why the density of monopoles should be larger or smaller will be scrapped. What will remain is evidence that somehow the physical processes that determine the abundance of monopoles in the early universe are somehow different from what we currently model them to be.

I should add, to be fair, that several other astrophysical arguments maintain that the cosmic abundance of magnetic monopoles might be far less than that which would make it relevant as dark matter. Perhaps the most solid statement is that of astrophysicist Eugene Parker, which has since become known as the "Parker bound." Parker pointed out that if the magnetic monopole abundance in the galaxy exceeded a certain level, then as they move in response to the galactic magnetic field, they will sap energy from this field, dissipating it away to zero on time-scales far too brief to be replenished to the value it is observed to have today. As long as the motion of magnetic monopoles in the galaxy is governed primarily by the galactic magnetic field, Parker's argument holds independent of the monopole mass, and it depends only on the magnetic charge on the monopole. For a monopole with the magnetic charge predicted by Dirac, the Parker bound limits the flux of magnetic monopoles at the earth's surface to be less than about 10^{-14} per square centimeter per second.

How does this compare with the flux of magnetic monopoles if they are the galactic halo dark matter? Using our magic number of 10^7 GeV per square centimeter per second for the mass flux in elementary particle halo candidates, we see that this results in a particle flux far in excess of the Parker bound for any monopole with a mass characteristic of the elementary particles we know and love, like the proton. But Grand Unified Theories not only predict the existence of magnetic monopoles, they fix their mass to be about one hundred times the mass scale at which unification occurs. Since present limits from the nonobservation of proton decay suggest that the GUT scale is in excess of 10^{15-16} GeV, this implies that GUT monopoles have masses in excess of 10^{17-19} GeV, or 17–19 orders of magnitude heavier than the proton. If the galactic halo were made of monopoles this heavy, their flux at the earth's surface would be less than 10^{-10} monopoles per square centimeter per second. The Parker bound is about 4 orders of magnitude smaller than this flux. However, for monopoles heavier than 10^{17} GeV, the Parker bound becomes less stringent, because such particles are so heavy that the galactic magnetic field could not compete with gravity in governing their motion; accordingly, they would not so effectively remove energy from this field as they move about the galaxy. Thus, although the Parker bound indicates that magnetic monopoles should be pretty rare beasts, for very heavy magnetic monopoles this upper limit is still compatible with a monopole-dominated galactic halo.

Aside from arguments about which is lower, however, both the Parker upper bound and the predicted flux of a GUT-mass halo density of monopoles are pitifully small. How can one hope to measure such a slight flux, which would be more than 20 orders of magnitude smaller than a flux of, say, GeV-mass galactic halo WIMPs?

Here one encounters the chief distinction between magnetic monopoles and the other dark matter candidates I have described. Though the flux of WIMPs or axions at the surface of the earth might be quite substantial, only a very small portion of this flux will actually register a signal in a detector at any one time. However, although the flux of monopoles would be orders of magnitude smaller, one can design experiments that should be sensitive to each and every monopole that crosses the detector volume.

This is not to say that monopoles might be discovered with ease, however. Even if every GUT monopole that enters a detector leaves a signal, the monopole flux is likely to be so small that any normal-sized detector volume would be lucky to register one event per year. If such a detector is also sensitive to other stuff, such as the flux of particles contained in the cosmic rays, which continually bombard the earth, any monopole signal could easily get lost.* Any detector of monopoles obviously must be based on a distinct signature not produced by any other particles.

This was the idea behind Cabrera's detector. One of the fundamental predictions of the laws of electromagnetism, at the basis of all electric power generators, is that changing the strength of a magnetic field near a wire causes a current to flow in the wire. Such a change can be caused by moving a small permanent magnet around inside a loop of wire; this is what happens inside an electric generator. If a magnetic monopole passes through a loop of wire, it too will cause a current to flow, because as the monopole travels through the loop the changing magnetic field due to the moving monopole will induce a current, by Maxwell's laws. It turns out that the total current induced by the moving monopole is independent of its speed; it depends only on its magnetic charge.

Normally the induced current produced by a monopole moving through a loop of wire would not be measurable, because, depending on the electrical resistance in the wire, the current could be extremely minute, and in any case it would be quickly dissipated. However, at very low temperatures, certain materials become superconducting, and this changes everything. As noted earlier, in a superconductor all electrical resistance disappears and currents can flow indefinitely. Cabrera, a low-temperature physicist, whose previous research had been in

---

*"Cosmic ray" is the generic name for the flux of elementary particles emanating from outside the earth's atmosphere. Cosmic rays contain standard particles such as electrons and protons, from a wide variety of sources, some exotic and some not so exotic. These include local origins such as solar flares on the sun's surface, and faraway, poorly understood sources such as distant supernovae. The flux of such particles has been measured and provides an unavoidable background for any particle detection experiment on the surface of the earth.

the field of superconductivity, was the first to recognize that a superconducting detector could be designed that would give an unambiguous signal if a single monopole passed through, but would not register the same signal if any other particle passed through.

However, unless a way could be found to reduce the earth's ambient magnetic field signal in the detector, then the change induced when a monopole entered would be unnoticeable. Here Cabrera used a trick based on superconductivity. Because of the "Meissner effect," superconductors expel ambient magnetic fields that attempt to enter their surface. By fashioning a "balloon" made of superconducting lead, which is malleable, Cabrera could begin with an uninflated balloon, inside of which would be some remnant of the earth's magnetic field that was trapped when the lead was cooled and became superconducting. Then, by "inflating" the balloon, he could reduce the strength of the field significantly, since the superconducting lead surface would keep out any more stray magnetic fields, and the trapped magnetic field strength inside would be greatly diluted by the increase in volume. By inserting a new uninflated balloon inside the first and inflating this second balloon and then repeating this process several times, the magnetic field in the region inside the lead shield could be reduced to a value far below that which would result from the traversal of a single monopole through the volume. (Remember that magnetic monopoles are predicted to be so heavy that they are not easily stopped, and they can travel large distances through ordinary matter.) Moreover, the lead balloon would shield the region inside from any fluctuations in the outside magnetic field which could otherwise accidentally mimic the signal that is predicted to result from a single monopole traversing the detector inside the shield.

Having thus carefully shielded this volume, Cabrera could monitor the current flowing through a superconducting wire placed inside it. Because of the particular form of the equations of electromagnetism, only a magnetic monopole that passed through the surface spanned by the wire loop would produce a current. Similarly, because the induced current requires a nonzero magnetic charge, any normal cosmic ray particle, for example, which is not magnetically charged, would produce no signal. Finally, because the magnitude of the induced current depends only on the magnitude of the monopole charge, and not on its velocity or mass or anything else, the resulting signature from a monopole traversing the detector would be unambiguous. When a monopole passed through the wire loop, the current in the wire, which we can assume would be initially zero, would jump quickly to a fixed nonzero value. This would be a calculable multiple of the magnetic charge on the monopole. This situation is shown schematically in figure 12.1.

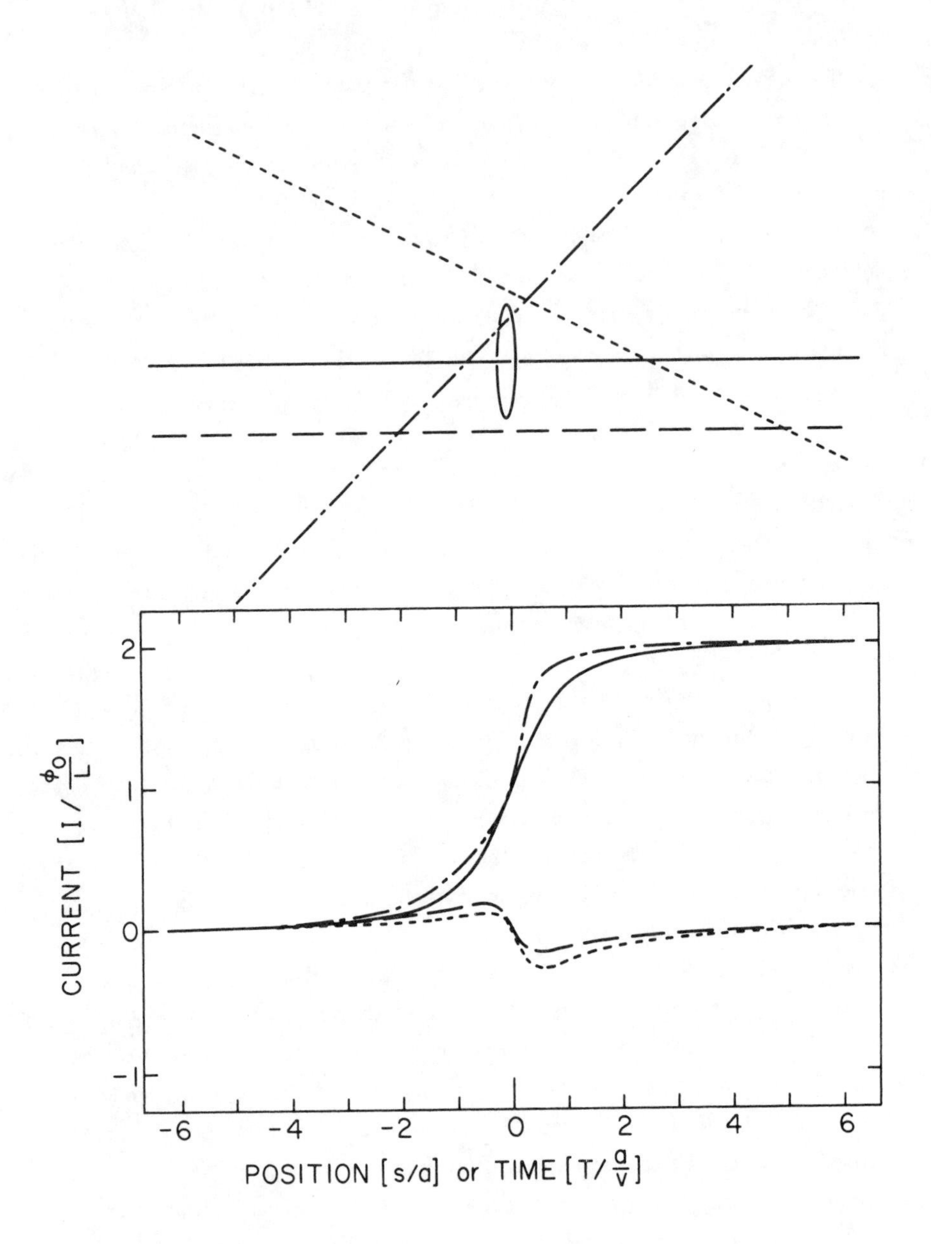

FIGURE 12.1

Four different trajectories of a magnetic monopole near a superconducting wire loop are shown in the upper half of the figure, two of which actually go through the loop. In the lower half is shown the calculated response of the current (normalized in terms of the expected response to a Dirac monopole) inside the loop as a function of monopole position (or time) for each trajectory. Only in the case of the trajectories going through the loop is the net current in the loop permanently affected. In this case, the amount by which the current changes is proportional only to the magnetic charge on the monopole, and not to its mass or speed. (Courtesy of Blas Cabrera.)

Cabrera built this detector and left it running in a laboratory located in a subbasement of the Stanford University physics building. With a chart recorder, he could check the history of the current flow in the wire periodically to see whether a monopole had passed through. Moreover, he could calibrate the detector by turning on an electromagnet contained inside the volume of the detector to confirm the response of the superconducting loop current. On the day after Valentine's Day in 1982, Cabrera found the strip recorder output depicted in figure 12.2.

This recording strip appeared to register a completely clean signal from a passing magnetic monopole, with precisely the charge first predicted by Dirac. By carefully checking all possible noise sources, Cabrera could not reproduce the signal that had been observed in his detector on that Valentine's Day. Moreover, the cosmic monopole flux, which would be inferred from observing this one event in the time since the detector had first been turned on, was about 10^{-9} per square centimeter per second. Using our magic mass flux formula, we would arrive at a halo density of monopoles with this flux for a monopole mass of around 10^{16} GeV, exactly where the simplest GUT monopoles might be expected to lie.

Regrettably, as we noted previously, no other experiment since then, including ones done by Cabrera's group, has observed a similar event. Given the running time of his original detector, and the size of detectors built later, the probability that cosmic monopoles with the flux quoted here should have resulted in

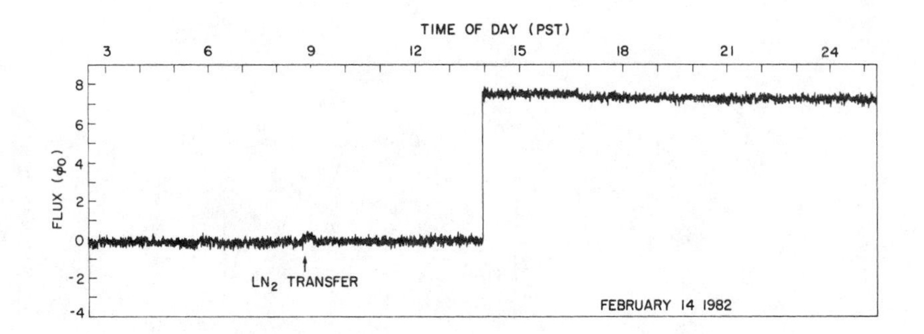

FIGURE 12.2

The actual output of Cabrera's strip recorder for the period of February 14, 1982. The small bump around 9:00 A.M. is due to a transfer of liquid nitrogen in tanks involved in cooling the experiment. The "monopole-like" event at around 2:00 P.M. is clearly seen. For comparison, the very small shift in the measured signal at around 5:00 P.M. is the response of the detector being jarred by hand. (Courtesy of Blas Cabrera, from *Physical Review Letters* 48 [1982].)

an event in the original detector and no other events since is about one in a thousand. Although no one has successfully explained what caused the original event, we must admit that the odds are against it having been due to a magnetic monopole with that mass and flux.

Because some random fluctuation in a single superconductor might accidentally, with very low probability, mimic the signal from a monopole, Cabrera and other groups around the world who build superconducting detectors altered the design to remove this possibility. The alteration is very simple: use more than one loop. For example, consider the configuration of six independent loops arranged around the sides of a box, as in figure 12.3.

Because of the geometry of the arrangement, it is impossible for a monopole to pass through the box without going through two and only two loops. The signal that one would expect, then, would be a simultaneous jump, *with the correct magnitude,* in the current observed to flow in two of the loops, with no similar change in any of the other loops. Any other bizarre process, such as a rare thermal fluctuation, a cosmic ray that strikes one of the loops and excites some local current, or a power surge in the whole detector that excites all the loops, or anything else we can imagine, will not excite just two and only two loops by exactly the right amount at the same time.

All superconducting monopole detectors now have such a "coincidence"-sensing geometry of one sort or another. The largest such detector in the world at present was built by Cabrera's group; it involved a total area spanned by superconducting loops of about 10 square meters. Even without all the magnetic shielding

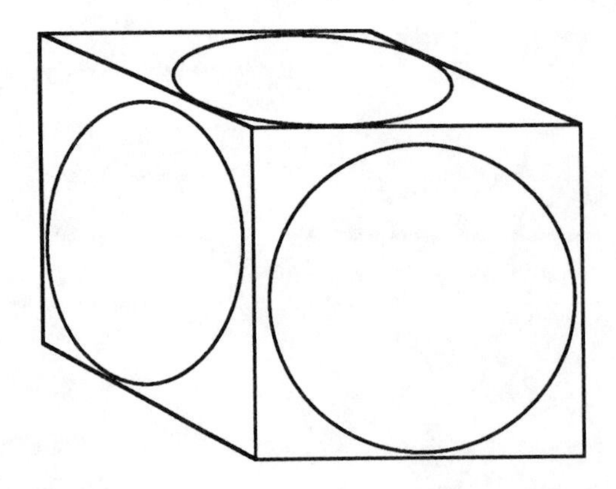

FIGURE 12.3

If a monopole passes through a cube with six independent loops arranged around the sides, the currents in two and only two of the loops will be permanently affected.

required, as well as the refrigeration apparatus necessary to keep the whole thing superconducting, it is truly a gargantuan device. The combined running time of Cabrera's and other detectors built since about 1983, during which no signal has been seen, can be used to place a statistical upper limit on the total flux of cosmic monopoles impinging on the earth's surface.

Present observations suggest that with a probability of better than 90 percent the monopole flux at the earth's surface is less than about 4×10^{-14} monopoles per square centimeter per second. This would rule out a halo density of monopoles with masses less than almost 10^{21} GeV. A single monopole this heavy would be better measured in grams. In this system of units, monopoles weighing less than about 100 milligrams are disqualified as dark matter candidates. Nevertheless, notice that even after the concerted detection effort that has so far been devoted to monopoles, the limits obtained from superconducting detectors were still almost 2 orders of magnitude larger than the Parker upper bound for monopoles less massive than 10^{17} GeV. Therefore, we might not have expected, if this limit is correct, to have detected fluxes this large in any case, if the monopole mass is less than 10^{17} GeV. To obtain sufficient sensitivity to probe for monopoles having a flux at the earth below the Parker limit, it is obvious that the size of monopole detectors must be greatly expanded. The superconducting detectors of today, given their limited area, would have to run for well over a century before they could produce limits in this range.

Because of the cost and size of such a huge monopole detector, we probably need more evidence before physicists or funding agencies will consider developing such a device. On the other hand, some general-purpose large detectors (without superconducting loops) are also believed to be sensitive to magnetic monopoles. Building mammoth devices is nothing new in particle physics. One need only consider the vast underground water detectors used to look for proton decay, which detected the neutrinos from Supernova 1987A. As I have described, the capacity of these cubic detectors, up to 60 meters on a side, exceeds 50,000 tons of water, which is continuously monitored for the light emitted when very fast charged particles travel through matter (see figure 6.5).

If one uses a general-purpose detector to search for monopoles, one must also attempt to reduce the cosmic ray backgrounds. As I have indicated, at the surface of the earth these would be so great as to swamp any monopole signal. For this reason, the large particle detectors now being developed to seek monopoles, as well as other particles of extraterrestrial origin such as neutrinos from supernovae, are placed very deep underground, where the flux of ordinary particles in cosmic rays is severely depleted. The miles of rock overhead stop all but the most energetic particles. One such detector, called MACRO (Monopole Astrophysics and Cosmic Ray Observatory), was constructed in a huge underground laboratory connected to the automobile tunnel at Gran Sasso, in Italy. This detector, completed in 1990, has roughly 1,200 square meters of detection area, over 2 orders of magnitude greater than all the previous monopole detectors combined. This impressive detector was designed to have nine identical modules, each of which can operate independently (see figure 12.4).

The MACRO detector is similar to a conventional cosmic ray particle detector in that it is designed to detect the ionization caused when a particle passes through the detector with enough energy to ionize atoms, but it is much larger and should also be sensitive to monopoles. When various particles, including charged particles and monopoles, traverse matter, they loose energy by colliding against atoms and ionizing them. When the electrons in the material recombine

FIGURE 12.4
The Monopole Astrophysics and Cosmic Ray Observatory (MACRO) detector in the Gran Sasso Tunnel in Italy. (Photo courtesy of Laura Petrizii.)

with the ionized atoms, light of a well-defined frequency is emitted. In certain clear liquid materials, this "scintillation" light can propagate out of the material and be measured. The amount of light emitted tells how much energy was deposited in the detector. This special detector is called an LSD (liquid scintillation detector). The MACRO detector contains many volumes filled with LSDs sandwiched between blocks of absorbing rock and other charged particle detectors. Any normal particle would be stopped by such a mass of material. For example, for an electron to pass through the entire detector, it would have to possess an initial energy in excess of about 20,000 GeV, or about 20 million times the energy associated with its mass. A magnetic monopole weighing 10^{20} GeV, on the other hand, has an energy of motion far in excess of 20,000 GeV. Even if it is moving very slowly, it packs quite a kick. Thus, even if a monopole *could* interact as strongly as an electron with detector material, it would have no problem passing through the entire detector volume.

Conventional LSDs are insensitive to monopoles that might come from the galactic halo because these objects travel very slowly. Normally energetic particles in cosmic rays move at or near the speed of light. Thus, as they traverse detectors, they deposit their energy in the span of millionths of a second. A galactic monopole, however, which may be moving with a speed of only 1 part in 1,000 of the speed of light, will take much longer to pass through the detector. The light from the ionization caused by such a monopole would then take orders of magnitude longer to accumulate, perhaps requiring a span of thousandths of a second. Such prolonged faint signals are usually vetoed in conventional detectors in order to reduce the background noise signal. At the depths of the MACRO detector, and in other LSD monopole detectors, the noise is sufficiently weak that the electronics designed to probe for such long signals is swamped. The only uncertainty that remains is whether in fact a slowly moving monopole will ionize sufficiently to produce enough scintillation light to measure. Calculations suggest that it should, but this is still an area of some uncertainty.

Previous LSD detectors ran with effective areas in excess of 10 square meters for more than a year. To the extent that one accepts the claims of sensitivity for slowly moving monopoles, the limits were at least 1 order of magnitude better than from the superconducting detectors. MACRO has already improved the limits on magnetic monopoles in the galaxy to well below the Parker upper bound. This is the region that is most interesting in terms of physics and in which a positive signal would probably spur funding of a large detection device dedicated to monopoles alone. Such a device could then verify a signal seen by MACRO.

Whether or not MACRO discovers a galactic halo of monopoles, the search for monopoles is not in vain. The technology developed for monopole detection has applications in a variety of other areas in cryogenics. For example, one spinoff of the low-temperature physics technology developed for monopole detection has been the ability to measure extremely small magnetic fields, such as might be produced by currents inside the brain. This technology promises noninvasive diagnosis and treatment for brain tumors and other related problems.

The excitement generated by the single event observed on 14 February 1982 bestowed a lasting legacy on theoretical physics. With the possibility that magnetic monopoles, and thus GUTs, were real, theoretical research in many areas was spurred. Among the beneficiaries was cosmology. Remember that the inflation theory was developed in direct response to the question of the existence of monopoles in Grand Unified Theories. The stir caused by the possible discovery of a magnetic monopole inspired people to delve deeper into inflation theories and other astrophysical consequences of GUTs and monopoles. Imagine what excitement and developments might ensue if a verifiable discovery of magnetic monopoles were to take place at MACRO under the mountains of Eastern Italy.

CHAPTER THIRTEEN

Of Thermometers and Radios

Monopole detectors are in some sense an established art. The recognition that cosmic WIMPs or axions too might yield to the intrepid experimentalist has been more recent. This chapter will show that despite the considerable technological challenges of building working WIMP or axion detectors, the race to find dark matter may at last be entering the home stretch.

WIMPS: THE SEARCH CONTINUES . . .

By far the most active area of experimental research in the matter of dark matter detection involves weakly interacting massive particles, WIMPs. New results are constantly emerging from the most surprising places, a vast range of detectors devoted solely to WIMPs is under development at dozens of laboratories around the world, and a whole new field of applied technology is springing up. With the current level of research activity, we soon may be able to probe most of the parameter range that would be expected if WIMPs are in fact dark matter. This effort, in combination with new accelerators, will, among other things, determine whether supersymmetry governs nature at energies accessible to experiments on human time-scales. These developments are all the more exciting when one considers the short history of this field. Before 1985 the possibility of directly detecting WIMPs as dark matter was not even considered viable.

To my mind, the event that started things rolling was the publication in 1985 of a theoretical paper by one of the most well known theoretical physicists today, Edward Witten, primarily known for his mathematical work associated with string theory, and his student Mark Goodman at Princeton. Goodman and Witten deserve credit for recognizing and publicizing a deceptively simple fact that may have been known earlier in other contexts, as well as for pointing out its significance for dark matter detection.

Anyone who has ever considered the interactions of the only weakly interacting particle we know exists for certain, the neutrino, realizes that the strength of the interactions of neutrinos with ordinary matter increases as the neutrino energy rises. For this reason, high-energy neutrinos produced in modern accelerators are detectable, while the low-energy cosmic background of light or massless neutrinos which we know must exist interacts so feebly with matter that no one has ever proposed a viable method for detecting it. Now, if a very heavy new 2 GeV-mass neutrino were to make up the dark matter in the galactic halo, then, given an expected mean velocity of about 300 kilometers per second, the mean energy of motion of these objects would be about one-millionth of the energy associated with their mass, or about a thousand eV. If we blindly insert this energy value in the standard neutrino interaction calculations, we would find that such objects would be so weakly interacting that they could pass through, on average, about *100 million light-years of rock without interacting once.* If this were true, we would have no hope of detecting these neutrino WIMPs. What Goodman and Witten pointed out, however, is that the interaction strength of very heavy neutrinos is *not governed by the standard equations.* Instead of growing as the square of this energy, the interaction strength grows as the square of their mass. For a 2 GeV-mass neutrino moving at 300 kilometers per second, the ratio of the enhanced interaction strength compared to the naive estimate is a factor of 10^{12}, and this difference would be enormous. Instead of requiring 100 million light-years of rock before the average neutrino WIMP interacts once, one would only need about 10^{13} centimeters of rock, about 100,000 times the size of the earth.

When framed this way, the gain may not seem worth crowing about, but, in fact, it pushes WIMPs like massive neutrinos from the realm of the hypothetical into the realm of the observable. A simple calculation of probabilities illustrates this. If you have a detector that is 10 centimeters on a side instead of 10^{13} centimeters, the probability of stopping a given WIMP is about 1 part in 10^{12}. Thus, if 10^{12} WIMPs pass through your detector, on average, one of them will interact. If you can detect this interaction, you are in business. Now, how long should it take for 10^{12} WIMPs to pass through your detector? Since the flux of 100-GeV halo WIMPs is, from our magic formula (see chapter 12), about 10^5 per square centimeter per second, and a detector that measures 10 by 10 by 10 centimeters has a surface area of about 100 square centimeters, it should take about 100,000 seconds, or about one day, before a WIMP scatters in your detector! (This calculation is very rough, and factors of 2 and π can lower the actual estimates.)

What actually happens when a WIMP interacts in a detector? Normally this interaction implies that a WIMP particle merely "bounces" off of the nucleus of some atom in the detector. Because the masses predicted for WIMPs are comparable to the masses of atomic nuclei, what happens resembles the collision of two billiard balls. The WIMP bounces off in a new direction and the nucleus gets a kick and recoils (see figure 13.1).

In the process, the WIMP can deposit at most an amount of energy equal to its initial energy of motion. This would happen, for example, in a head-on collision with a nucleus of mass equal to the WIMP mass, just as when you hit a billiard ball head-on your cue ball can stop and the target ball can proceed onward with the same velocity that the cue ball had initially (see figure 13.2).

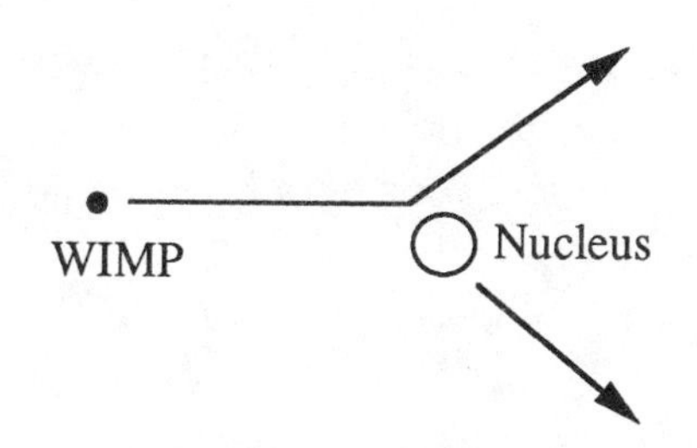

FIGURE 13.1

In a WIMP collision with an atomic nucleus, the two objects behave essentially like billiard balls, with the WIMP changing speed and direction and the nucleus recoiling.

FIGURE 13.2

In a head-on collision of a WIMP and an atomic nucleus, both of the same mass, the WIMP can transfer all of its energy of motion to the nucleus, which will recoil with the same velocity as the WIMP initially had, just as for billiard balls of equal mass.

Normally, collisions are glancing and the WIMP will not lose all of its energy. On average, the energy deposited is usually less than half of the incident energy of motion carried by WIMPs. Since a WIMP having a mass of a few 100 GeV, moving at 300 kilometers per second, has an energy of motion of a few hundred thousand eV (1,000 eV = 1 keV), a WIMP detector must be capable of measuring energy deposits smaller than this amount, in volumes in excess of a few kilograms, which occur less frequently than about once per day or so. Therein lies the rub.

As a practical matter, it is worth noting that the interaction strengths estimated by Goodman and Witten for heavy neutrino-like WIMPs are some 3–6 orders of magnitude larger than those of light neutrinos emitted by nuclear reactors, which are themselves "routinely" detected. Of course, the flux of such neutrinos near a reactor is much larger than the predicted cosmic WIMP flux, but the increase in interaction strength for WIMPs largely compensates for this. The major factor that works in favor of detecting reactor neutrinos and against detecting WIMPs is the fact that the energy of motion of reactor neutrinos is about 3 orders of magnitude greater than the energy of motion for halo WIMPs, and thus larger energy deposits per interaction are possible. The greater the energy deposit per event, the easier it is to detect it. In fact, it is this small energy available in WIMP scattering that provides the biggest challenge for modern detectors, as we shall see. Nevertheless, the fact that particles with weaker interactions than WIMPs have been detectable in experiments beginning as early as about 1956 gives us some confidence in the practical possibilities for WIMP detection.

Surprisingly, while Goodman and Witten, and, independently, Ira Wasserman at Cornell, concerned themselves with investigating the possibility of the direct detection of dark matter, the immediate impact of their ideas inspired an explosion of theoretical work which led to experiments that could *indirectly* probe for galactic halo WIMPs. Logically, we can frame this development as follows. Given the interaction length of about 10^{10}–10^{13} centimeters for WIMPs, we have two choices if we wish to detect them: (1) build a manageable detector that makes use of probabilistic effects to detect a small fraction of the WIMP flux, or (2) use a detector big enough to detect almost all of them. In the second case, the detector would have to be at least as large as the size of the earth, far beyond human capabilities. But nature may have already provided us with such a detector—the sun!

The story of how the sun can be used indirectly as a WIMP detector demonstrates the rapidity with which new cosmological developments have occurred. In science, as in life, fortune favors the prepared mind, and when the scientific community is primed, sparks can fly with amazing speed.

About the time when Goodman and Witten's work was being done, Blas Cabrera, Frank Wilczek, and I had just proposed a new type of device to detect solar and reactor neutrinos. Witten happened to visit us at the Institute of Theoretical Physics in Santa Barbara and discussed his ideas about direct detection with us. It was apparent that the device we were proposing could also be used to detect the WIMP events referred to by Witten and Goodman. Because Wilczek and I had been considering noise backgrounds in our detector, and signatures that might allow one to extract a small signal from these backgrounds, we spent an afternoon thinking of various signatures for dark matter in such a detector as well. One idea, which we soon discarded, was the possibility that most of the WIMPs would be stopped by the earth so that a detector on one side of the globe would see more events when that side was turned in the same direction as the sun's motion through the galaxy, as opposed to when that side was in the earth's "shadow." The idea was abandoned because we calculated that most of the WIMPs would not be stopped by the earth, so there would be no "shadow."

A few weeks later, upon my return to Harvard, I learned of the findings of two colleagues at the Center for Astrophysics. They had rediscovered an interesting fact that had been noted several years earlier by John Faulkner and colleagues at a time when it seemed of little application. If there existed in the sun a population of exotic objects with masses in the range of 5–10 GeV, constituting only 1 part in 10^{-12} of the sun's total mass, then under certain conditions these objects could change the conditions inside the stellar core in such a way as to reduce the temperature in the very center of the core. This would reduce the observed flux of neutrinos from the sun, bringing theory in line with observation, without grossly affecting any of the other successful features of the standard model for the solar interior. Unfortunately, arbitrarily positing such an abundance of exotic particles, which they called *cosmions,* in the sun made the idea seem contrived, which is probably one reason why the idea had not taken off earlier.

I noticed, however, that the required interaction strengths for these hypothetical particles upon scattering in collision with the hydrogen in the sun were somewhat larger than, but close to the WIMP interaction strengths calculated by Goodman and Witten. Our work on neutrino detectors in Santa Barbara came to mind, and I tried to determine whether WIMPs might be stopped by the sun instead of the earth. The radius of the sun, at about 10^{11} centimeters, is much closer to the 10^{13} centimeters of the approximate mean free path of WIMPs in matter. A quick estimate suggested that a fair number of WIMPs not only might scatter at least once in the sun, but also would then lose sufficient energy to be gravitationally captured inside the sun afterward. When I worked out what the present abundance of dark matter particles inside the sun would be today if all of those from the halo which were incident on the sun since its formation had been captured, I found a remarkable result: they would produce a net abundance of about 1 part in 10^{12} by weight—almost what had been required for the "cosmion" mechanism to work. If WIMPs were actually the dark matter, here was a potential way to have enough particles of the right specifications inside the sun to reduce the solar neutrino flux today to levels compatible with observations. Perhaps the dark matter problem and this solar neutrino problem (see chapter 3) had the same solution! Rarely have I been so excited by a calculation. It really seemed as if the dark matter might actually be made of WIMPs.

My exuberance was short-lived. Shortly after performing my calculations, I contacted my colleague and was reassured that he and his student had independently arrived at similar conclusions, based on the assumption that their hypothetical cosmions formed the galactic halo dark matter. We met soon after that and discussed our complementary results. They had done more detailed estimates of gravitational capture in the sun, but since cosmions had no real physics associated with them they could not realistically estimate scattering processes. I had investigated the significance of collisions and scattering of actual WIMPs off of hydrogen versus helium in the sun, for example. In the midst of our enthusiasm, however, two realizations put a damper on the whole business, at least as far as WIMPs were concerned.

A notable aspect of WIMPs is that their presumed remnant abundance today is due to the notion that WIMP and anti-WIMP annihilation in the early universe should dynamically "freeze out," as I noted earlier, so as to leave a nonzero abundance of these objects. This not only implies that there is a definite connection between their mass, interaction strength, and abundance today: it means in general that not only WIMPs would be captured by the sun, but that their antiparticles—also present in the galactic halo—would be captured as well. As the WIMP abundance builds up inside the sun, eventually WIMPs will encounter anti-WIMPS and they will begin to annihilate. The night after our discussions, I calculated that under very general circumstances annihilation would quickly take precedence over capture to determine the remnant abundance of WIMPs in the sun today. This remnant abundance could then be shown on general grounds to be at least 4 orders of magnitude less than one would predict in the absence of annihilation, and thus at least 4 orders of magnitude too puny to affect solar neutrino emission rates. Moreover, using detailed capture estimates the actual interaction rates of WIMPs—that is, neutrinos and LSP's; were probably 2–4 orders of magnitude too small to do the job in any case, even without annihilations. One could try to get around this, but as far as WIMPs were concerned, this problem did not bode well for any proposed solution to the solar neutrino problem.

Because of these obstacles, it was decided that I would not publish my manuscript proposing solar capture of WIMPs. Their parallel proposal, based on hypothetical cosmion capture, was submitted for publication. Within a month or so we would combine our results in a joint publication illustrating the problems and possible solutions. This would allow the possibility of a dark matter–solar neutrino connection to be pointed out independent of the grim realities of particle physics. (In retrospect, this was probably a mistake. It led to some confusion about the idea's problems and about any connection, or lack thereof, between "cosmions" and WIMPs.)

In any case, some clouds have silver linings. Although annihilation of WIMPs in the sun appeared to kill immediate hopes of using WIMPs to solve the solar neutrino problem, their annihilation turned out to be the key that allowed a new generation of experiments to probe for their existence indirectly. A year or so earlier, Mark Srednicki at Santa Barbara and his collaborator, Joseph Silk at Berkeley, had analyzed the rate at which one specific WIMP candidate, the photino, might be detected by the photons or other energetic cosmic rays that might be produced by annihilations of photinos in the galactic halo itself, independent of any buildup in astrophysical objects. Stimulated by the proposal that dark matter might be captured in the sun, Srednicki and collaborators also then examined annihilation in the sun. Besides confirming my results about the effect of annihilations on stellar abundance, they went on to point out an important potential signature of such annihilation. They argued that among the particles produced when photinos annihilated were likely to be some good old-fashioned light electron-type or muon-type neutrinos. These neutrinos could easily escape the sun. Moreover, they would carry enormous energies characteristic of the mass of the photinos that had annihilated. It was suggested that these might be observable on earth as a new high-energy neutrino background. Wilczek and I had independently been exploring similar ideas, and we continued our investigation with Srednicki of the different possible signatures from WIMP annihilations in the solar system. We all concluded that this neutrino signature—from annihilations of WIMPs captured either in the sun or possibly in the earth—seemed the most likely to be detectable.

How could such a background of high-energy electron or muon neutrinos at the earth's surface be detected? Once again, the huge underground water detectors designed to search for the decay of a proton turned out to have a broader utility than originally expected. I think this is a specific example of a more general occurrence in science. Whenever a bold new project is undertaken, unexpected applications well beyond the original design are often found. Nothing could be more true than in the case of proton decay detectors. These colossal devices which I discussed briefly in chapter 6, originally produced to examine an esoteric prediction made by particle theorists about physics at energies orders of magnitude beyond that which we can experience directly on earth, are also serving as valuable astrophysical observatories. They already created a minor revolution by detecting for the first time neutrinos during Supernova 1987A. The new work I have described here hinted that they might help discover dark matter as well.

If a proton decays, it must decay into something lighter. The total energy that can be released in the decay is limited to that associated with the mass of the proton, about 1 GeV. Since the proton is charged, at least one charged object must be emitted in the decay, so physicists decided to probe for the trails left by charged particles in a substance that could be amassed in large volumes and monitored continuously. Recall from my discussion in chapter 6 that large volumes were necessary because theory suggested that at best only one decay or so would take place each year among 10^{30} protons. To minimize the ever-present cosmic ray backgrounds that contain energetic charged particles, these devices have to be placed far below the surface of the earth. Several enterprising groups of experimentalists around the world have built spectacular cavernous water tanks underground. Thousands of light-sensing detectors are located at their peripheries, each scanning for the light emitted when a charged particle travels through the water, exciting atoms along the way.

One of the chief cosmic ray backgrounds limiting the efficacy of proton decay detectors comes from what are called "atmospheric neutrinos." These neutrinos are produced when cosmic ray protons or other particles smash into the earth's atmosphere. The neutrinos that result from these collisions are not stopped by the miles of intervening rock and can pierce the depths at which proton decay detectors are located. If such a high-energy neutrino, with energy in the range comparable to the proton's mass, interacts in a water detector, it will produce an energetic charged particle, such as an electron, or its heavier cousin, the muon, which will be detected. Such a single charged track beginning in the detector can closely mimic some of the possible proton decay modes for which the experiments were searching. (In fact, this "atmospheric neutrinos" background has itself proved to be remarkably useful. Recent observations of this background by the Superkamiokande detector in Japan have provided the first solid evidence that at least certain neutrinos have small, but nonzero, masses.)

The early predictions, later refined by various groups, all indicated that WIMPs could produce a neutrino background with characteristics similar to the atmospheric neutrino background. Suddenly a nasty background had become a useful signal. If the actual neutrino background could be measured in proton decay detectors, and compared with the predicted atmospheric background, any excess could be due to WIMP annihilations. And the lack of such an excess could allow limits to be placed on WIMPs.

Since that realization, all backgrounds in proton decay detectors have been analyzed in detail, and no WIMP signal has yet been seen. From this absence, limits have been placed on WIMPs that annihilate in the sun and earth. It turns out that the photinos, which initially inspired these investigations, actually would produce a background below that which is measured, and so escape these bounds. However, one hypothetical WIMP, the supersymmetric partner of the neutrino, called the *sneutrino,* was actually ruled out for masses in excess of 5–6 GeV.

In the intervening decade there have been a number of important developments that have affected the search for the indirect annihilation signature of WIMPs. A number of interesting new detectors sensitive to such a signal have come online, or are about to. Even the MACRO detector has been configured to search for the neutrino and muon signatures from WIMP annihilations. One of the most exciting new possibilities has been the creation of truly gargantuan underground water detectors, up to 1,000 times larger in volume than the special purpose detectors built for proton decay and solar neutrino detection. These detectors will instrument either sea water or Antarctic ice in order to detect neutrinos and muons, some of which might come from WIMP annihilations in the sun and earth. Two proposed detectors include the Deep Undersea Muon and Neutrino Detector (DUMAND), to operate off the coast of Hawaii, and the Antarctic Muon and Neutrino Detector Array (AMANDA), to operate, obviously, in Antarctica. It turns out that a kilometer below the surface the Antarctic ice shelf is incredibly transparent to light. By using a "drill" made from a high velocity stream of warm water, one can drop photosensitive detectors deep inside the ice and monitor huge volumes of underground ice for the signatures of high energy particles traversing the volume.

At the same time, constraints on SUSY models, due to accelerator measurements, have shifted the range of interesting WIMP masses from an original range of 5–50 GeV to a range of 50–500 GeV. For these heavier masses, the earth becomes a more interesting target, because of abundance of heavy elements such as iron, whose masses are better matched to the WIMP masses from the point of view of energy transfer. Calculations by Andrew Gould at Ohio State University, and others, now suggest that for a certain range of masses, searching for WIMP annihilations in the earth can provide stronger constraints on SUSY parameter space than can the sun.

One recent result that has increased interest in indirect detection of WIMPs has been the recognition that there may be a second WIMP population in the solar system. While I was lecturing on WIMPs at CERN in Geneva, the eminent French theoretical physicist and expert in General Relativity, Thibault Damour, was in the audience. He was interested in whether WIMPs captured in the solar system might affect calculations of the orbits of the planets relevant for testing General Relativity. We determined that they would not affect things at the level of accuracy of planned experiments, but his questions after the lectures spurred us to reconsider WIMP scattering in the sun: some of these WIMPs might not be captured but might instead move into bound orbits about the sun. A year later we had completed our calculations showing that under certain circumstances—in fact precisely those circumstances relevant for WIMPs that might be detectable directly in terrestrial detectors—such a new population should exist, with a density comparable to the background-halo WIMP density, but with very different velocities with respect to the sun. While we first focused on the implications of this new population for direct detection experiments, it became clear that this could also have an impact on indirect detection experiments. This new population could be captured by scattering in the earth and could increase the expected neutrino or muon signal coming from the center of the earth as a probe of captured WIMPs, as we have calculated with Lars Bergstrom and his colleagues in Sweden.

To date no experiment has yet seen any such signal, and the indirect experiments have already ruled out a corner of supersymmetric parameter space. To probe further, and if we really want to confirm or reject WIMPs as dark matter, however, we must be prepared to detect them directly in a controlled experiment.

Goodman and Witten provided experimentalists with estimated rates that appeared to be detectable. Of course, this alone is a long way from guaranteeing in practical terms that they are observable. The introductory pages of chapter 12 described the potential challenges and difficulties to be overcome in attempting to measure rates as small as one event per day in a kilogram of material containing upward of 10^{25} atoms and involving energy deposits less than a few keV—energies smaller by orders of magnitude than those produced in most radioactive decays, for example.

Goodman and Witten had been inspired to make their estimates by a detection scheme earlier proposed for detecting solar and/or reactor neutrinos, a task that poses similar problems. This proposal, by Leo Stodolsky and André Drukier in 1984 at the Max Planck Institute in West Germany, involved a significant departure from the standard particle detection technique of looking for either the ionization or the light produced when particles scatter off the atoms in a detector. They pointed out that if one could design a detector made of millions of very tiny superconducting granules, then even depositions of very small amounts of energy produced when a nucleus recoils, for example, might cause sufficient heating in a single grain to raise its temperature until it was no longer superconducting. Because superconducting grains will expel an external magnetic field, when a grain "went normal," it would suddenly allow this field to enter and become magnetized. With very sensitive magnetometers, they hoped that such a change could be measured. It has turned out that such detectors have not been successful for this purpose. Nevertheless they did suggest that one might consider detection techniques for dark matter by registering not ionization but heat. The possibility of using heat to detect particles, first proposed in principle in the 1970s by T. Ninikowski at the European Center for Nuclear Research, has proved very fertile indeed.

Before launching into a discussion of this exciting new area of technology, it is worth considering whether any previous particle detectors meet the specifications required to detect WIMPs as dark matter. Once again, an ambitious experiment, originally designed for another esoteric purpose, comes close to filling the bill. Earlier I said that one of the chief problems encountered in detecting WIMPs was the fact that even the smallest contamination of radioactive materials can produce unacceptable backgrounds. To gain familiarity with this drawback, physicists looked to those experiments that had already conquered such backgrounds.

The most successful experiments involve the search for the rarest known nuclear decays in nature. Since the beta decay of the neutron into a proton, an electron, and an antineutrino was first discovered early in this century, we have realized that a much more exotic, but related, process must take place in the nuclei of certain atoms. If the energy levels in the nuclei are just right, it becomes energetically favorable for two neutrons in the nucleus to beta decay simultaneously, producing a new nucleus, twice removed from the original nucleus. In the process, known as "double beta decay," two electrons and two antineutrinos are released. Because two simultaneous mediations of the weak interaction are involved, double beta decay is the rarest known radioactive process that has been calculated. The lifetimes for the nuclei that might undergo this decay exceed 10^{22} years in some cases.

Another argument has encouraged researchers to investigate double beta decay. As I stated, two antineutrinos normally are emitted in this process. However, it may be possible, if neutrinos have a mass, for these two antineutrinos mutually to annihilate. In this case, called "neutrinoless" double beta decay, the observations of double beta decay would change in character. All that would be seen to be emitted from the decaying nucleus would be two lone electrons, each taking up half of the available energy of the decay. Now, since particles can only annihilate with their antiparticles, this implies that the neutrino would not be distinct from its antiparticle, the antineutrino. This alone is not so special. The photon of electromagnetism, for example, has no distinct "antiphoton." But the nature of a neutrino mass that could allow the neutrinos to annihilate with each other *is* special. What makes neutrinoless double beta decay so interesting to particle physicists is that the only way the electron neutrino can have a mass that would allow this process to take place is if there exists new physics beyond the standard model. Indeed, the most likely place for such a mass to originate is in the physics associated with the GUT scale. Thus, a very sensitive experiment at normal energies in the laboratory could probe for new physics at scales 12 orders of magnitude larger than presently accessible in the most powerful accelerators in the world.

Searching for decays associated with lifetimes of 10^{22} years, or longer in the case of neutrinoless double beta decay, in a laboratory at the earth's surface is no mean feat. The materials that can undergo double beta decay are not as easy to manage or obtain as water, so proton decay-sized detectors would be out of the question. Samples of 1 to 5,000 grams or so of material such as germanium are used instead. Given a kilogram of material containing 10^{24} atoms, a lifetime of 10^{24} years would correspond to only *one decay in the whole sample every year*. Meanwhile, the experimentalists must contend with cosmic rays bombarding their detector at a rate of about one particle per second, and with radioactive materials located in their detector. If these materials have a lifetime less than about 10^{14} years, as most do, they must have an abundance in the sample smaller by the ratio of this lifetime to the double beta decay lifetime—a factor of about 10^{10} or so—if their decay rate is not to overwhelm the process under investigation.

To date, the lowest background double beta decay experiments involve the element germanium, for several reasons. In the first place, because of its importance—along with silicon—in semiconductors, materials essential to modern computers and other high-tech electronics, large quantities of this material are available with purities far in excess of those available for other solids. During a process called "zone refining," crystals of germanium or silicon are melted at a single location, and the zone is then slowly dragged across the crystal. The impurities move to just in front of this melted zone and are thus drawn away from the crystal with great efficiency, leaving chemically active impurities at levels of less than 1 part in 10^{12}. Second, because of the chemical properties of germanium which make it an ideal semiconductor, a germanium block can not only be used as the source for double beta decays, it can act as its own detector as well. The ionization produced during radioactive decays inside the germanium can be monitored by putting an electric field on the crystal and measuring the flow of charge. From the amount of ionization, the energy of the decay can be measured with notable accuracy and great sensitivity.

Pure germanium alone, however, does not make a background-free detector. While the actual germanium source itself might weigh in the neighborhood of a few hundred grams in the largest detectors, this source must then be surrounded by a shield against background radioactivity, coincidence counters to detect cosmic rays, and finally all the cryogenic paraphernalia associated with keeping the system at supercold temperatures in order to reduce "noise" in the electronics. It is a tribute to the ingenuity of experimenters that they were able to reduce the backgrounds to the required levels to perform the experiment. Indeed, this whole business of low background measurement is as much an art as a science. Before this work, no one had ever required such low backgrounds, so each step was a shot in the dark. The most innocuous things, from solder used on the electronic leads to the 99.999 percent pure lead blocks used in shielding, sometimes turned out to be "hot" with radioactivity, at least on the scale of the signal to be detected. Sometimes different samples of the same material, from the same manufacturer, would vary tremendously in their level of radioactivity.

Other problems of low-energy backgrounds had to be overcome before germanium detectors could achieve the extreme levels of sensitivity in energy and rate to probe for WIMPs. In the first place, the sensitivity of the electronics at the lowest energies accessible in the keV range had to be reduced. It is this electronic "noise" that fixes the minimum sensitivity of germanium ionization detectors. One more kink had to be fully resolved. As we have already noted, when a dark matter particle such as a WIMP collides elastically with the nucleus of an atom, it can transfer at most an energy equal to its initial energy of motion. If it does this successfully, it will kick the nucleus out from its resting position and cause it to move slowly among the crystal lattice for a short distance. Now, the less energy imparted to the nucleus, the more slowly the nucleus will move. But the more slowly it moves, the less strongly it will ionize, because the electrons that normally exist in neutral combinations around the atoms on the lattice will be able to keep up with the nucleus as it travels. In this case, if the ionization is reduced, there is less energy to detect in an ionization detector. Eventually the nucleus will cease to ionize altogether below a certain energy.

Unfortunately, few experiments had been performed to show how much ionization is produced by very slow-moving nuclei in germanium or silicon. Moreover, these are extremely difficult to do; until recently, there seemed little need. Using the one set of measurements obtained previously from nuclei moving in germanium with an energy in excess of 10 keV, and a theoretical calculation that seems to fit the data well, however, experimenters have shown these nuclei do ionize atoms about 30 percent as efficiently as electrons of the same energy. New experiments have since been performed to verify this result.

If this value is used, and the data from present experiments analyzed, these experiments can probe a vast WIMP parameter range directly, by looking for the energy deposits that would be induced by WIMPs scattering in the detector. Not seeing any such deposits can in turn rule out WIMPs with a given mass and interaction rate. From the absence of any such signal for energy deposits in ionization in excess of about 3 keV, experimenters provided limits that appear to rule out any heavy neutrino galactic halo WIMP with mass between 12 and almost 10,000 GeV.

The reason why neutrino-WIMPs with mass less than about 12 GeV were not ruled out by these detectors is that they do not transfer enough energy in their collisions with germanium. Just as the elastic collision of two billiard balls allows the most efficient transfer of energy between them when they both have the same mass, the same holds true for WIMP collisions with atoms. The mass of a germanium nucleus is about 75 GeV. Hence, germanium is optimized, at least as far as energy deposits are concerned, to detect 75-GeV WIMPs. To be sensitive to a wider mass range, experimenters must consider using detectors containing materials with lighter nuclei. Since silicon is electronically as easy to work with as germanium, but weighs only about 28 GeV per nucleus, a silicon ionization detector would be able to probe down to WIMP masses of about 5 GeV.

In fact, there is now no reason to probe for heavy neutrino WIMPs below 12 GeV because of a wonderful symbiosis between direct WIMP detectors and accelerators. After the discovery at CERN in Geneva of the W and Z particles that mediate the weak interactions, a new accelerator was built in order to explore the properties of these particles with great precision. The Large Electron-Positron (LEP) collider produced millions of these particles, and their lifetime and decay mechanisms were explored. Since neutrinos couple to the W and Z particles, these particles can in turn decay into neutrino–antineutrino pairs, at least as long as the neutrinos are light enough. Since the Z particle weighs about 90 GeV (90 times the mass of the proton), as long as the neutrino in question was less massive than about 45 GeV, a neutrino–antineutrino pair could be produced via Z decay. While the neutrinos could not be observed directly, this additional possible decay channel would affect the Z particle lifetime in a calculable way. By very carefully measuring this lifetime, the experimenters at CERN determined that there were precisely three types of neutrinos lighter than 45 GeV, the well-known electron, muon, and tau neutrinos, all of whose masses were known to be far less than 1,000 times smaller than that required to be WIMPs. This result alone did not preclude the existence of heavier neutrino WIMPs. However, when combined with the limits from direct detectors, the entire neutrino WIMP-phase space was ruled out. This combination points out the great utility of not putting all of your eggs in one basket. It is important for the health of particle physics and particle astrophysics to keep an active accelerator and nonaccelerator physics program intact.

There is another feature to this story that I think gives an important lesson. When I and others did the calculations that demonstrated that heavy neutrino WIMPs were ruled out, we did so *after* the accelerator limits on neutrino species were announced. There was nothing about the calculations themselves that required us to wait for the announcement in order to derive the consequences of it. While I cannot speak for the other groups, the reason I had not derived the result before is that I never thought seriously about the implications of the experiment until it was performed. I have found this to be a general result. It is remarkable how until one is presented with actual data it is difficult to motivate oneself to think deeply enough about all the possible implications of future results. There is a very healthy interplay between theory and experiment, which is why, after all, physics is an *empirical* science.

In any case, if we are to proceed to explore for WIMPs other than neutrinos, ionization detectors alone are limited. Because less than 30 percent of the energy deposited in a collision of a WIMP with a nucleus ends up in ionized particles, this detection technique is inefficient at best. Moreover, the fact that nuclei below a certain energy ultimately will no longer ionize means that there is a theoretical limit below which no practical technological improvements in this detection methodology will be possible. Of course, as theoretical arguments in favor of heavier WIMPs increase the need to push down to lower energies becomes less severe. In any case, to probe efficiently for WIMPs, one must build a dedicated WIMP detector. Several different technologies are being employed for this purpose, and they can be roughly separated into two categories: cryogenic detectors, searching for "heat" produced via WIMP collisions, and "non-cryogenic" detectors searching for other signatures of WIMP collision with nuclei.

The principles behind the first generation of cryogenic WIMP detectors were the same as the principles at work when a thermometer is placed in a roast in a hot oven. The equipment was just a little fancier. When a WIMP scatters off a nucleus in a crystal, it jars the nucleus, and the nucleus recoils. This, in turn, jars the rest of the crystal, producing a wave of vibrations that travel outward and eventually dissipate. More than 70 percent of the energy deposited in the crystal by the initial scattering event ends up in such vibrational waves. On a macroscopic scale, it is these increased vibrations of the crystal lattice which we measure as a rise in the temperature of the bulk sample.

These "acoustic" waves of vibrations in a crystal lattice can behave very much like particles in systems that are supercold. Because of this similarity, solid-state physicists give these waves a name that recalls names given to particles: *phonons*. To create a single energetic phonon excitation in a crystal of silicon or germanium may require only about one-thousandth of an electron-volt (eV). Contrast this with the energy needed to ionize atoms in a crystal, which may require several eV per ion pair. Thus, not only does more of the energy of low-energy scatterings in solids end up in phonons than in ions, many more of them are produced as well. If somehow we could detect each energetic phonon wave as efficiently as we can detect ions, having many more to register would mean that both the accuracy and the sensitivity of such a detector might greatly exceed that of an ionization detector.

In fact, however, a detector of phonons does not always have to discriminate between individual waves. After all, a thermometer does a pretty good job of measuring the total thermal energy due to the random vibrations of particles inside a roast, and no one ever talks about phonons. When the same process is done in a laboratory, using a fancy thermometer to measure the total amount of heat energy generated when an external incident particle scatters in a detector volume, the procedure is given a fancy new name: *bolometry*.

Earlier I outlined the essential problems to be solved in detecting WIMPs, axions, or other weakly interacting particles, such as neutrinos from reactors. How can one measure (1) very small energies, (2) deposited very infrequently, (3) in macroscopically large volumes? When Cabrera, Wilczek, and I began to think about this problem, along with several other groups, it seemed to cry out for bolometry, and also for a detector made of silicon or something like it. We later learned that Ninikowski a decade earlier had reached similar conclusions. One of the many bolometric "miracles" of silicon that impressed all of us as we sought a new way to detect neutrinos was the way a tiny energy deposit could be converted into a macroscopic measurable signal. This marvel depends crucially on the properties of materials as they are cooled to very low temperatures.

Physicists have defined a quantity that refers to the amount by which the temperature of an object changes when we add a specified amount of energy into it. This quantity, called the *specific heat* of the material, is the reason why some things stay hot longer, or appear to be cool even on hot days. For example, I sometimes burn my tongue on the raisins in toasted raisin bread. Why do the raisins remain hotter much longer than the toast itself? One reason is that they have a larger specific heat. It takes more energy to heat up a raisin in a toaster to the same temperature as the bread. This is because there are many more internal configurations that can be excited and that store thermal energy in the material of a raisin, compared to the material of the bread. Once out of the toaster, if the raisin and the bread both radiate energy at the same rate initially, then the loss of a given amount of energy may cool the toast dramatically, while the raisin remains hot, because at its initial temperature it had stored much more heat energy.

What qualities make silicon and similar materials so attractive for use in bolometry? At very low temperatures there are almost no internal degrees of freedom to be excited. This means that the specific heat of silicon becomes infinitesimally small. This, in turn, implies that even a very small energy deposit could raise the temperature of a macroscopic amount of silicon by a measurable amount: exactly what the doctor ordered.

This appealing idea had been previously tested in a number of experiments. A group associated with D. McCammon at the University of Wisconsin and Harvey Moseley at the Goddard Space Center in Maryland, in trying to develop a new X-ray detector for use on space missions, had demonstrated in 1984 that bolometers could provide energy-sensitivity levels unattainable with other technologies. Using a small silicon wafer of about 10^{-5} grams cooled to a temperature of about one-tenth of a degree above absolute zero (or about 460 degrees below zero on the Fahrenheit scale), they were able to measure energy deposits due to X rays with energies of a few thousand eV, with a resolution of better than 30 eV.

Based on these results, we did some simple scaling calculations. If the specific heat of silicon continued to drop as it had down to 0.1 degrees Kelvin, then an energy deposit of 1 keV in 100 grams of silicon, cooled to several *thousandths* of a degree above absolute zero, would raise the temperature of the entire sample by a few thousandths of a degree. We envisaged a detector that would consist of one or several small blocks of silicon, cooled to very low temperatures and fitted with some sort of thermometer. When a neutrino or a dark matter WIMP bounced off one of the nuclei in the silicon, it would deposit energies in the range of 1 keV. We estimated that for the course of one-thousandth of a second or so, this energy would produce a sharp temperature peak and fall as measured electronically in the thermometer (see figure 13.3).

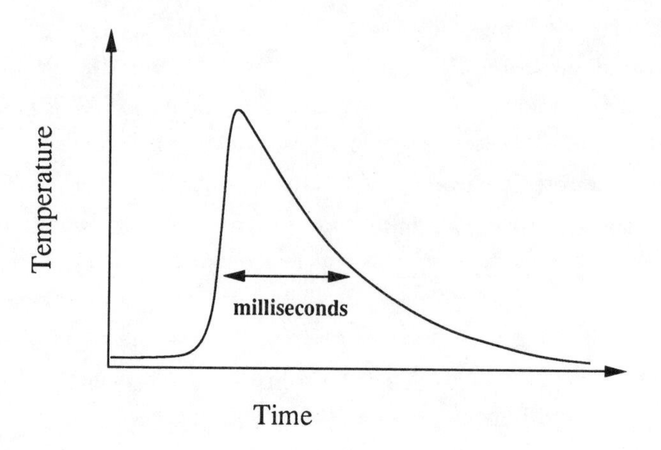

FIGURE 13.3
A schematic representation of the thermal pulse that might be expected to be measured for a small deposit of energy in silicon by the collision of a WIMP with a silicon nucleus.

Actually performing the experiment and describing it are two different things. Many questions of principle have to be tested first. In this case, no one had ever cooled bulk samples of silicon down to such low temperatures to see whether the specific heat actually dropped as fast as one hoped: some arguments suggested that it might not. Next, efficiently cooling large volumes to temperatures of, say, 0.1 degree above absolute zero sounds fantastic, but it is actually straightforward, although each hundredth of a degree requires a lot of sweat and effort. After all, the relative difference between one-thousandth of a degree above absolute zero and one degree, which is already pretty darned cold, is greater than the difference between one degree and room temperature. The first time I saw a refrigerator capable of cooling a sample to twenty-thousandths of a degree above absolute zero, this relative difference was dramatically brought home to me. The builder of the device talked about a certain chamber as being at "room temperature," and I realized she was referring to a region that was cooled to the temperature of liquid helium, about 450 degrees below zero Fahrenheit.

Finally, there is the issue of "thermometry." This seems innocuous at first, but it is one of the most challenging features of the whole business. Traditionally, very low temperatures are measured by using resistance. A piece of wire whose internal resistance varies greatly in the temperature range of interest is hooked up to the sample, and to a small-voltage current as well. As the temperature in the wire, and hence its resistance, varies, the current that flows in the wire will vary.

By measuring this current in a properly calibrated wire, the experimenter can determine the temperature of the system. Now consider the following. The specific heat of a kilogram block of silicon in the region of a few thousandths of a degree above absolute zero is so small that if I attach a thermometer containing, say, a millionth of a gram of metal to the block, the *thermometer itself* will dominate the specific heat of the combined system. In addition, how can we couple the vibrational phonon energy to the thermometer? At low temperatures, the acoustic waves that will eventually be detected as "heat" travel relatively unimpeded over long distances and also do not cross barriers between materials easily. Therefore, one might expect problems both with coupling the heat energy from the silicon to the thermometer and also with actually *thermalizing* the energy, which begins with a few very energetic phonons, into a thermal spectrum of waves characterized by some well-defined temperature.

In spite of these practical difficulties, several teams of experimenters have worked on bolometric devices and other cryogenic tools designed specifically to detect dark matter.

It has become clear that the spectrum of phonons being measured is not exactly thermal, but nevertheless, calibrations can be done so that the total energy input can be well measured by observing the phonon signal. At the present time, at least two groups are actively involved in WIMP detectors using bolometric-related techniques. Cabrera's group at Stanford has been collaborating with Bernard Sadoulet, who played a key role in the discovery of the W and Z particles and is now at Berkeley, where he helped set up the Center for Particle Astrophysics. Along with colleagues from Santa Barbara, the Soviet Union, Fermilab, and my own department at Case Western Reserve University, they have been collaborating on the Cryogenic Dark Matter Search Experiment (CDMS) experiment, (see figure 13.4A) using germanium targets, currently located at Stanford University. Ettore Fiorini's group in Milan, Italy, has been involved in developing an array of telurium oxide crystals as bolometers for a WIMP search.

Techniques have developed dramatically in the fifteen years since the early cryogenic WIMP proposals were first made. As preferred WIMP masses have increased, heavier elements such as germanium have come to be preferred over target materials like silicon. Moreover, phonon detection schemes have themselves changed. In the case of germanium, for example, a layer of germanium itself doped with impurities on the outside of a crystal can act as an effective thermometer, which also allows good heat flow between the bulk sample and the thermometer.

Of course, strict bolometers do not measure individual phonons, but rather a

FIGURE 13.4A

Shown are the components of a WIMP detector being assembled by the Cyrogenic Dark Matter Search experiment now running in California. The silver-colored disks are made of high-purity germanium crystals cooled to ultra low temperatures allowing detection of minute temperature changes resulting from a collision between a WIMP and a single germanium nucleus. (Photo courtesy of Dan Akerib)

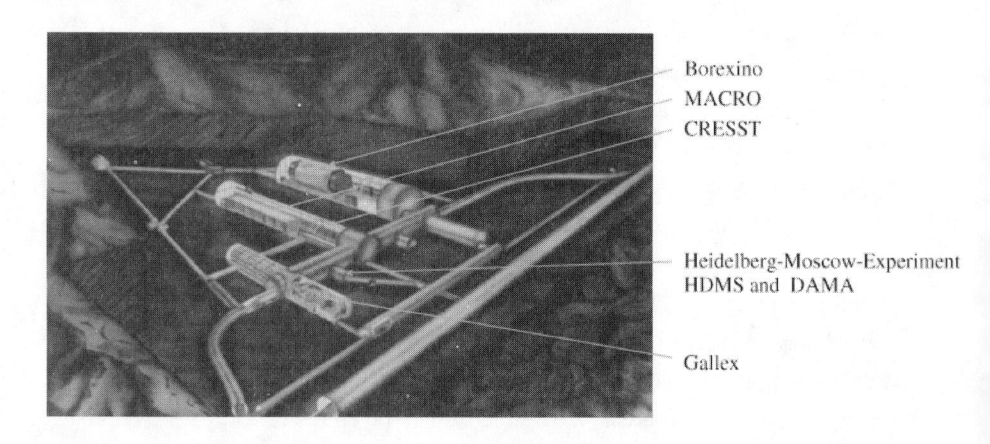

FIGURE 13.4B

Schematic view of the Gran-Sasso Underground-Laboratory. Presently three big halls are in use. The locations of some important experiments are marked with arrows. The Genius Experiment is a large proposed WIMP detection experiment consisting of a tank, approximately 16 meters in diameter and 18 meters in height containing liquid nitrogen. In a first step 40 HPGe-detectors will be immersed into the center of the tank, corresponding to approx. One hundred kg of nat Ge-target-material in the search for dark-matter WIMPs. (Photo courtesy of Hans Volker Klapdor Kleingrothaus)

FIGURE 13.4C:
Picture of apparatus in the Boulby Mine, in England, using Sodium Iodide Scintillation detectors to search for WIMPs. In this experiment, WIMP collisions with atoms in the scintillator cause them to emit light, which is detected by photosensitive detectors. (Photo courtesy of Neil Spooner.)

collective thermal ensemble. Over the past decade different attempts have been made to move cyrogenic detectors more in the direction of being "phonon" detectors rather than "heat" detectors. An interesting recent development has involved the use of thin superconducting tungsten wires bonded to the surface of the detector. These are kept quite close to their superconducting transition. Moreover, by careful application of external voltages and choice of materials, one can design "traps" that concentrate the phonon energy coming from the bulk sample to thin regions near the wires. This extra energy then drives the wires from being superconducting to normal. The resulting change in current is easily measured. Moreover, it turns out that the length of time these wires remain in their normal phase before returning to their superconducting phase is proportional to the total energy deposited in the signal, with an accuracy of better than 1 keV.

A key advantage of techniques such as this is that the response is very fast. Collecting the initial phonon energy before it thermalizes means these detectors have a response of a few millionths of a second. In this way, one can literally determine the time difference between when the signal arrives at different wires, and in so doing attempt to localize the deposited energy. This is very important for removing events from radioactive backgrounds, for example, which tend to occur near the surface of the detector, where contaminants are likely to lie.

Perhaps the most striking development of all, however, centers around not a technique for measuring the WIMP signal, but rather a technique for distinguishing it from backgrounds. In our initial work on bolometry in silicon, Cabrera, Wilczek, and I mused about whether one might, in the end, design phonon detectors as "particle" detectors. Namely, could it be that different particles would produce distinct signatures in such a detector, even if they deposited the same amount of energy? One possibility was clear immediately. WIMPs scatter off of nuclei, producing slow moving particles that deposit most of their energy in phonons. Other particles, such as electrons and photons, tend to deposit more of their energy in ionization. If one could in principle measure *both* ionization energy deposits and phonon energy deposits, then one could separate out possible WIMP signals from other backgrounds.

It turns out that the predicted WIMP event rates in detectors are so small that one cannot gain by going to larger and larger detectors unless one can systematically reduce the backgrounds which would otherwise limit the sensitivity of detectors. Shielding alone may not be sufficient. However, if one can remove backgrounds on an event-by-event basis, then it will be possible in principle to explore even exceedingly low predicted signal rates. To do this in cyrogenic detectors, one should aim for an ability to reject at least 995 out of 1,000 background events, even in deep underground detectors.

The CDMS group has now demonstrated an ability to instrument their detectors for both ionization and phonons. They can distinguish events due to incident neutrons (which produce WIMP-like energy deposits) from incident gamma rays from radioactive sources with the same energies and the hoped-for rejection sensitivity. In their next, deep site, in a mine in Minnesota, the ambient cosmic ray and neutron backgrounds should be small enough so that, when combined with the ability to distinguish backgrounds from signals, one could probe most of the available SUSY WIMP phase space with a kilogram-sized detector.

Another proposed cryogenic scheme is similar in spirit to ballistic phonon detection, although not yet as experimentally advanced. A group at Brown University has considered the use of superfluid liquid helium as a possible WIMP detector. Instead of a local energy deposition producing phonons, as is the case in a crystal, microscopic vortices or eddies in the liquid helium are caused. These excitations are called *rotons* by solid-state physicists. If a roton detector could be developed at the surface of the liquid helium, it might serve the same function as a phonon detector on the surface of a crystal. These ideas remain to be demonstrated as a practical detector with good energy resolution. One possible advantage of this scheme, however, is that liquid helium is the purest substance in the world: when it liquefies, all impurities condense out, leaving pure helium behind. Thus, the radioactive background from the sample itself would be minimal.

Of much greater surprise to me than the developments in cryogenic detectors, however, is the developments in other possible WIMP detection schemes. Here two different thrusts have produced results that are to date competitive with and, in certain regions of parameter space, exceed those previously obtained from bolometry. The first involves a return to the original germanium-ionization detection schemes used for double-beta decay experiments that were used in developing the first direct constraints on WIMPs. Remember that these detectors were limited in their ability to reduce the energy of the detected signals due to the small percentage of the total energy deposited in ionization by WIMPs. However, as the favored mass range of SUSY WIMPs has increased, the expected energy deposits in a heavy target like germanium have also increased, thus removing one of the disadvantages of this technique. Since there is a rich history of experience working with these detectors, experimenters have also been able to address problems of radioactive backgrounds. One particularly ambitious proposal from the Heidelberg group, under the direction of Hans Klapdor-Kleingrothaus, involves the proposed use of 1 ton of germanium surrounded by a vast quantity of liquid refrigerant such as nitrogen or xenon, which can also serve to screen out radioactive backgrounds. This detector would be operated in the Gran Sasso Laboratory, located next to an automobile tunnel under a mountain in Italy (see figure 13.4B).

Another research front, explored by groups in the United Kingdom under the direction of Peter Smith, one of the longstanding leaders in exploring all facets of WIMP detection, and a group at Sheffield University under the direction of Neil Spooner, and also of a Paris-Italian collaboration working at Gran Sasso, involves the use not of phonons or of ionization, but of light itself as a probe of WIMPs (see figure 13.4C). Certain traditional particle detectors at accelerators such as sodium-iodide (NaI) detectors are transparent and produce what is called "scintillation light" when incident particles deposit energy by collisions with atoms in the detector. The initial ionization is transferred to light when the electrons knocked out of atoms get reabsorbed, and because these materials are transparent, the light can be collected elsewhere. From the intensity of the measured light, one can determine the energy deposited.

Again, of great surprise to me, NaI detectors have already been operating to produce important constraints on SUSY WIMPs. While radioactive backgrounds are a problem, these materials are sufficiently cheap so that large amounts can be used in detectors, and one can hope to disentangle signals from backgrounds by statistical means.

With at least a dozen active groups operating WIMP detectors in deep sites from California to Russia, the perspectives for directly observing WIMPS has changed dramatically since I wrote the first edition of this book a decade ago. That is the good news. On the other hand, it is worth noting that we are still talking about "limits" and "constraints" rather than positive signals! Nevertheless, it is clear that we are just beginning to approach interesting regions for SUSY WIMPs, and the next decade may provide the first direct signal for nonbaryonic WIMP dark matter in the universe.

Of course, as a theorist, it is worth thinking ahead and imagining what might happen when such a signal is claimed to be seen. How will we be sure it is due to WIMPs and not to some unknown radioactive background? After all, the expected signal is simply a few excess rare events at low energy in such detectors. A WIMP signal would appear as a little extra "noise," just as the first observation of the Cosmic Microwave Background was first mistaken as noise resulting from pigeon droppings inside an antenna! It is true that the expected energy dependence of the signal should be characteristic of the velocity distribution of WIMPs in our halo, but many, many events would be needed before this could be mapped out. Instead, two other features of the predicted signal are being explored.

The first involves the well known fact that the earth is going around the sun. Thus, as the sun moves through the galactic WIMP background, for half the year the earth travels in the same direction as the sun, and for half the year it travels in the opposite direction. This little extra velocity kick means that for half the year WIMPs collisions should deposit a little more energy detectors, and for half the year a little less. This translates into a predicted annual modulation in both the energy and the overall event rate in detectors. The modulation is extremely small, however, since the motion of the earth around the sun is small compared to the velocity of the sun through the galaxy. Predicted variations are at the 3–6 percent level, requiring many hundreds of events at a minimum before such a signal could be extracted, even ignoring the fact that the specific features of the signal depend on the unknown features of the galactic halo WIMP distribution. Nevertheless, it is worth pointing out that the DAMA NaI detector group in Italy has already claimed to observe such an annual modulation in their detector, consistent with what one would expect from a WIMP signal! Unfortunately, however, it is also consistent with what one might expect from a radioactive background such as radon, which might also be produced with different rates over the course of the year.

The present debate over the veracity of the claimed DAMA signal illustrates why annual modulation is not likely to be that revealing. A far better signature would be to probe the direction of the recoiling nuclei hit by WIMPs. If you are running in the rain, you will expect to get wetter in the direction facing the rain, rather than vice versa. Moreover, if your direction changes over the course of the storm, you can track this direction by examining which direction relative to you the rain is coming from. In principle, *if we had the ability to measure the direction of the recoiling nuclei from WIMP scattering,* this would be a powerful and relatively unambiguous indicator that any claimed signal was indeed WIMP-related. With colleague Craig Copi and student Junseong Heo at CWRU, I have demonstrated that in spite of our present uncertainties about the intrinsic features of the galactic halo, only 25–30 events with directional sensitivity would be required to disentangle a WIMP signal from a direction-independent radioactive background. This is very exciting. Unfortunately, however, while there are several general ideas floating around for a possible directional WIMP detector, no working prototypes yet exist.

Some or all of these speculative ideas may be vital if WIMPs are to be detected as dark matter. Extracting the WIMP signal may be very challenging indeed. While I have not stressed it here, the estimated rate for WIMP scattering is in the range of one to one hundred events per kilogram per day only for the most optimistic cases. Supersymmetric particles, such as a the LSP, might have scattering rates that are suppressed by several orders of magnitude. Instead of scattering coherently off the protons and neutrons in a way that would increase the interaction rate as the square of the number of particles in the nucleus (as is the case for neutrinos, for example), the LSP might instead only couple to the spin of a nucleus. This means that they would not scatter off nuclei that have no net spin; moreover, their net scattering rate could be vastly smaller.

Of course, we must first detect a WIMP signal before we worry about its characteristics. Nevertheless, if such a signal is eventually seen, the very characteristics of the signal may tell us a great deal about the nature of the galactic halo and, in turn, the formation of the galaxy itself. In the meantime, in the absence of the "ultimate" WIMP detector, experimentalists would be more than content to see any positive signal, even an ambiguous one. The good news is that if WIMPs are out there, such a signal is likely to be seen in the near future.

The technologies that I have discussed here for WIMP detection involve state-of-the-art techniques from many areas, including cryogenics and electronics. Whenever advanced technologies are combined, commercial applications often arise. This may be true of WIMP detectors. I have already mentioned particle identification as potentially useful in medical work, for example. As WIMP detectors are developed they might also have important side benefits.

One application involves neutrino detection. Commercial nuclear power reactors produce neutrinos in abundance. If such neutrinos could be detected with rates high enough to make detectors practical to install near such reactors, they could be used as safety monitors. Neutrinos emerge from inside the reactor volume at the speed of light and thus give instantaneous feedback about power production. Traditional monitoring usually involves temperature measurements near the reactor core. But it takes time for power fluctuations to manifest themselves as temperature rises. The extra time gained by monitoring the neutrino signal instead might offer an extra margin of safety.

Another potential application is relevant to the computer industry. As computer memory cores become smaller and smaller in new supercomputers, a single radioactive event can alter the state of a memory location. It thus becomes extremely important to monitor the microchips for even very small radioactive backgrounds. Today's silicon and germanium WIMP-detection devices provide the most sensitive measures of intrinsic radioactivity in these materials in existence: the technology used in their fabrication may have direct application in monitoring supercomputers.

One can muse about other applications, but I have discovered from experience that before any businessperson will take any of this seriously, he or she must be convinced that a working prototype exists. As a discriminating reader, you too may feel the same about WIMP detection. Nevertheless, developments are proceeding at an amazing pace, and we may soon know for sure whether WIMPs are anything more than a catchy name.

AN AXION RADIO

I have saved the axion "radio" for last because it is such a nifty idea. When "invisible" axions were first proposed, and even when it became clear that they might make up the dark matter in the universe, no one imagined they would be detectable. The reason for this pessimism was simple. Invisible axions would couple to matter approximately 10^{10} times more feebly than even neutrinos, whose detection already pushes the limits of modern technology. How can we hope to detect them? The surprising answer relies on two features associated with cosmic axions: first, the nature of the axion background itself, and second, the nature of axion couplings to matter.

To get a proper perspective, let me propose a thought experiment. I have said earlier that gravity is by far the weakest force in nature, some 40 orders of magnitude weaker than electromagnetism, for example. The coupling of a *graviton,* the particle thought to convey gravitational forces, to matter is so negligible that it makes axion couplings seem extravagant. Nevertheless, you can perform the following experiment in your home: (1) jump up, (2) observe what happens.

Presumably you came back down to earth, so it is safe to assume that the earth's gravity somehow managed to alter your movement. How? The answer is simple yet subtle. While the gravitational attraction of each atom in the earth for each atom in your body is small beyond belief, the attraction of all of the atoms in the earth on all of the atoms in your body can add up coherently to produce a macroscopic effect. If this were not so, we would not have any idea that gravity exists.

Could we use the same trick with axions? It was Pierre Sikivie who first had the bright idea. In the language of Feynman diagrams, we can draw our jumping experiment as shown in figure 13.5.

The mass of the earth acts as a coherent source of gravitons, which I absorb and which then change my momentum from up to down. Now, it is one of the universal properties of any axion model that axions can couple, albeit extremely

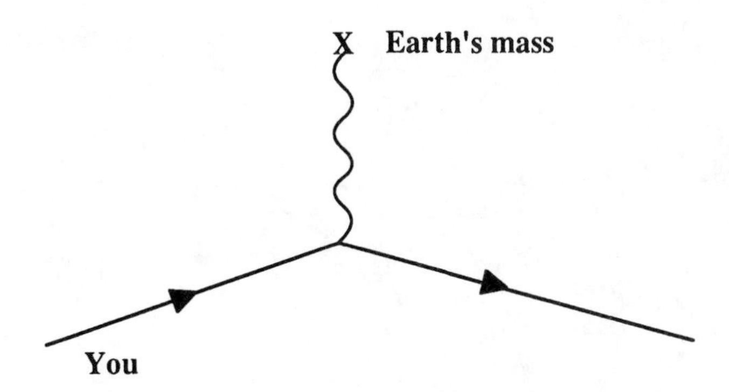

FIGURE 13.5
A Feynman diagram schematically representing the interaction with the earth which causes you to change your motion (that is, come down) after you jump up.

weakly, to photons. Because of this interaction, one might hope coherently to couple axions to, say, a background magnetic field. What would happen in this case? If we draw a diagram analogous to what we drew earlier for the gravitons, but this time we show the axion interaction with photons, the result is shown in figure 13.6.

Here, the background magnetic field acts as a coherent source of photons, which are absorbed by the axion. If we could coherently couple an incoming axion to a coherent background magnetic field, the axion might be converted into an outgoing real photon. How wonderful.

As dazzling as this possibility is, however, were it not for another property of cosmic axions, it would not be very useful. Recall that remnant cosmic axions in the early universe were not produced in a thermal configuration, but rather resulted out of an initial condensate of particles in the lowest energy ground state of matter. Thus, to a first approximation, all axions in this background had exactly the same energy, equal to the mass of the axion. Moreover, their initial state was "coherent," that is, the individual background axions were coupled so that the whole axion background over a horizon volume behaved uniformly. These two facts have two momentous implications. First, the photons into which the axions might convert must, by conservation of energy, have the same energy as the incident axions. Since these incident axions all possess approximately the same energy, the outgoing photons will all have the same energy, and thus the same frequency. This is very important if we are to distinguish the miniscule axion conversion signal from other backgrounds. If the axion background was

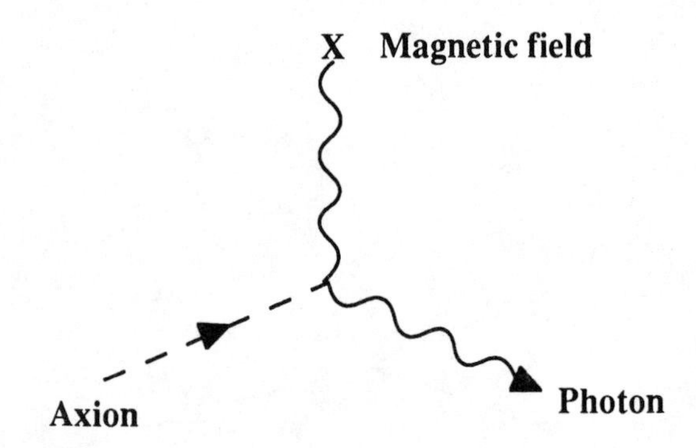

FIGURE 13.6

A Feynman diagram showing how an axion can convert into a photon in the presence of a background magnetic field. In spirit this is similar to the gravitational process displayed in figure 13.8, in that the interaction can take place coherently over a large volume. Hence it might be macroscopically measurable.

thermal, so that it had a large range in energy, the photons into which they could convert in the presence of a background magnetic field would also have a broad range in frequency. A signal of a given magnitude spread out over a large frequency range is much harder to detect than a very narrow spike, provided we know where to look for the spike.

Next, because the axions in the background were produced in a "coherent" configuration and one that was, to first approximation, spatially uniform, the photons that are produced at different points from individual axions will have their electromagnetic fields coupled together in a well-defined way. This is also very significant for detection. If the photons produced at different points were uncoupled, then the signal could not build up over time. It would wash out. However, the individual axions "pump" the electromagnetic fields in unison, as a child pumps on a swing. The net result is that the photon signal in principle can grow, just as the amplitude of the child's swinging can grow.

Now, neither property of the axions as I just described them is strictly true. When the axion background around what is now our galaxy first collapsed under its own gravitational attraction, the axions picked up some nonzero velocities, and hence some nonzero energy of motion, in addition to the energy associated with their mass. As a consequence, the background began to vary spatially to a small degree, and also the axion energies were no longer all identical. These effects are small, however. If the axions have a mean velocity like everything else in the galaxy, namely, about 300 kilometers per second, then their energy of motion is about 10^{-6} as large as the energy associated with their mass. Similarly, one can show that the galactic axion background would remain spatially uniform and coherent today over distances of about 20 meters or so, at least for axions with the "suggested" mass of 10^{-5} eV, that is, those which naive arguments suggest would close the universe today. This is far beyond the size of any detector likely to be built.

We may thus hope to build an axion "radio" to detect cosmic axions. We would construct a large microwave cavity that is "tuned" to precisely the frequency of the photons that would be produced from axion conversion, and then place this cavity in a large background magnetic field. Axions would then enter the cavity and convert to photons at the resonant frequency of the cavity. The photon signal could then build up for some finite time, to a level where we might hope to detect it with a very sensitive receiver. The proposed axion radio is shown schematically in figure 13.7.

Now, this works beautifully in theory, but several practical problems must be overcome. In the first place, no one knows exactly what the axion mass is supposed

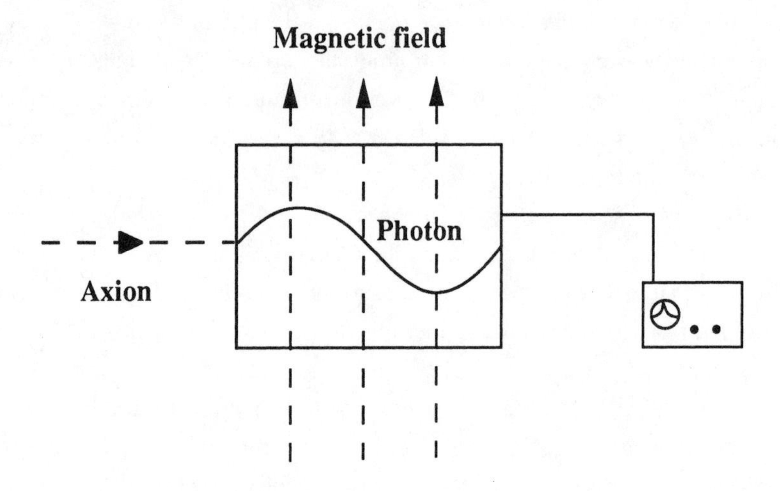

FIGURE 13.7
If cosmic axions enter a properly tuned microwave cavity inside of which a large magnetic field is maintained, then these axions may be converted into photons that will resonate inside the cavity, to be detected by a receiver connected to the cavity.

to be. Thus instead of a single fixed cavity, one must have a tunable cavity, whose resonant frequency can be varied to probe for cosmic axions in different mass ranges. This is not very easy. Large cavities give larger signals (the axion photon conversion rate is proportional both to the magnetic field we place in the cavity and also to the volume spanned by this field), but large cavities are hard to tune. Second, to allow the photon signal from axions to build up for the maximum allowed time, the "reflectivity" inside the cavity must be extremely good, so that no losses occur. This means that we probably must resort to superconducting cavities. Next, the receivers and amplifiers that allow detection of the photon signal must be exceedingly sensitive. Even microscopic smidgens of electrical noise in these devices will swamp the signal. Finally, even after all the splendid, coherent effects are added up, the axion signal is still very tenuous. Frank Wilczek, John Moody, and I, along with Donald Morris of Lawrence Berkeley Laboratory, calculated how small the signal would be. In a cavity 1 cubic meter in size, in which photons can bounce back and resonate for more than a million cycles before being absorbed, and with the largest known magnetic field that can span such a volume, the power converted from axions into photons would be only 10^{-24} watts. Even though in theory a cosmic axion field, if discovered, could solve the world's energy problems by acting as a continuous source of microwave energy, one would need a converter the size of the sun to power a single light bulb!

Nevertheless, experimentalists are a hardy bunch, and a group at Lawrence

FIGURE 13.8
Shown is the housing for Microwave Cavity for the Livermore-Brookhaven-Florida Axion Search experiment. (Photo Courtesy of Pierre Sikivie.)

Livermore National Laboratory, in collaboration with Sikivies group at Florida, and a group at Fermilab, using a huge Livermore magnet, have already completed a ground-breaking experiment that has achieved the sensitivity required to begin to probe for cosmic axions. Of course, because the axion parameter space is large and somewhat uncertain, and because different techniques have to be used to probe for different masses, we will have to be patient before closing the books on axions, or rejoicing over their discovery (see figure 13.8).

Given time, patience, and money, an axion radio may someday resonate with the pure sweet tones of an axion field that has waited more than 10 billion years for us to hear its music.

CHAPTER FOURTEEN

Searching for Nothing?

I cannot conclude this book without returning for at least a brief discussion of the possible physics that might be associated with the most surprising potential discovery of the century: that empty space may dominate the energy of the universe, govern its ultimate destiny, and swamp all matter, even dark matter, in ultimate cosmic importance. By such a discussion we are going full cycle, back to the aether and quintessence of Aristotle. But now we are no longer satisfied merely to postulate such a background energy; we want to detect it directly! Having devoted significant space to the remarkable search for dark matter, it would be unfair not to discuss the challenges associated with detecting vacuum energy directly, and also the theoretical arguments that might shed light on its nature and origins.

In fact, the first proposal to shed light on the possible direct experimental detection of vacuum energy came from none other than novelist and social critic George Orwell, who stated in 1946, "To see what is in front of one's nose requires a constant struggle!" These words aptly describe the workings of modern cosmology. The universe is all around us—we are a part of it—yet scientists must sometimes look halfway across it to understand the processes that led to our own existence on earth. As we have seen, while we believe that the underlying principles of nature are simple, unraveling them can be subtle.

Orwell's adage is doubly true, however, for it provides the key to how we could directly probe a nonzero vacuum energy in the laboratory today. Remember that the effect of a nonzero vacuum energy is to cause space to expand and for this expansion to accelerate. Objects moving away from us will therefore speed up. We can then ask the following question: How far away will objects get before they are traveling away from us faster than the speed of light?

This may sound like a silly question. After all, Einstein told us that nothing can travel faster than light! However, when he developed General Relativity, the wording of this law had to be revised. Nothing can travel through space faster than the speed of light. However, space can do whatever it wants! It can expand or contract faster than the speed of light, and objects at rest in an expanding space will be carried away from us along with the space. Thus, the effect of a vacuum energy will be to cause all objects farther than a certain distance away from us to be receding from us faster than the speed of light. Put another way, such objects could not be seen.

Now, let us ask ourselves how small the vacuum energy has to be so that we can indeed, à la Orwell, see the ends of our noses. When we put in the numbers, it turns out that in order to do merely that, the vacuum energy must be 70 orders of magnitude smaller than the estimate particle physicists first came up with for this quantity.

Thus, in a very real sense, you can do a cheap and very powerful experiment here on earth to probe for the existence of a vacuum energy: Look at the end of your nose! The results are very strong indeed! Unfortunately, however, while these results are strong, they also point out precisely why we will never, ever, be able to do an experiment in the laboratory that matches the sensitivity of astronomical probes for the possible existence of a cosmological constant.

In a laboratory you can see to the end of your nose, and even to the end of the room. However, in cosmology, we can see to the end of the universe! The fact that we can see remote objects moving at sublight speeds away from us as far as 10 billion light-years distant puts an incredibly tight constraint on the maximum possible value of the vacuum energy: about 125 orders of magnitude smaller than the naïve first estimate of particle physicists, and about 55 orders of magnitude smaller than you obtain by staring at your own nose. This limit is so very small that there is no direct way we can ever hope to measure such a quantity.

So much for laboratory probes of a cosmological constant. Nevertheless, as I have already described, cosmological observations of receding supernovae, combined with a host of other indirect probes, have now provided strong, if not yet undeniable evidence that the universe is dominated by something that behaves like vacuum energy. If this is the case, and if this energy density, while larger than the net energy density of all matter in the universe—including dark matter—is so very small compared to any reasonable expectations, what could possibly be happening to cause this?

The gut response to this question is that we have no idea whatsoever! That is what makes this possibility so exciting to physicists. Or rather, we have no really good ideas. Nevertheless, I feel it is incumbent upon me to at least introduce some of the speculations that are now bandied about in the absence of sound theoretical arguments. Perhaps one day speculation will turn to fact.

(1) Dirac's Large Number Hypothesis. Paul Dirac, of whom I have already spoken, was one of the most creative physicists of the first half of this century, next to perhaps Einstein and Werner Heisenberg. Dirac often made observations that were only properly appreciated years later. This might be one of them. He noted that there is one "natural" large number in the universe: the age of the universe. In terms of any fundamental atomic timescales, the 10-billion-year timescale of the evolution of our universe is truly astronomical (if you forgive the pun, and you shouldn't!). Dirac pointed out that if any physical quantity varied uniformly with time, either increasing or decreasing, then it could "naturally" become either very large or very small by the present epoch. Perhaps the vacuum energy of the universe is just such a quantity. If it evolved with time, this might naturally explain how it could be so minuscule today!

While this is, in principle, a great idea, it fails miserably, at least in the context of our current understanding of gravity. General Relativity requires, as one of its central features, that any true energy of the ultimate vacuum of nature remain precisely constant and unchanging, regardless of the vicissitudes of fate as far as ordinary matter is concerned.

This minor wrinkle has not intimidated theorists, however. A number have tried to explore how variations in general relativity might alleviate this constraint. Another group has explored a completely different direction, one that we have already explored in the context of our discussion of axions.

Recall that at early times, before the axion background "relaxed" into its ground state, the coherent axion background behaved precisely as if it were a temporary cosmological constant. Perhaps there is another background in nature, unlike axions, that has not yet relaxed but is in the process of so doing, leading to an ever decreasing vacuum-energy density in the observable universe.

This idea sounds good on the surface, but it is pretty ugly once you explore a little more deeply. In the first place, in order to not have relaxed by the present time, the parameters one has to apply to the particle physics models in question are absurdly small, making the theory appear quite unnatural, if perhaps not as unnatural as a bare cosmological constant. Secondly, and more damning from my point of view, such models do not address the real question, which is what is the ultimate energy of truly empty space? Such models require this energy to be either zero or very very small, so that this impostor field can mimic a cosmological constant and get away with it. The enormity of the cosmological constant problem remains.

Finally, there is another nagging operational question at stake here. If the field is varying very slowly on cosmic timescales, there is absolutely no observational distinction between such a field and a true cosmological constant, so we will not be able to empirically validate this hypothesis directly. On the other hand, if the field is varying enough for us to measure this by cosmological probes today, the question arises: Why is the field varying on a 10-billion-year timescale and not a trillion-billion-year timescale? Alternatively, we are led to ask why this is the first time in the history of the universe that we could make the distinction, if indeed we can.

All of these negatives have not stopped theorists from exploring this possibility, however. And as happens in Hollywood or New York, or in between, a catchy name gets you a long way. Physicists Robert Caldwell and Paul Steinhardt coined the name "quintessence" for such a field, possibly borrowed from my own use in discussing dark matter a decade ago. While this is in fact a much better candidate for a true "quintessence," in the spirit of Aristotle than perhaps dark matter, I find this no reason to latch onto it, at least at the present time. And preliminary observational data suggests no evidence whatsoever for a cosmic field varying on a 10-billion-year timescale.

(2) Are we too myopic? We have seen how, during an inflationary phase of expansion, the universe can get stuck in a "false" vacuum, which leads to exponential expansion exactly of the sort produced by a fundamental cosmological constant. In the case of inflation, this expansion ends when the true vacuum state is reached. Perhaps we are being too myopic to assume that we live in the ultimate ground state of the universe today. Perhaps there is another phase transition yet to complete, and the vacuum energy we observe is merely that stored in the false vacuum state.

Of course, if this is the answer, it is not a particularly satisfying one, at least as far as humanity is concerned, since in such a case, when the phase transition completes, it is quite likely that the properties of matter would dramatically alter, and with them the properties of the observable universe. It is unlikely we should survive such a transition!

Nevertheless, let us not let petty human concerns get in the way! We can even ask what kind of parameters would produce such an inflationary phase today, and the answer is quite intriguing. The energy stored in the vacuum, if we are correct in our cosmological inferences, has precisely the scale comparable to the mass scale recently proposed for one species of neutrinos based on the observations of atmospheric neutrinos at the Kamiokande detector. Is this a coincidence, or is the smell of grand synthesis in the air? I might be compelled to suppose the latter except for one thing. There is no sensible particle model at the present time that predicts such a transition to occur. But who knows? In any case, even in this proposed resolution, why the ultimate vacuum should have zero energy is still not addressed, so the cosmological constant problem itself remains unresolved in this picture.

(3) Maybe we are not myopic enough? Another proposal, recently argued by Steven Weinberg and colleagues at the University of Texas, invokes the last resort of cosmologists: the anthropic principle. They point out that if the observed universe is really one of perhaps an infinite ensemble of universes, each of which has different values for the fundamental physical parameters, then the cosmological constant would be too large and life would never evolve. Perhaps then the cosmological constant is small in our universe precisely because we happen to be able to live in only such a universe! Moreover, Weinberg and his colleagues have tried to quantify this argument, and they get numbers comparable to the value of the vacuum energy today claimed as "reasonable." Nevertheless, I think it is fair to say that many physicists do not find this a particularly attractive possibility, as the pursuit of physics has traditionally been based on explaining why the universe must be the way it is, rather than why it should in general be quite different.

I think this sample has given the proper flavor for the arguments. They are intriguing, if not compelling, and the resolution of this problem will no doubt require at some level a wholescale revision in our thinking of the microphysics of the universe. It seems to me likely that whatever physics results in a universe whose dominant form of matter is dark is probably not unconnected to this larger issue. Thus, the discovery of the true nature of dark matter, which will no doubt occur before we unravel the deeper mystery of the nature of empty space, will be a very important step on the road to an ultimate understanding of why we are so lucky as to be here today to ponder such cosmic questions at all.

EPILOGUE

The Best of Times?

Lately it occurs to me . . . what a long strange trip it's been.

—The Grateful Dead

Perhaps nothing could be more fitting after humanity's long intellectual journey than the discovery of a dark background of axions with an axion "radio." What could come closer, after all, to the music of the spheres of Pythagoras and Kepler? Still, some readers may wonder, will our axions, or WIMPs, or even the modern "quintessence" seem as quaint to physicists two thousand years hence as the "indefinite" of Anaximander seems to us today? Are we as far from the correct picture of what dominates the universe today as the ancient Greeks were? I think not. Science is a field in which objective progress can be made, even though it proceeds in small steps. The theories I have described here have the force of experiment behind them. No matter how they fit within our theories in the future, they will not vanish. So it is with dark matter. We have proved that it is out there, and we will keep advancing in our understanding of what it is not, until someday we discover what it is.

As I reflect on the ideas and developments I have described in this book I cannot help but marvel at our present situation. The picture we have of the universe today would have been indescribable in the first decades of this century: the ideas and language simply did not exist then. Now we have discovered that most of the universe is dark, and there is a real possibility that at last much of this darkness will reveal its identity within our lifetimes. Since the dawn of civilization, human beings have pondered the origin of the world we perceive, what it is made of, and what its future will be. It is mind-boggling that within less than a quarter-century we have come within striking distance of the answer to any one of these questions. It is also rather amazing that the answer to all three lies in determining what constitutes the dark matter and energy that astronomers have discovered surrounding all the astronomical systems we can see.

Some may question whether the search is worth the effort. Of what possible use can it be to the man or woman on the street whether the universe will end in some unfathomably long time with a bang or a whimper? Who cares whether a hidden supersymmetry may govern nature, or whether an axion solves a puzzle that has no impact whatsoever on our daily lives? I must say that it is not so much the answers that make the search worthwhile as the search itself. What raises us up from the tedium of mere existence if not our capacity to understand our place in the universe? And what makes life worth celebrating if not our ability to dream, to imagine? It is the progress of science and the arts which continually rekindles our imagination. If we cease to explore the universe, we will eventually cease to wonder about it.

I have no idea how many of the notions discussed here will survive intact through the course of time, even in the short term. Their greatest legacy may be the contribution they make to our collective imagination, their ability to spark our curiosity, and our awe. I view theoretical particle physics and cosmology as enterprises as valuable as art, music, and theater in their gifts to our imagination: they give us a better sense of ourselves and our place in the world. And while science may have led to an enhanced realization of the cosmic insignificance of humankind, it has at the same time demonstrated unequivocally the awe-inspiring wonder of the universe in which we live.

While there is every possibility that some of what I have described is simply wrong, there is also the possibility that much of it is right. More thrilling is the fact that in science we can distinguish between the two. We may never be able to prove that a theory is completely correct. A new experiment may always lurk around the corner to disagree with one's predictions. It is easy as pie to prove a theory wrong, however.

In this sense I find the story of the search for dark matter to be, in microcosm, the story of modern physics. The only thing more astounding than these wild and fantastic notions about nature is the fact that many of these ideas appear to describe correctly the world in which we live. No, I take that back. More remarkable still is the fact that we have the ability to tell the difference. Whether or not the dark matter is made of exotic new particles, most of the ingredients of the tale that I have spun here will survive the test of time, because they have survived the test of experiment. Even if much of the presently accepted wisdom about dark matter or energy in the universe and its role in the formation of structure turns out to require alteration—and we should know soon—we still win. In the end we will better understand the universe, and our place within it. We could do worse.

APPENDIX A

ORDERS OF MAGNITUDE AND THE SCALE OF THE UNIVERSE

In order to understand the universe properly, one should first try to gain an appreciation of the numbers needed to describe it. In this regard, it is important first to understand what is meant by an "order of magnitude." This in turn is related to the way scientists write down numbers. Their representation of numbers is called "scientific notation," for several reasons that I hope to make clear here.

Because physics deals with such a wide variety of scales, very large or very small numbers can occur in even quite simple problems. Now, to a physicist any number has two pieces. The first piece tells, to within a factor of 10, the overall magnitude of the number: that is, is it large or is it small? This part of the number, called its "order of magnitude," contains all the important information that sets the scale of any discussions involving this number. Once the overall scale has been determined, the second piece specifies more exactly what the value of the number is.

Scientific notation and the definition of an order of magnitude depend on a simple property of multiplication. If I multiply the number 10 by itself two times, I get the number 100. Because 100 is equal to 10 *squared,* I may write it as $100 = 10^2$. The number 10^2 then represents the number 1 followed by *two* zeros. In a similar fashion, I may define the number $10 \times 10 \times 10 \ldots \times 10$ (n times) $= 10^n$, which represents the number 1 followed by n zeros. We say that this number is "10 to the nth power." (The number 10 can thus also be written as 10^1, and the number 1 can be written as 10^0.) Similarly, I can write 1/10 as 10^{-1}, which signifies the number given by a 1 in the first place after the decimal point, that is, 0.1. In this way $10^{-n} = .0000 \ldots 1$ (nth place after the decimal point).

Any number can now be written as the product of a number between 1 and 10, multiplied by 10 to some power. For example, take the number 14,959,000,000,000. In scientific notation I could rewrite this as: 1.4959×10^{13}. Not only is this latter form more economical, but it is more meaningful, at least as far as physics is concerned. The power of 10 is explicitly displayed. It is this factor that gives the scale, or order of magnitude, of the number. Most often, to get the overall feeling for a physical quantity, one can forget the prefactor, which in this case is 1.4959, and say that this number is approximately equal to 10^{13}. This makes it clear that this number is (*a*) pretty big and (*b*) about 10 times larger than a number that is approximately equal to 10^{12}. In this book, if I refer to the order of magnitude of some quantity, that means I am not worrying about the exact value of any prefactor, and that, to within such a factor between 1 and 10, the quantity in question has a certain size. This is because I will usually be interested in the overall scale of quantities here, rather than their exact values.

When discussing "cosmic" size questions, as I do in this book, very large or very small numbers are the rule rather than the exception. For reasons of economy, therefore, I will have to write certain numbers in scientific notation. For example, the number of neutrinos emitted in a supernova explosion is about 10^{58}. It would be hopeless to try to write this number down the normal way, and if I did, no one would have any idea exactly how large it was. At least this way one understands that it is 58 orders of magnitude larger than 1, or that I would have to multiply 10 by itself 57 times to make such a large number.

Indeed, without scientific notation and the idea of an order of magnitude, it would be very difficult to get a feeling for the scale of the universe as a whole and the range of scales which define the phenomena that occur within it. With these ideas in hand, however, let me now describe those scales that are important for the discussions in this book.

Curiously, the scale of human existence falls about midway between the largest and the smallest scales about which we have any direct experimental knowledge. Consider the distance traveled by an Olympic marathoner in the course of a half-hour—about 10 kilometers. (I remind you that there are about 2 kilometers in a mile.) If we multiply that length by 10, and then by 10 again, and continue the process about 22 times (that is, 10^{23} kilometers), we come to the distance a light ray created in the primordial Big Bang explosion would have traveled before being measured today in an antenna as part of the cosmic black body background. If, on the other hand, we divide that original length by 10 about 22 times, we come to the distance separating the particles that will collide in the most powerful particle accelerators presently in existence. The 45 orders of magnitude separating these lengths represents the span of distance about which we have direct empirical knowledge.

The speed of light is about 3×10^{10} centimeters per second. There are about 3×10^{7} seconds per year. That means that light travels a distance of about 9×10^{17} (roughly equal to 10^{18}) centimeters in a year. Because this kind of distance is more typical of the scales we encounter in astronomy and cosmology, it is given the special name "light-year." Astronomers, to complicate the situation, also use a related quantity to define the scales they study. This quantity, called a "parsec," is equal to about 3.26 light-years. The "sec" in parsec has nothing to do with time, but rather refers to a "second of arc," which is an angular measure. (It turns out that a star at a distance of 1 parsec from the earth will shift in position across a background of much more distant stars by about 1 second of arc when viewed from the earth on opposite sides of its orbit around the sun.) In any case, the disk of our galaxy is about 30,000 parsecs from end to end, or about 99,000 light-years across. That means that it takes the light from the far end of the disk of our galaxy about 80,000 years to get here.

The distance to the nearest large galaxy similar to our own, Andromeda, is about 2 million light-years. The distance to the center of the large Virgo supercluster of galaxies is about 45 million light-years. Since the universe is between about 10 and 20 billion years old, the furthest distance we can see is about 10 billion, or 10^{10}, light-years. Since a light-year is about 10^{18} centimeters, this is how I obtained the estimate of the distance traveled by the light forming the cosmic photon background from the Big Bang explosion.

Finally, the Hubble constant, which sets the rate at which the universe is expanding, and whose uncertainty sets the scale for our uncertainty in measuring the distance to faraway objects, takes a value between 60 and 80 kilometers per second per million parsecs. This means that on average a galaxy one million parsecs distant from us will be moving away from us with a speed of between 60 and 80 kilometers per second, while a galaxy 2 million parsecs distant will be moving away with a speed of between 120 and 160 kilometers per second, and so on.

It turns out that cosmological theory allows the Hubble constant to be related to the present age of the universe (in a way that depends in detail upon the mean density of matter in the universe, which in turn depends on how much dark matter there is today). Without knowing any theoretical details, however, but using some arithmetic manipulations, we can demonstrate this relation. The units in the Hubble constant are kilometers per second per megaparsec (one million parsecs), or kilometers/(seconds × megaparsecs). Since a megaparsec is a unit of distance, we can write megaparsecs in terms of kilometers. In this system of units the Hubble constant would be kilometers/(seconds × kilometers). Canceling out the kilometers in the numerator and denominator, one would get the unit 1/seconds. Thus, in this system of units, the Hubble constant has the units of the inverse of time. Hence, the inverse of the Hubble constant has the units of time. If we plug in the numbers, the inverse of a Hubble constant of 100 kilometers per second per megaparsec in these units takes the value 3×10^{17} seconds, which is almost equal to 10 billion years. Hence, the Hubble constant not only sets the scale of the expansion of the universe, it turns out to set the magnitude for the age of the universe as well. This in turn sets the scale for the size of the visible universe. Thus, if you want to remember one number that best describes the universe we live in, remember the Hubble constant.

APPENDIX B

A *REALLY* BRIEF HISTORY OF TIME

By solving the equations for an expanding universe, and extrapolating backward using the known properties of matter and radiation, we can describe how properties such as scale and temperature of the universe evolved over time in the standard Big Bang model. During the period when the energy density of the universe was dominated by relativistic particles (that is, radiation), its scale size increased as the square root of time. Once it became dominated by nonrelativistic matter, this proportionality changed, so that the scale of distances increased as the two-thirds power of time. (The expansion is faster in a universe dominated by matter because the energy density of matter falls less quickly as the universe expands than does the energy of radiation. Since it is this energy density that drives the expansion, if the energy density decreases more slowly, the expansion rate will remain larger for a longer time.) Finally, throughout most of the period of the expansion of the universe, the temperature of radiation fell inversely as the scale size grew. Thus, once we know how the scale size varies with time, we automatically know how the temperature varies with time.

With these relations in mind, I present here two "histories" of the universe, showing graphically how both the temperature and the radius of what is now the observable universe are predicted to have changed over time according to the equations of the Big Bang expansion, along with showing some temporal milestones (see figures B.1 and B.2).

I have used a scale that conveniently allows me to show ever smaller times with equal detail. However, it takes some getting used to. Each tick mark in time indicates a period *10 times later* in the history of the universe than the tick before it. As a result, the whole period since the formation of the earth has transpired between the arrow marking the present time (c) and the tick mark on the graph just before that line. Similarly, the period covering the origin of life on earth up to the present is contained within the width of the line marked "today" (c)! I remind you (see appendix A) how the position of this line is determined: the present age of the universe is roughly 10 billion (10^{10}) years. There are about 30 million seconds in a year, thus making the universe almost 10^{18} seconds old today.

I have shown explicitly in the figures the recombination time, at around one million years into the expansion, after which matter and the cosmic photon background were no longer in thermal contact. Earlier, at about 100 seconds, you see the time when Big Bang nucleosynthesis began in earnest. Theoretical arguments suggest that at yet earlier times two other important milestones occurred: the transition from a

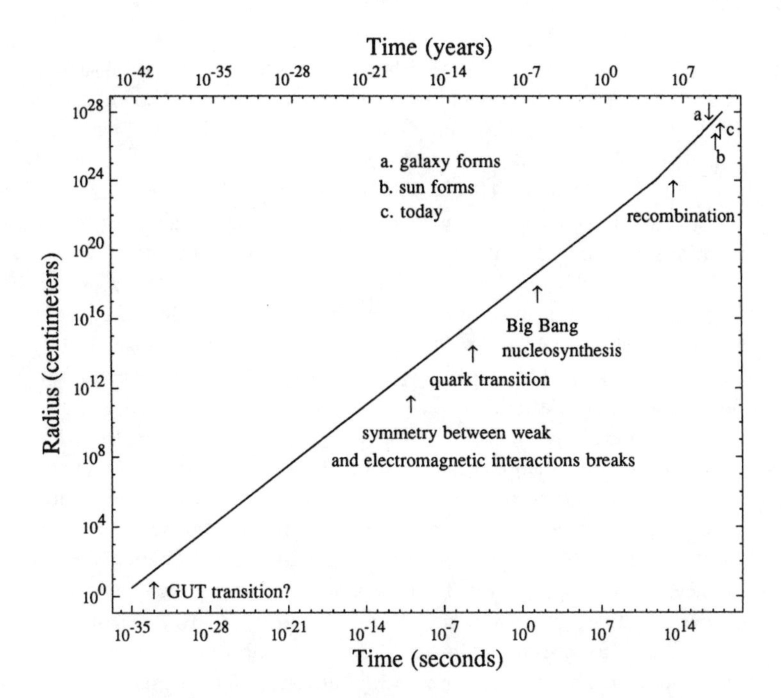

FIGURE B.1 HISTORY OF THE RADIUS OF THE UNIVERSE

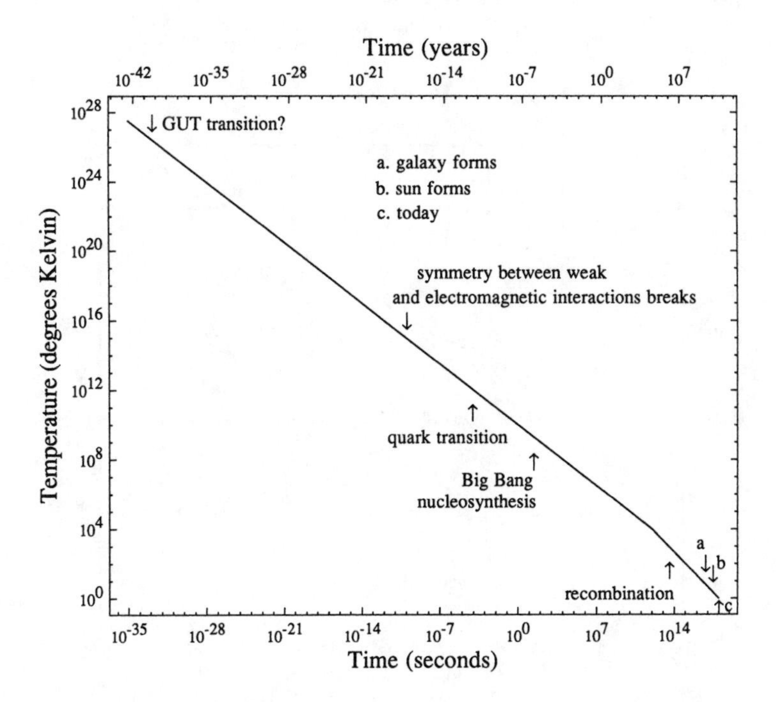

FIGURE B.2 HISTORY OF THE TEMPERATURE OF THE UNIVERSE

dense quark gas to a gas of baryons and other particles which we can observe in the laboratory, and before that, the time after which the symmetry between electromagnetism and the weak interaction was no longer manifest. Finally, much earlier, I label the time when the strong, weak, and electromagnetic interactions may have become unified in a single Grand Unified Theory (GUT). There could have been a lot of important action around this time. There may have been an inflationary period during which the scale size–time relation may have departed drastically from the behavior shown in figure B.1. In addition, it could have been during this time that the excess of protons and neutrons over antiprotons and antineutrons, which eventually left behind everything we can see in the universe, was dynamically established. However, the entire period of universal history shown before the time of weak symmetry breaking is at present pure speculation.

Returning to much later times, the "break" in both the temperature-time and radius-time curves shortly before recombination signals the onset of matter domination. It is roughly at this time that density fluctuations in a massive neutrino background could begin to grow. This is to be compared with the time when density fluctuations in normal baryonic matter can grow, that is, the recombination time, which occurred somewhat later. In figure B.3 I display with a dashed line the distance that could have been spanned by a light ray at any given time. This "horizon" distance grows linearly with time and today encompasses (by definition) the observable universe, so that it meets the solid line, representing the radius of the observable universe, today.

In figure B.3 I have blown up that period of the universe from about one year up to the present time. I have also drawn lines that display how large the region would

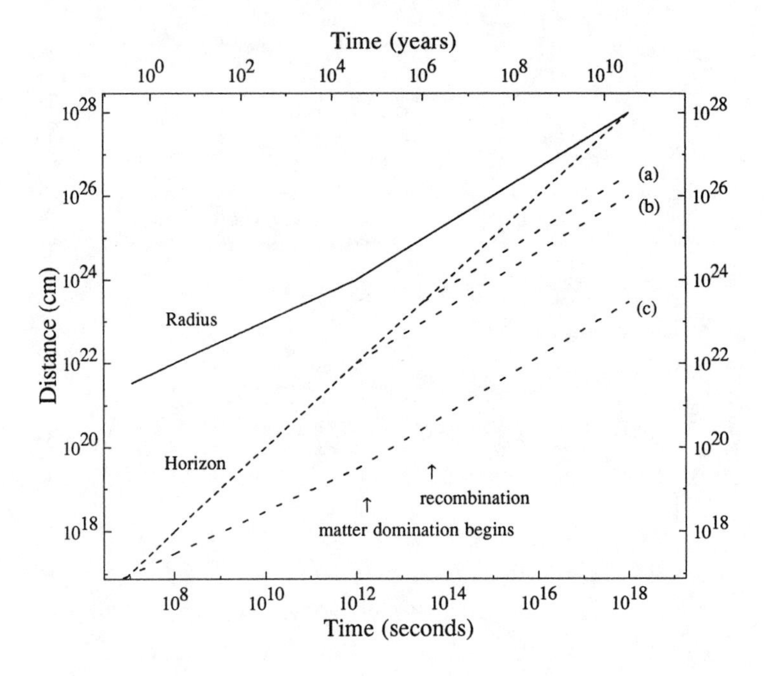

FIGURE B.3 HISTORY OF LARGE-SCALE STRUCTURE

be today that would have been encompassed inside the horizon at (*a*) the recombination time, (*b*) the time when the universe became matter dominated, and (*c*) the time when the universe had a temperature of about 10 million degrees Kelvin. These scales represent the first scales on which the growth of structure would generally be expected to occur in (*a*) a baryon-dominated universe, (*b*) a universe dominated by light neutrinos, and (*c*) a universe dominated by material that is just on the verge of being classified as cold dark matter. Only in this last case does growth begin first on galaxy-sized scales, which seems to be necessary based on the results of the numerical simulations of large-scale structure described in part 3.

NOTES

CHAPTER ONE: MAKING SOMETHING OUT OF NOTHING

1. Lucretius, *The Poem on Nature,* trans. C. H. Sisson (Manchester: Carcanet New Press, 1976).
2. *Enuma Elish: The Seven Tablets of Creation,* trans. L. W. King (London: Luzac, 1902); for a later translation, see Alexander Heidel, *The Babylonian Genesis: The Story of Creation* (Chicago: University of Chicago Press, 1967).
3. *Hymns from the Rigveda,* trans. A. A. Macdonell (Madras: Diocesan Press, 1966).
4. Epiphanus Wilson, ed., *The World's Great Classics: Sacred Books of the East* (New York: Colonial Press, 1900); for a later translation, see *The Rig Veda: An Anthology,* trans. Wendy Doniger O'Flaherty (Middlesex: Penguin, 1981).
5. E. A. Wallis Budge, ed. and trans., *Legends of the Gods: The Egyptian Texts* (London: Kegan, Paul, Trench, Trubner, 1912).
6. J. M. Plumley, "The Cosmology of Ancient Egypt," in Carmen Blacker and Michael Loewe, eds., *Ancient Cosmologies* (London: George Allen and Unwin, 1975); reprinted in J. H. Weaver, ed., *The World of Physics,* vol. 1 (New York: Simon & Schuster, 1987).
7. Plato, *Theaetetus,* in G. S. Kirk, J. E. Raven, and M. Schofield, *The Presocratic Philosophers,* 2d ed. (Cambridge, Eng.: Cambridge University Press, 1983).
8. Aristotle, *Metaphysics,* trans. Hippocrates G. Apostle (Bloomington: Indiana University Press, 1969).
9. Simplicius, *Physics,* in C. H. Kahn. *Anaximander and the Origin of Greek Cosmology* (New York: Columbia University Press, 1960); see also Kirk, Raven, and Schofield, *The Presocratic Philosophers,* and J. Barnes, *The Presocratic Philosophers* (London: Routledge and Kegan Paul, 1979).
10. Lao-Tzu, *Tao Te Ching,* trans. D. C. Lau (Middlesex: Penguin, 1963).
11. Aristotle, *De caelo,* in Kirk, Raven, and Schofield, *The Presocratic Philosophers.*
12. Quoted in Stephen Toulmin and June Goodfield, *The Architecture of Matter* (New York: Harper & Row, 1962), p. 83.
13. Anaxagoras fragment from Simplicius, *Physics,* in Barnes, *The Presocratic Philosophers.*
14. Kahn, *Anaximander,* p. 94.

15. Fragments from Empedocles in R. E. Allan, ed. and trans., *Greek Philosophy: Thales to Aristotle* (New York: Macmillan, 1966); reprinted in Weaver, *The World of Physics*. For additional Empedocles source references, see Kirk, Raven, and Schofield, *The Presocratic Philosophers*, and J. Barnes, *Early Greek Philosophy* (Middlesex: Penguin, 1987).
16. Toulmin and Goodfield, *The Architecture of Matter*, p. 99.
17. Christian Huygens, *Traité de la Lumière* (1690), facsimile reprint (London: Dawsons of Pall Mall, 1966), p. 3. I have translated into English the excerpts quoted here.
18. Ibid., p. 10.
19. Ibid., p. 9.
20. Sir Isaac Newton, *Opticks* (1730) (New York: Dover, 1952), pp. 348–49.

CHAPTER TWO: FILLING THE VOID

1. Steven Weinberg, *The First Three Minutes: A Modern View of the Origin of the Universe* (New York: Basic Books, 1976).

CHAPTER FOUR: BEYOND OUR ISLAND IN THE NIGHT

1. Richard Feynman, *The Character of Physical Law* (Cambridge, Mass.: MIT Press, 1967).

CHAPTER SIX: THE TIP OF THE ICEBERG

1. For a more detailed discussion of inflation theory, see Alan H. Guth and Paul J. Steinhardt, "The Inflationary Universe," *Scientific American* 250 (May 1984): 116–28; and Guth, "The Inflationary Universe," (New York: Addison-Wesley, 1997).

CHAPTER TWELVE: THE MUSIC OF THE SPHERES?

1. Aristotle, *De caelo*, quoted in R. E. Allan, ed., *Greek Philosophy: Thales to Aristotle* (New York: Macmillan, 1966).

INDEX